Coal

Information Report

INTERNATIONAL ENERGY AGENCY

2, RUE ANDRÉ-PASCAL 75775 PARIS CEDEX 16, FRANCE

The International Energy Agency (IEA) is an autonomous body which was established in November 1974 within the framework of the Organisation for Economic Co-operation and Development (OECD) to implement an International Energy Program.

It carries out a comprehensive programme of energy co-operation among twenty-one* of the OECD's twenty-four Member countries. The basic aims of IEA are:

   i) co-operation among IEA Participating Countries to reduce excessive dependence on oil through energy conservation, development of alternative energy sources and energy research and development;

   ii) an information system on the international oil market as well as consultation with oil companies;

   iii) co-operation with oil producing and other oil consuming countries with a view to developing a stable international energy trade as well as the rational management and use of world energy resources in the interest of all countries;

   iv) a plan to prepare Participating Countries against the risk of a major disruption of oil supplies and to share available oil in the event of an emergency.

IEA Member countries: Australia, Austria, Belgium, Canada, Denmark, Germany, Greece, Ireland, Italy, Japan, Luxembourg, Netherlands, New Zealand, Norway, Portugal, Spain, Sweden, Switzerland, Turkey, United Kingdom, United States.

Pursuant to article 1 of the Convention signed in Paris on 14th December, 1960, and which came into force on 30th September, 1961, the Organisation for Economic Co-operation and Development (OECD) shall promote policies designed:

   – to achieve the highest sustainable economic growth and employment and a rising standard of living in Member countries, while maintaining financial stability, and thus to contribute to the development of the world economy;

   – to contribute to sound economic expansion in Member as well as non-member countries in the process of economic development; and

   – to contribute to the expansion of world trade on a multilateral, non-discriminatory basis in accordance with international obligations.

The Signatories of the Convention on the OECD are Austria, Belgium, Canada, Denmark, France, the Federal Republic of Germany, Greece, Iceland, Ireland, Italy, Luxembourg, the Netherlands, Norway, Portugal, Spain, Sweden, Switzerland, Turkey, the United Kingdom and the United States. The following countries acceded subsequently to this Convention (the dates are those on which the instruments of accession were deposited): Japan (28th April, 1964), Finland (28th January, 1969), Australia (7th June, 1971) and New Zealand (29th May, 1973).

The Socialist Federal Republic of Yugoslavia takes part in certain work of the OECD (agreement of 28th October, 1961).

TABLE OF CONTENTS

Part I

SUMMARY OF
WORLD COAL DEVELOPMENTS

1. INTRODUCTION

Coal has re-emerged in the last decade as an important fuel with a promising future. Particularly for the OECD, coal is an attractive energy source because some OECD countries possess both vast reserves of coal and the potential for a large expansion in export capacity. In addition, coal is economically competitive with oil and natural gas in several uses, principally in electric power generation and in some industrial applications. Moreover, coal can also compete with nuclear in certain cases, depending on the location, structure and richness of the deposits.

Several studies in recent years have examined the technical and economic potential for expanded coal production, trade and use in the world for the rest of the century. In 1978 the International Energy Agency published *Steam Coal Prospects to 2000**. That study was initiated when OECD total oil imports were continuing to increase and delays in nuclear development programmes were stalling the displacement of oil-generated electricity by nuclear generation. Encouraging about coal's future prospects, the study projected that OECD total solid fuels demand would rise from a 1976 level of 708 Mtoe[1] to 1160-1199 Mtoe in 1990 and 1313-1472 Mtoe by 2000 if existing member government policies were maintained. The study suggested further that if governments adopted certain policies to increase coal use, an additional 153 Mtoe of coal in 1990 and 353 Mtoe in 2000 could be substituted for oil.

Another examination of coal prospects was published in the IEA's *World Energy Outlook*** in 1982, in which OECD coal demand by the year 2000 was projected to range between 1400 and 1600 Mtoe.

In 1979, IEA Ministers adopted a number of Principles for IEA Action on Coal designed to increase stability and transparency in international coal trade and to create conditions that would maximize the growth of coal to displace oil.*** In

---

*   *Steam Coal Prospects to 2000* (IEA/OECD, Paris, 1978)
**   *World Energy Outlook* (IEA/OECD, Paris, 1982)
*** These Principles called, among other things, for a periodic review of IEA countries' coal prospects and policies. The first such review was held at the end of January 1982 and results were published in early summer, 1982 (Coal Prospects and Policies in IEA Countries, 1981 Review, IEA/OECD, Paris 1982).

1.   Million tons of oil equivalent . Throughout this report historical data presenting interfuel relations and all projections are expressed in a common energy unit, tons of oil equivalent (toe). For futher information, see Chapter 1 in Part II of this report.

order to develop continuing dialogue among and advice from those actively involved in the mining, moving and use of coal, IEA governments established the Coal Industry Advisory Board (CIAB) in 1979.

In December 1980, IEA Ministers agreed that publication of more comprehensive information on projections for coal use, production and infrastructure development would help reduce uncertainties constraining coal's role in IEA energy systems. Based on recommendations from the CIAB, identification of information needs and review of available information, a decision was reached in December 1981 to implement a comprehensive IEA Coal Information System. The IEA Coal Information System is intended to provide both OECD member country governments and those employed in all facets of the coal industry with information of the present world coal market and prospects for coal in the rest of the century. *The present report is the first result of the IEA Coal Information System* and includes historical data and forecasts of coal reserves, production, trade, demand, prices and transportation infrastructure.*

Part I of this report integrates these parts of the world coal market to provide a total view of coal's prospects. It is based on historical data of OECD energy supply and demand collected by the IEA, data of other world regions supplied by outside sources and forecasts of trends in solid fuels supply, demand and trade supplied by the individual OECD Member countries. The forecasts of future trends are subject to major uncertainties. They may reflect policy goals rather than the "most likely case" forecasts. Moreover, the projections may well have been prepared some time ago and could, therefore, reflect a more optimistic view than would now be taken of the prospects for coal.

Part II gives a more detailed and comprehensive picture of coal developments and future prospects for the OECD by regions and countries. The historical data in Part II are all from IEA/OECD energy statistics based on annual submissions from Member countries. The projections in Part II are based on submissions to the IEA from OECD Member countries. A more detailed review of sources used are given in Part II.

---

*   Information on coal resources and transportation infrastructure is provided by the World Coal Resources and Reserves Data Bank Service and the Economic Assessment Service of the IEA Coal Research, London. These organisations were established under the auspices of the IEA to fulfill R&D information needs identified by coal specialists in a number of IEA countries. They are, however, separately funded from and managed independently of the IEA and are solely responsible for the accuracy of the information they have provided for this report.

2. COAL USE

2.1. Historical Overview

Between 1960 and 1973, demand for coal remained virtually constant in the OECD area, increasing from 682.4 Mtoe in 1960 to 702.2 Mtoe by 1973 (see Table 2.1) despite a rapid rise in total primary energy requirements (TPER). The relative importance of coal accordingly declined from more than 37% of TPER in 1960 to less than 20% by 1973. This overall decline is explained largely by greatly reduced coal demand for thermal use by non-electricity industries and for heating use in the residential sector, primarily in Europe, which was partly offset by coal demand growth in the electricity sector, mainly in the United States.

Table 2.1 **OECD Historical Coal Use[1]**
(Mtoe)

	1960	1973	1978	1981
OECD TOTAL	682.4	702.2	730.1	825.8
NORTH AMERICA	249.1	337.4	391.0	435.8
PACIFIC	68.0	86.0	82.7	104.3
EUROPE	365.3	278.7	256.4	285.7

Source: IEA/OECD Energy Balances

1. Total Primary Solid Fuels Requirements

Within this overall picture, there are considerable regional differences. Coal use in North America increased by 35.5% and in the Pacific countries by 26.6%, while coal use in OECD Europe declined by 23.7% from 1960 to 1973. As a result, North America accounted for almost 50% of total OECD coal use by 1973, compared with only 36.5% in 1960.

Even in the five years following the sharp oil price increases of 1973, total OECD coal demand did not grow significantly. Between 1973 and 1978, coal demand increased by only 4% and the share of coal in TPER decreased slightly from 20.0% in 1973 to 19.3% in 1978. During this period, the demand growth of coal for electricity generation in each OECD region was partly offset by the decline in coal use for coke ovens in North America and Europe. However, the oil price increases following the Iranian revolution boosted demand for coal through fuel switching and conversion. Between 1978 and 1981, coal demand increased by 13% largely due to the demand growth of coal in electricity generation in each OECD region and as a result, the share of coal in TPER in the OECD increased to 22.6% in 1981.

Although total OECD coal consumption increased only slightly from 1973 to 1978, coal use in electricity generation increased by 20% during this period. However, the share of total electricity produced from coal declined in all regions of the OECD and in many individual countries, as seen in Table 2.2. This trend was sharply reversed in the OECD from 1978 to 1981, when the share of coal-fired generation increased in all OECD regions. Of the countries shown in Table 2.2 only France exhibited a different trend. The share of coal-fired generation decreased in France from 1978 to 1981, coinciding with a large expansion of nuclear capacity.

Coal use in industry declined in absolute terms from 1973 to 1978 in all OECD regions (see Table 2.3). However, from 1978 to 1981 the share of coal in total energy inputs increased in all regions except North America, where industrial coal use in the United States declined by 14%. Japan showed a significant increase in industrial coal use, mainly in the cement industry, during this same period.

The continuous, and probably irreversible, decline in coal use in the residential sector is mainly due to the inconvenience of handling coal and the environmental problems associated with it.

Table 2.2. **OECD Coal Use in Electricity Generation**

	1973		1978		1981	
	Mtoe	% Share[1]	Mtoe	% Share[1]	Mtoe	% Share[1]
OECD TOTAL	382.7	36.6	461.2	34.4	555.4	39.3
NORTH AMERICA	229.8	43.3	282.1	39.4	347.9	46.7
PACIFIC	26.5	15.8	32.8	15.6	38.1	18.2
EUROPE	126.4	38.4	146.3	37.8	169,4	40.5
Canada	8.2	12.9	12.3	15.0	15.0	16.1
United States	221.6	47.2	269.9	42.9	332.9	51.4
Japan	12.2	7.9	12.8	5.5	15.2	8.1
France	10.7	19.9	17.4	28.5	17.4	21.8
Germany	48.0	64.5	52.7	57.5	58.5	61.9
Italy	1.6	4.5	2.7	6.6	5.7	12.4
United Kingdom	50.5	61.9	48.5	65.4	51.8	74.5

Source: IEA/OECD Energy Balances

1. Share of electricity generation produced from solid fuels

Table 2.3. OECD Coal Use in Industry

	1973		1978		1981	
	Mtoe	% Share[1]	Mtoe	% Share[1]	Mtoe	% Share[1]
OECD TOTAL	220.7	20.1	189.6	17.8	191.1	19.7
NORTH AMERICA	91.9	18.1	87.1	17.1	78.6	17.0
PACIFIC	44.4	25.4	39.1	21.2	47.1	28.2
EUROPE	84.4	20.5	63.4	17.0	65.5	19.0
Canada	5.8	12.0	10.1	18.4	12.2	19.5
United States	86.1	18.7	76.9	16.9	66.4	16.6
Japan	35.8	23.3	29.9	18.9	38.2	27.0
France	12.3	20.2	9.3	16.0	8.6	16.0
Germany	24.1	25.0	16.3	20.5	19.0	24.9
Italy	4.6	9.0	4.5	9.7	5.4	12.2
United Kingdom	17.1	23.8	10.7	18.2	7.5	15.9

Source: IEA/OECD Energy Balances

1. Solid fuels share of total energy use in industry, including non-energy use of petroleum products.

2.2. Current Situation

Preliminary data for 1982 (see Table 2.4) indicate a clear break in the steady growth in total OECD coal use observed since 1978, both in absolute terms and in share of total primary energy requirements. Overall coal use declined by 2.3% from 825.8 Mtoe in 1981 to 806.7 Mtoe in 1982, while the share of coal in total primary energy requirements remained relatively constant (22.8% in 1982 compared with 22.6% in 1981). As a matter of reference, total primary energy requirements (TPER) declined by 3.5% from 1981 to 1982.

Table 2.4. OECD Current Coal Use [1]

	1981		1982*	
	Mtoe	% Share [2]	Mtoe	% Share [2]
OECD TOTAL	825.8	22.6	806.7	22.8
NORTH AMERICA	435.8	21.9	413.0	21.8
PACIFIC	104.3	23.1	108.9	24.3
EUROPE	285.7	23.5	284.8	24.0
Canada	27.0	12.0	29.4	14.2
United States	408.9	23.1	383.7	22.7
Japan	69.1	19.1	68.2	19.3
France	32.7	16.7	30.5	16.4
Germany	86.4	32.8	86.0	33.7
Italy	14.4	10.3	14.7	11.0
United Kingdom	67.8	34.8	63.8	32.9

Source: IEA/OECD Energy Balances

*   Preliminary data
1.   Total primary solid fuels requirements.
2.   Share of solid fuels in total primary energy requirements.

This total picture does, however, cover some significant differences in the regional and country picture as seen in Table 2.4. The decline in the absolute level was most significant for North America with the share of coal use remaining relatively constant, while the Pacific region experienced an increase in coal use both in absolute and relative terms. The absolute level of coal use in Europe remained practically constant with a slight increase in the relative share of coal in total primary energy.

2.3. Future Prospects

Table 2.5 presents projections of OECD coal demand by region. The figures for 1985 and 1990 are based on 1982 government submissions to IEA.

Table 2.5.   **OECD Present and Projected Coal Use**[1]
(Mtoe)

	1981	1982*	1985	1990
OECD TOTAL	826	807	976	1176
NORTH AMERICA	436	413	535	640
PACIFIC	104	109	130	161
EUROPE	286	285	311	374

Sources: IEA/OECD Energy Balances, IEA Country Submissions (1982).

*   Preliminary data
1.   Total Primary Solid Fuels Requirements

The use of coal in electricity generation in the OECD is projected to increase by 6% from 1981 to 1985 and then by an accelerated 22% from 1985 to 1990 (see Table 2.6). However, in all regions except the Pacific region, the share of electricity generated by coal is expected to decrease slightly during the 1980's and the share in the total OECD will remain almost constant at about 40 per cent. These declines are probably due to projected large expansions of nuclear capacity in many OECD countries. In particular, France and the United Kingdom both forecast declines throughout the 1980's in amounts of coal used for electricity generation, as well as in the share of electricity generated by coal. However, Japan projects a steady growth in coal use for electricity generation in the 1980's and Italy forecasts a sharp increase in coal-fired generation from 1985 to 1990. These two countries now depend heavily on oil for electricity generation. Both plan, however, to construct new coal-fired units.

Coal use in industry is anticipated to increase from 1981 levels by 32% in 1985 and 54% in 1990 and the share of coal in energy use in this sector will rise from 20% to 25%. This would imply an increase of coal use of about 160 Mtoe over the 1980's (see Table 2.6). In particular, Canada, the United States, France and the United Kingdom project significant increase in coal's share of energy inputs to the industrial sector. These projections probably underestimate the economic and technical potential but may still be optimistic in view of the investment difficulties and institutional problems that need to be overcome.

Table 2.6. **OECD POTENTIAL COAL USE BY SECTOR** [1]

	1981		1985		1990	
	Mtoc	% Share[1]	Mtoe	% Share[1]	Mtoe	% Share[1]
Electricity Generation						
OECD TOTAL	555.4	39.3	587.1	38.5	718.8	39.4
NORTH AMERICA	347.9	47.7	371.7	46.7	445.3	46.9
PACIFIC	38.1	18.2	52.9	21.7	73.3	24.1
EUROPE	169.4	36.3	162.5	33.0	200.3	34.6
Canada	15.0	16.1	22.7	17.3	24.3	15.6
United States	332.9	51.4	349.0	51.6	421.0	52.3
Japan	15.2	8.1	26.3	12.1	40.3	14.6
France	17.4	21.8	11.3	14.4	8.5	9.2
Germany	58.5	61.9	52.0	51.1	60.0	50.8
Italy	5.7	12.4	6.7	11.1	22.4	31.8
United Kingdom	51.8	74.5	47.0	71.6	44.0	65.9
Industry						
OECD TOTAL	191.1	19.7	299.6	24.0	346.6	24.7
NORTH AMERICA	78.6	17.0	152.8	23.6	171.7	23.7
PACIFIC	47.1	28.2	59.5	30.7	70.6	31.5
EUROPE	65.5	19.0	87.3	21.4	104.3	23.0
Canada	12.2	19.5	16.8	23.5	20.7	24.4
United States	66.4	16.6	136.0	23.9	151.0	23.7
Japan	38.2	27.0	47.8	29.5	57.9	30.6
France	8.6	16.0	13.6	22.5	18.8	27.3
Germany	19.0	24.9	16.0	18.2	17.0	18.7
Italy	5.4	12.2	7.1	14.0	7.6	13.8
United Kingdom	7.5	15.9	11.0	20.4	12.0	20.7

Sources: IEA/OECD Energy Balances and IEA Country Submissions (1982)

1. For electricity generation, this represents the share of total electricity generated by solid fuels. For industry, this represents solid fuel's share in total inputs.

Demand for coking coal has historically tended to fluctuate around 250 million metric tons since the mid-1960's despite the expansion in crude steel output. This has been largely due to the modernization of steel production and the growing use of fuel oil as a supplemental heat source.

In the OECD countries, about 371 Mtoe of oil (including non-energy use) was used in 1981 in the industrial sector. A large share of this oil was used for process steam, direct heat and space and water heating. These are the areas where coal could gradually replace oil. The use of solid fuels in the industrial sector, including iron and steel, was about 191 Mtoe in 1981. A recent study by the CIAB* suggests that the technical potential for converting from oil to coal in industry is significantly greater than had previously been assumed and that steam coal use in the industrial sector could reach as much as 490 Mtoe by 2000. However, the full technical potential is not likely to be reached because of financial, environmental and other difficulties in converting from oil to coal which can delay decisions to convert. Based on other industry estimates, an increase of steam coal use in the industrial sector to about 245 Mtoe by 2000 (50% of the CIAB potential) appears more in line with the High Demand Scenario in the WEO**.

Synthetic fuels from coal are not expected to be developed on any commercial scale in the 1980's but could have some impact on coal demand in the 1990's if oil prices increase significantly. The current slackening in world oil prices exerts a dampening effect on synthetic fuel projects. Although these projects are not all cancelled some of them are shelved for an indefinite period.

Coal use in the residential sector is projected to remain at about the current level of 41 Mtoe. The direct use of solid fuels in the residential/commercial sector is anticipated to decrease toward the end of the century, but the development of district heating could add some coal use in this sector. Increasing solid fuel use for the residential sector is most likely to occur indirectly through electricity generation.

---

*   CIAB, *The Use of Coal in Industry*, (IEA/OECD, Paris 1982).
**   *World Energy Outlook* (IEA/OECD, Paris 1982).

3. COAL PRODUCTION

3.1. Historical Overview

Coal production trends have varied considerably between regions as indicated by Table 3.1. While world total coal output grew by 16.6% between 1960 and 1973, total OECD production increased only slightly over the same period (2.9%). In contrast, coal production in the Centrally Planned Economies (CPE) grew by 23.3% and production in the rest of the world increased by 66.0%. Thus, virtually all of the growth in world coal production between 1960 and 1973 took place outside OECD. The CPEs were by 1960 and remain today the world's largest coal-producing group, accounting for almost 55% of total world coal production, while the share of OECD coal production in world total coal output declined from 41.3% in 1960 to 36.4% in 1973.

From 1973 to 1981 total world coal output grew by 25%. Coal production in the OECD during this period showed a comparable increase of 26.6 %, a significant acceleration from earlier trends. Thus, the share of OECD coal production in world total coal output has remained relatively constant around 36% since 1973.

Table 3.1. **World Total Coal Production by Regions**
(Million metric tons)

	1960	1973	1978	1981
World Total	2600.5	3032.8	3494.8	3791.2
OECD	1073.2	1104.4	1185.9	1398.4
CPE	1420.2	1750.6	2068.7	2081.3
All others	107.1	177.8	240.2	311.5

Source: IEA/OECD Coal Statistics and U.N. Energy Statistics.

From 1960 to 1973, total coal production in North America increased by 34.3%, while coal production in Europe fell by 27.0%, mostly because of the progressively reduced competitiveness against oil of high-cost European coal. During this period, coal production in the United Kingdom was almost halved, while German production declined at a slower pace. For similar reasons coal production in Japan was also reduced sharply. In Australia there was strong growth, with output more than doubling. This expansion was due partly to coking coal exports to Japan.

Since 1973, OECD coal production has increased progressively and was in 1981 30.3% above its 1960 level or 26.6% above its 1973 level. Japanese output has been stable at around 18 million tons for the last few years (about 30% below

1973 level). European production has increased in the last few years with 1981 production some 8% above 1973 level, due to increases in brown coal tonnage production which more than offset declines in European hard coal production. Output in the United States and Canada has increased at annual average rates of 2.8% and 8.7% respectively. Australian coal production also has continued to expand at a high rate of 5.1% per year since 1973.

The overall coal production trends as indicated by Table 3.1 cover both Hard Coal and Brown Coal. A more detailed picture is given in Tables 3.2 and 3.3 for Hard Coal and Brown Coal respectively.

Table 3.2. **Hard Coal Production by Regions/Countries[1]** (Million Metric Tons)

Regions/Countries	1960	1973	1978	1980	1981
TOTAL WORLD	1966.9	2183.0	2545.7	2718.0	2750.2
OECD [2]	935.8	910.5	940.9	1093.3	1097.9
Canada	6.4	12.3	17.1	20.2	21.7
United States	391.5	529.6	576.8	710.4	698.1
Australia	21.9	55.5	71.8	76.6	92.1
Germany	148.0	103.7	90.1	94.5	95.5
United Kingdom	196.7	130.2	121.7	128.2	125.3
Other OECD	171.3	79.1	63.3	63.6	65.1
NON-OECD [3]	1031.1	1295.1	1604.8	1634.8	1642.3
AFRICA	39.7	68.4	95.9	121.7	136.9
Botswana	—	—	0.3	0.4	0.4
South Africa	38.2	62.4	90.6	116.0	131.2
Zimbabwe	—	3.5	3.1	3.1	2.9
Other Africa	1.5	2.5	2.0	2.2	2.4
ASIA	486.6	557.1	756.9	767.6	782.2
China	420.0	430.0	593.0	594.0	593.0
India	52.6	77.9	101.5	109.1	123.1
Other Asia	14.0	49.2	62.3	64.5	66.1
USSR	355.9	461.2	501.5	492.9	510.0
EAST EUROPE	141.5	196.7	233.0	233.7	202.6
Czechoslovakia	26.2	27.8	29.2	28.3	27.2
Poland	104.4	156.6	192.6	193.1	163.0
Germany (East)	2.7	0.8	0.1	—	—
Other East Europe[4]	8.1	11.5	11.1	12.2	12.4
CENTRAL & SOUTH AMERICA	7.3	11.6	17.5	19.0	20.6
Colombia	2.6	3.0	4.8	5.3	5.3
Mexico	1.8	4.3	6.8	7.0	8.1
Venezuela	—	0.1	0.1	—	—
Other C & S America	2.9	4.3	5.9	6.6	7.2

1. Hard coal includes anthracite and bituminous coal. For further information, see Principles and Definitions in Part II.
2. Source: IEA/OECD Coal Statistics.
3. Source: United Nations Energy Statistics.
4. Includes Bulgaria, Hungary, Rumania and Yugoslavia.

OECD hard coal production fell by 2.7% from 1960 to 1973. Production then increased by 3.2% from 1973 to 1978 but accounted for less than 10% of the total world production increase during this period. However, in the years since 1978, OECD production has increased by 17%, representing about 75% of the total world increase between 1978 and 1981. Substantial production increases after 1973 were recorded in Australia and the United States which more than compensated for the reduction of two major OECD European producers (Germany and the United Kingdom). The bulk of the increase in non-OECD production from 1973 to 1981 was assumed by China (163 Mt), the U.S.S.R. (49 Mt), India (45 Mt) and the Republic of South Africa (69 Mt). Together they represent about 50 per cent of the total world hard coal production increase between 1973 and 1981 (567 Mt).

The increase in world brown coal production was more modest than for hard coal. Out of a total increment of 214 million metric tons about half was assumed by the OECD area. Another third of this increase occurred in Eastern Europe and the USSR with the balance in China and other Asian countries.

Table 3.3 **Brown Coal/Lignite Production by Regions/Countries** [1] (Million Metric Tons)

Regions/Countries	1960	1973	1978	1980	1981
TOTAL WORLD	633.7	827.3	949.1	968.1	1041.0
OECD [2]	137.4	194.0	245.0	281.4	300.5
Canada	3.3	8.1	13.3	16.5	18.3
United States	2.5	13.4	31.2	42.3	46.4
Australia	14.2	24.1	30.5	32.8	33.0
Germany	98.0	118.7	123.6	129.9	130.6
Spain	1.8	3.0	8.3	15.5	20.9
United Kingdom	—	—	—	—	—
Other OECD	17.5	26.7	38.2	44.4	51.3
NON-OECD [3]	496.3	633.3	704.1	723.6	740.5
AFRICA	—	—	—	—	—
ASIA	4.6	12.9	41.6	46.0	48.9
China	—	—	24.9	26.0	27.0
India	—	3.3	3.6	4.5	6.0
Other Asia	4.6	9.6	13.1	15.4	15.9
USSR	134.2	153.5	162.9	159.9	160.0
EAST EUROPE	357.4	466.9	499.7	517.6	531.6
Czechoslovakia	58.4	81.2	94.9	94.9	95.2
Poland	9.3	39.2	41.0	36.9	35.6
Germany (East)	225.5	246.2	253.3	258.1	267.0
Other East Europe[4]	63.9	99.4	110.5	127.8	133.8
CENTRAL & SOUTH AMERICA	0.1	0.1	—	—	—

1. Brown coal represents lower grade coal and includes lignite. For further information, see Principles and Definitions in Part II.
2. Source: IEA/OECD Coal Statistics.
3. Source: United Nations Energy Statistics.
4. Includes Bulgaria, Hungary, Rumania and Yugoslavia.

The distinction between coking coal and steam coal is important in projecting coal demand and for related production and trade forecasts, because the big producers of coking coal are limited in number and its big consumers are major steel-producing countries. Of total OECD hard coal production of 1097.9 million metric tons in 1981, steam coal accounted for 76% or 834.3 million metric tons (See Table 3.4). Coking coal production was 263.6 million metric tons.

From 1978 to 1981, coking coal production increased by 12.2% while steam coal production increased by 18.2%.

Table 3.4 **OECD Hard Coal Production By Type** (Million metric tons)

	Coking Coal			Steam Coal		
	1978	1980	1981	1978	1980	1981
OECD TOTAL	234.9	260.9	262.6	705.9	832.5	834.3
Canada	13.8	14.1	15.1	3.4	6.0	6.6
United States	87.7	117.7	114.5	489.1	592.7	583.6
Australia	43.4	42.1	50.5	28.4	34.4	41.6
New Zealand	0.2	0.2	0.3	1.8	1.8	1.8
Japan	8.7	6.9	6.2	9.9	11.1	11.5
Germany	52.2	56.0	57.2	37.9	38.5	38.4
United Kingdom	15.1	10.1	7.9	106.6	118.2	117.4
Other Europe	13.9	13.6	11.8	28.9	29.8	33.4

Source: IEA/OECD Coal Statistics.

3.2. Current Situation

Preliminary data for 1982 (see Table 3.5) indicate that total OECD coal production increased by 0.7% from 808.2 Mtoe in 1981 to 813.7 Mtoe in 1982.

Table 3.5. **OECD Current Coal Production** [1]

	1981		1982*	
	Mtoe	% Share [2]	Mtoe	% Share [2]
OECD TOTAL	808.2	100.0	813.7	100.0
NORTH AMERICA	488.8	60.5	490.0	60.2
PACIFIC	84.1	10.4	89.3	11.0
EUROPE	235.3	29.1	234.3	28.8
Canada	27.9	3.5	29.0	3.6
United States	460.9	57.0	461.1	56.7
Australia	69.1	8.5	74.3	9.1
Japan	12.9	1.6	12.8	1.6
France	14.8	1.8	13.7	1.7
Germany	93.0	11.5	92.9	11.4
United Kingdom	73.2	9.1	71.5	8.8
Other OECD	56.4	7.0	58.4	7.2

Source: IEA/OECD Energy Balances.

*   Preliminary data
1.   Indigenous production of total solid fuels.
2.   Share of total OECD production.

Following the reduction in OECD coal demand observed in Section 2.2 and an increase in net imports (see Chapter 4) the OECD area experienced a significant stock build during 1982.

The regional and country specific distribution of 1982 coal production as shown in Table 3.5 further confirms some of the major historical trends observed above. This implies a further increase in the share of total OECD coal production in the Pacific region, notably in Australia, a further decline, even though small, in the European share of total production and a more or less constant share of about 60% for North America.

3.3. Future Prospects

Given the amount of accessible coal in significant coalfields both within the OECD area as well as in non-OECD coal-producing areas, ultimate availability of coal is very large. Ample supply potential exists in existing and new mines in Australia, Canada and the United States which could meet expanded coal demand. However, massive development of new coal production capacity and construction of the necessary infrastructure for coal transportation requires long lead times and without assured or firm expectations of substantial continuing growth in coal demand, the necessary investments may not be made. Therefore, actual coal production capacity will largely be determined by demand for coal in the OECD area and in particular in the electricity sector. Keeping pace with the estimated gradual growth in demand for coal, total coal production in the OECD area is estimated to increase by 14% in 1985 from the 1981 level. Between 1985 and 1990 OECD total coal production is projected to grow by another 20% (see Table 3.6).

During the next ten years, the *United States* will continue to be the largest single coal producer providing 60% of total OECD coal output. The country has available coal reserves large enough to support rapidly expanding domestic and export needs.

Australia's share could also increase from 7% of OECD output in 1981 to 12% by 1990, providing 25% of incremental coal production. Ample potential for expanded production from existing and new mines exists with the commercial groups involved seeking long-term contracts. Currently, a number of new mine development plans are under way in New South Wales and Queensland. The major constraint on developing new mines is the current reluctance of consumers to enter into long-term commitments with adequate pricing arrangements which would provide the basis for the investment to develop new mines and associated infrastructure.

Canada is projected to increase total coal production (including lignite) from 21 Mtoe in 1981 to 56 Mtoe in 1990. Coal resources are estimated to be adequate to supply an increasing domestic demand and an expanded export market. A major factor is that the main coal resources are located in the middle west of the country, separated from current domestic markets in the east by long distances and from western exporting terminals by difficult terrain underscoring the importance of mastering future transport cost development.

Table 3.6. **OECD Projected Hard Coal Production**
(Mtoe)

	1978	1981	1985	1990
Coking Coal				
OECD TOTAL	153.81	173.57	196.58	202.34
Canada	9.23	9.62	18.10	20.00
United States	56.10	73.23	84.00	84.00
Australia	28.13	32.74	37.00	40.00
New Zealand	.15	.23	.45	1.00
Japan	6.33	4.56	4.40	4.60
Germany	35.68	39.04	37.00	37.00
United Kingdom	8.69	4.53	4.40	4.40
Other Europe	9.50	9.62	11.23	11.34
Steam Coal				
OECD TOTAL	449.10	529.43	586.86	711.72
Canada	2.25	4.18	7.20	19.90
United States	312.76	373.18	417.00	514.00
Australia	18.44	26.97	44.80	53.70
New Zealand	1.42	1.28	1.80	2.65
Japan	7.23	8.37	9.00	9.30
Germany	25.86	26.22	24.00	25.00
United Kingdom	61.29	67.52	64.00	62.00
Other Europe	19.85	21.71	19.06	25.17

Sources: IEA/OECD Coal statistics and IEA Country Submissions (1982)

Note: In many cases projections of coking and steam coal are IEA Secretariat estimates. For further details, see Table B1 for individual countries in Part II.

Total coal production in *OECD Europe* is projected to show a moderate increase from 213 Mtoe in 1981 to 231 Mtoe in 1990, as Spain, Greece and Turkey foresee a substantial expansion of output, mainly of lignite. Most other countries are estimated to maintain their current production level. *Japan*'s production level is estimated to be around 13 Mtoe over the next ten years.

A number of *non-OECD* countries have the potential capability to supply future coal demand. The Republic of South Africa has large reserves available at competitive cost in the international market and is expected to increase its exports. Colombia is developing a coalfield near the Atlantic coast and is a potential big supplier to OECD Europe in the 1990's. Botswana has geographically favourable coal seams and could enter the international market toward the end of the century, depending on solutions for the transportation issue. In Poland, however, investment programmes for increases in capacity have stopped and this will have severe effects on future export capability. The exports of China, mainly to Japan, are likely to remain at a relatively low level at least for the current decade. The USSR's exports are also expected to remain marginal and continue to be mainly inside the Eastern bloc. India is developing an ambitious coal programme but the export capability would be small because of large projected domestic needs.

4. COAL TRADE

4.1. Historical Developments

In 1960, most coal trade was intra-regional. Over half of the world's total coal trade in 1960 remained within Europe and the USSR, as shown in Table 4.1. Germany was the major exporter to the OECD Europe countries while Poland and the USSR were the major suppliers of the Eastern European countries. There was also a considerable coal flow from the United States to Canada. The only significant seaborne coal trade moved from the United States to OECD Europe and from the United States to Japan. OECD coal trade accounted for about 75 % of a total world trade of 132 million metric tons. A substantial share of non-OECD trade was intra-regional trade in Eastern Europe and the U.S.S.R. (28 million metric tons).

Table 4.1. **World Coal Trade in 1960; Total Solid Fuels**
(Million Metric Tons)

To:	North America	OECD Europe	Japan	OECD Total	East Europe	All Others[1]	Total Export
From:							
Canada	0.3	—	0.6	0.9	—	—	0.9
United States	11.9	15.0	5.2	32.1	0.5	2.2	34.8
Australia	—	—	1.4	1.4	—	0.3	1.7
Germany	—	28.2	—	28.2	0.5	0.4	29.1
United Kingdom	0.1	6.6	—	6.7	—	—	6.7
South Africa	—	0.1	—	0.1	—	0.9	1.0
Poland	—	8.1	—	8.1	11.9	0.9	21.0
Czechoslovakia	—	1.5	—	1.5	2.7	0.1	4.3
USSR	—	3.8	0.5	4.3	10.3	0.1	14.7
China	—	—	—	—	—	—	—
All Others	—	12.6	0.8	13.4	1.9	2.5	17.8
Total Imports	12.3	75.9	8.5	96.7	27.8	7.4	132.0

Source: UN Energy Statistics.

1.   Includes statistical difference.

Between 1960 and 1973, OECD gross imports of hard coal increased by 77% from 75.2 to 133.3 million metric tons (see Table 4.2), while total OECD coal exports increased by 63% from 65.8 to 107.5 million metric tons. Thus the OECD area as a whole increased its net imports from 9.4 million metric tons in 1960 to 25.8 million metric tons in 1973. Intercontinental coal trade became increasingly important to OECD countries, largely as a result of the growing imbalance between coal demand and supply in Europe and Japan and the growing production in North America and Australia. Most of the coal internationally traded between 1960 and 1973 consisted of coking coal. The United States was the world's largest exporter of coal by 1973, accounting for 40% of total OECD gross exports.

Between 1973 and 1980, OECD gross imports of hard coal increased further by 47% to 195.4 million metric tons. Australian exports expanded rapidly, from 28.1 to 42.9 million metric tons, but the United States continued to be the largest single coal exporter. Mainly due to the net import requirements in OECD Europe, 57.6 million metric tons of OECD gross imports were from non-OECD countries in 1980, in particular from Poland and the Republic of South Africa.

Table 4.2. **OECD Historical Hard Coal Imports and Exports**
(Million metric tons)

	1960		1973		1980	
	Import	**Export**	**Import**	**Export**	**Import**	**Export**
OECD TOTAL	75.2	65.8	133.3	107.5	195.4	160.2
North America	11.4	35.2	15.2	59.9	16.9	98.5
Oceania	—	1.2	—	28.1	—	42.9
Japan	8.7	—	58.0	—	68.6	0.1
Europe	55.1	29.3	60.1	19.4	109.9	18.8

Source: IEA/OECD Coal Statistics.

4.2. Current Trade Patterns

The current trading pattern of OECD countries has emerged largely in response to inter-regional transfers of coking coal which still dominates OECD coal trade. In 1981, however, steam coal represented 43% of total OECD gross hard coal imports, compared with 35% in 1978 (see Table 4.3).

Table 4.3 shows that the entire OECD was a net exporter of coking coal and a net importer of steam coal in 1981. Australia and the United States are the dominant OECD exporters of both coking and steam coal. Japan accounts for more than half of OECD coking coal imports and OECD Europe dominates steam coal imports.

In the United States, the share of steam coal in total production is particularly high, 84% of total coal production, due to the strong domestic demand from electric utilities. A small but growing portion of steam coal production is exported to Canada, Europe and other countries representing 42% of total coal exports from the United States in 1981. The United States doubled exports of coking coal and quadrupled exports of steam coal between 1978 and 1981.

About half of the United States coking coal output is exported, mainly to Europe and Japan. Preliminary 1982 figures indicate a slight regression in United States coal exports, attributable to lower steam coal exports reflecting both a slowdown in steam coal demand and strong competition resulting from surplus availability elsewhere.

Table 4.3. **OECD Hard Coal Trade by Type**
(Million metric tons)

	Coking Coal				Steam Coal [1]			
	Imports		Exports		Imports		Exports	
	1978	1981	1978	1981	1978	1981	1978	1981
OECD TOTAL	95.3	116.8	87.1	124.7	52.2	88.7	24.9	68.0
NORTH AMERICA	5.3	5.4	40.1	73.0	11.5	10.4	10.8	44.8
Canada	5.3	5.4	13.0	13.8	8.8	9.5	1.0	1.9
United States	—	—	27.1	59.2	2.7	0.9	9.8	42.9
PACIFIC	50.9	65.3	33.3	41.0	2.0	12.6	5.4	10.2
Australia	—	—	33.3	40.8	—	—	5.4	10.2
New Zealand	—	—	—	0.2	—	—	—	—
Japan	50.9	65.3	—	—	2.0	12.6	—	—
OECD EUROPE	39.1	46.1	13.7	10.7	38.7	65.7	8.7	12.9
Belgium	3.5	4.2	0.1	0.3	3.5	5.8	0.1	0.5
Denmark	—	—	—	—	6.1	10.9	—	—
France	8.3	10.2	—	—	15.1	17.3	0.4	0.7
Germany	0.7	0.7	13.4	9.2	6.2	10.1	5.4	2.8
Italy	10.0	10.8	—	—	2.5	7.7	—	—
Netherlands	2.9	3.2	—	—	2.2	4.7	0.5	0.9
Spain	3.1	3.9	—	—	0.4	3.2	—	—
Turkey	—	0.7	—	—	—	—	—	—
United Kingdom	1.4	2.6	0.1	1.0	1.0	1.7	2.1	8.1
Other Europe	9.2	9.8	0.2	0.2	1.7	4.3	0.2	—

Source: IEA/OECD Coal Statistics.

1.   Includes Anthracite

Canada exports significant amounts of coking coal to Japan and imports steam coal from the United States. For the first time in 1981, Canada was a net coal exporter. In Australia coal production is divided 45% and 55% between coking coal and steam coal respectively. Most of the steam coal is consumed domestically for electricity generation. Coking coal still comprises the bulk of the exports but steam coal has succeeded in increasing its share from about 12% in 1979 to about 25% in 1982. In OECD Europe, less than 20% of total coal production is coking coal and OECD Europe is a net importer of both coking coal and steam coal. The major European importers are France, Italy, Germany, Belgium, Denmark, the Netherlands and Spain. Germany exports some coking coal to neighbouring countries. The United Kingdom is basically self-sufficient, exporting a small amount, mainly steam coal, to other European countries.

The current world coal trade pattern shown in Table 4.4 is significantly different from that of 1960. Total world coal trade has more than doubled. Australia and South Africa have evolved from insignificant traders in 1960 to major exporters in 1981 and for 1982 this tendency is confirmed. Japan, which in 1960 represented only 6 per cent of total world imports, consumed almost 30 per cent of traded world coal in 1981 and is currently the largest coal importer.

Table 4.4. **World Coal Trade in 1981; Hard Coal**
(Million Metric Tons)

To:	North America	OECD Europe	Japan	OECD Total	East Europe	All Others[1]	Total Export
From:							
Canada	0.2	1.7	10.5	12.4	—	3.3	15.7
United States	14.8	48.2	23.8	86.8	2.4	12.9	102.1
Australia	—	8.5	34.6	43.1	0.4	7.5	51.0
Germany	—	11.5	—	11.5	0.1	0.4	12.0
United Kingdom	—	9.5	—	9.5	—	- 0.4	9.1
South Africa	0.7	20.2	4.3	25.2	—	4.0	29.2
Poland	—	6.6	0.1	6.7	7.1	1.2	15.0
Czechoslovakia	—	1.1	—	1.1	1.9	—	3.0
USSR	—	1.9	1.4	3.3	18.7	—	22.0
China	—	0.2	2.8	3.0	—	3.5	6.5
All Others	—	2.5	0.5	3.0	1.9	1.1	6.0
Total Imports	15.7	111.9	78.0	205.6	32.5	33.5	271.6

Source: IEA/OECD Coal Statistics/UN Energy Statistics/Secretariat Sources.

1.   Includes statistical difference.

About 76% of the internationally traded hard coal in 1981 was imported by OECD countries, mostly OECD Europe and Japan. Intra-European trade within the OECD amounted to about 23.0 million metric tons in 1981, mainly originating from Germany and the United Kingdom. Apart from the regional coal trade, OECD Europe mainly imported coal from the United States, South Africa, Poland and Australia. Japan imported 65.4 million metric tons of coking coal and 12.6 million metric tons of steam coal in 1981. While Japan's coking coal imports increased slightly from 1980 (5%), steam coal imports showed a significant increase from only 2.7 million metric tons in 1979 to 6.3 million metric tons in 1980 and to 12.6 million metric tons in 1981. A preliminary figure for 1982 indicates a further increase to about 14.4 million metric tons. Australia, the United States and Canada were the biggest suppliers to Japan. Canada imported both steam and coking coal from the United States. Of the remaining 60 million metric tons of internationally traded coal, about 25 million metric tons represent the traditional intra-Eastern European and U.S.S.R. trade. About 19 million metric tons are exported by Australia to South East Asia and by the United States to South America, South East Asia and North Africa.

On the export side, the United States is the largest exporter. In 1981 it exported 59.2 million metric tons of coking coal and 42.9 million metric tons of steam coal. Overseas shipments of steam coal soared from only 3.1 million metric tons in 1979 to 31.7 million metric tons in 1981 to compensate for the sudden

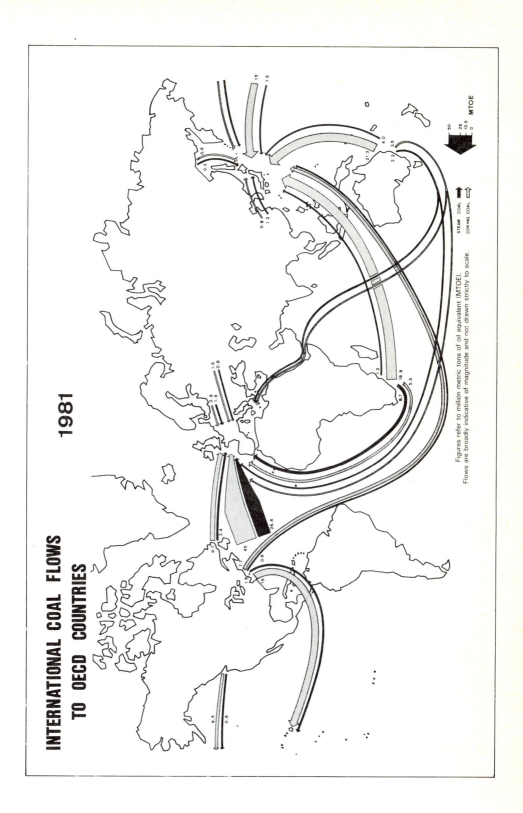

INTERNATIONAL COAL FLOWS
TO OECD COUNTRIES

1981

Figures refer to million metric tons of oil equivalent (MTOE).
Flows are broadly indicative of magnitude and not drawn strictly to scale.

MTOE

STEAM COAL
COKING COAL

decrease of Polish exports to OECD Europe. In Australia, total coal exports increased by 19% to 51.0 million metric tons in 1981, of which coking coal accounted for 80% (40.8 million metric tons). Its main client has been Japanese steel makers but currently many projects are under way or under negotiation to meet the steam coal demand from the cement industry and utilities in Japan and Europe. Canada exported 15.7 million metric tons of coal in 1981 of which almost 70% was sent to Japan. Outside of the OECD, the Republic of South Africa exported a record level of 29.2 million metric tons, slightly above the 1980 level but almost 30% above the 1979 level. On the Polish side, exports declined sharply from 40.6 million metric tons in 1979 to 30.9 million metric tons in 1980, due to reduced production by the temporary introduction of a 5-day working week and priority given to the domestic market. During 1981, the situation deteriorated further to an export level of only 15.0 million metric tons and traditional importers of Polish coal in OECD Europe were forced to search elsewhere. During 1982 Polish exports showed a substantial comeback with a total hard coal export level of about 28 million metric tons. As total world exports of coking and steam coal increased only slightly, the consequence was either stagnation or modest downard results for other traditional exporters. USSR's trade was mainly in the Eastern bloc and has declined by about 10% since 1979.

4.3. Current Situation

Table 4.5. **OECD Current Coal Trade** [1]
(Mtoe)

	Imports		Exports		Net Imports	
	1981	1982*	1981	1982*	1981	1982*
OECD TOTAL	154.57	150.95	134.45	124.84	20.12	26.11
NORTH AMERICA	10.55	11.20	76.02	72.06	- 65.47	- 60.86
Canada	9.63	10.69	10.02	9.76	- 0.51	0.93
United States	0.92	0.51	66.00	62.30	- 65.08	- 61.79
PACIFIC	57.01	56.77	34.56	32.75	22.45	24.02
Australia	0.02	—	33.09	32.33	- 32.07	- 32.33
New Zealand	—	—	0.16	0.22	- 0.16	- 0.22
Japan	56.99	56.77	1.30	0.20	55.68	56.57
OECD EUROPE	87.01	82.98	23.87	20.03	63.14	62.95
Belgium	7.91	7.78	1.11	0.74	6.80	7.04
Denmark	6.61	5.97	0.03	0.03	6.58	5.94
France	21.00	17.27	1.16	1.09	19.84	16.18
Germany	9.25	9.46	13.35	11.08	- 4.10	- 1.62
Italy	14.60	14.29	0.52	0.37	14.08	13.92
Netherlands	6.32	6.52	1.18	0.94	5.14	5.58
Spain	5.27	5.46	0.01	0.01	5.26	5.45
United Kingdom	2.54	2.65	6.20	5.49	- 3.66	- 2.84
Other Europe	13.51	13.58	0.31	0.28	13.20	13.30

Source: IEA/OECD Energy Balances.
*   Preliminary data.
1.   Trade in total solid fuels.

Preliminary data for 1982 (see Table 4.5) indicate a 2.3% reduction in OECD total coal imports from 1981 to 1982 and a 7.1% reduction in total coal exports. This brings net OECD coal import up to a level of 26.1 Mtoe in 1982 from 20.1 Mtoe in 1981 but still below the 1980 level of 32.5 Mtoe. It mainly reflects the current come back of Polish coal in international trade.

4.4. Future Prospects

Trade in steam coal is a relative novelty in international trade, since for many years only coking coal was traded regularly. Although coking coal trade is expected to increase in absolute terms over the 1980's, sluggish activity in the world steel industry is expected to produce a decline in its overall share of coal trade, while the attractiveness of coal for other energy purposes will increase. Future growth in coal trade will, therefore, be dominated by the expansion in steam coal.

Coal imports into the OECD are projected to grow substantially from 143.5 Mtoe in 1981 to 238.7 Mtoe in 1990 and could increase to about 500 Mtoe in 2000*, reflecting the growth in demand of steam coal in electricity generation in OECD Europe and Japan. During this period, the United States and Australia will continue to be the largest exporters. Projected OECD coal trade for 1985 and 1990 is shown in Table 4.6. These projections are based on submissions by Member countries. They are clearly subject to an even wider margin of uncertainty than the projections of coal consumption and production.

United States coal exports are projected to grow by 60% to 1990. Coking coal exports are, however, expected to decline slightly during this period, which means, in effect, that steam coal will account for the large increase in exports. Projections indicate that the United States will continue its role as the largest OECD exporter of both coking and steam coal to 1990. The immediate obstacle to such rapid expansion of coal exports is the status of coal handling installations.

Australia's exports are projected to reach 63 Mtoe by 1990. Coking coal exports are projected to increase only slightly but steam coal exports are expected to expand fivefold from 1981 to 1990. Coal production will not limit export expansion, as capacities of operating mines will be expanded and many new projects in New South Wales and Queensland should come on-stream by 1990. The main area of concern is the timely development of export handling capacities as well as their smooth operation, which was disturbed by strikes over the last two years.

The Canadian export coal market has developed in response to the Japanese demand for coking coal. Coking coal exports are projected to increase to

---

*   *World Energy Outlook* (IEA/OECD, Paris 1982)

Table 4.6.   OECD Projected Coal Trade
(Mtoe)

	Imports		Exports		Net Imports	
	1981	1990	1981	1990	1981	1990
Coking Coal						
OECD TOTAL	83.36	97.11	80.39	91.66	2.97	5.45
NORTH AMERICA	3.40	4.70	46.61	53.60	-43.21	-48.90
Canada	3.40	4.70	8.76	17.60	-5.36	-12.90
United States	—	—	37.84	36.00	-37.84	-36.00
PACIFIC	47.76	59.00	26.59	30.36	21.17	28.64
Australia	—	—	26.43	30.00	-26.43	-30.00
New Zealand	—	—	0.13	0.36	-0.13	-0.36
Japan	47.76	59.00	0.03	—	47.73	59.00
OECD EUROPE	32.20	33.41	7.20	7.70	25.00	25.71
Belgium	2.95	2.10	0.22	—	2.73	2.10
Denmark	0.01	—	—	—	0.01	—
France	7.15	6.20	—	—	7.15	6.20
Germany	0.45	1.00	6.30	7.00	-5.85	-6.00
Italy	7.97	8.00	—	—	7.97	8.00
Netherlands	2.24	4.10	0.02	—	2.22	4.10
Spain	2.70	3.10	—	—	2.70	3.10
Turkey	0.40	2.37	—	—	0.40	2.37
United Kingdom	1.50	1.40	0.60	0.60	0.90	0.80
Other Europe	6.83	5.14	0.06	0.10	6.77	5.04
Steam Coal						
OECD TOTAL	60.18	141.61	43.29	89.40	16.89	52.21
NORTH AMERICA	6.62	6.70	28.66	53.10	-22.04	-46.40
Canada	6.02	6.10	1.21	11.10	4.81	-5.00
United States	0.60	0.60	27.45	42.00	-26.85	-41.40
PACIFIC	9.23	38.90	6.65	32.90	2.58	6.00
Australia	—	—	6.61	32.90	-6.61	-32.90
New Zealand	—	—	0.04	—	-0.04	—
Japan	9.23	38.90	0.01	—	9.22	38.90
OECD EUROPE	44.32	96.01	7.98	3.40	36.34	92.61
Belgium	4.09	5.50	0.33	0.50	3.76	5.00
Denmark	6.52	7.90	—	—	6.52	7.90
France	12.10	11.60	0.49	—	11.61	11.60
Germany	6.92	11.00	1.89	—	5.03	11.00
Italy	5.72	24.00	—	—	5.72	24.00
Netherlands	3.32	6.70	0.61	0.50	2.71	6.20
Spain	2.23	6.40	0.01	—	2.22	6.40
Turkey	—	4.70	—	—	—	4.70
United Kingdom	0.97	1.00	4.64	2.40	-3.67	-1.40
Other Europe	2.45	17.21	0.01	—	2.44	17.21

Source: IEA/OECD Energy Balances and IEA Country Submissions (1982).

Note: Projections of coking and steam coal trade are in many cases IEA Secretariat estimates. For further details, see Table C1 for individual countries in Part II. Projections for the United Kingdom are based on more recent information from the United Kingdom authorities. Therefore, projections for the United Kingdom, OECD Europe and OECD Total do not correspond to previous estimates from the 1982 Country Submissions as published in the " Energy Policies and Programmes of IEA Countries, 1982 Review", IEA/OECD, Paris 1983. See Table 5/A3 for the United Kingdom in Part II.

17.6 Mtoe in 1990. Steam coal exports are, however, expected to expand from 1.2 Mtoe in 1981 to 11.1 Mtoe in 1990. Therefore, total coal exports could almost triple during the 1980's. The reserves of Canada's western surface coal deposits are more than adequate to supply anticipated demand. Coal throughput capacity at the port of Vancouver is expected to reach 35 million metric tons per year by the mid-1980's. Another new coal terminal is planned at Prince Rupert (B.C.) with an initial annual capacity of 12 million metric tons. With these expansions in place the port capacities should be more than adequate to handle the projected export levels by 1990. Other initiatives are under way to expand the rail capacity to handle the increased coal, grain, mineral and other commodities. There is also some interest in the concept of using slurry pipelines to transport coal to western Canadian ports for export markets.

Imports will continue to dominate OECD Europe coal trade. Germany will continue to be the only significant coking coal exporter for intra-European trade. In general, coking coal import levels are expected to remain relatively flat to 1990. However, steam coal imports to several European countries are projected to increase dramatically. Italy forecasts an increase of about 300 per cent in steam coal imports in the 1980's, as some of its predominantly oil-fired electricity generation system is planned to be converted to coal and new coal-fired power plants are planned. If these forecasts prove to be correct, Italy will become OECD Europe's largest steam coal importer. This would require a major effort in port expansion through the 1980's. Current plans for the 1980's indicate that development of new port capacity could handle the projected additional trade, if no delays in planned port expansion occur (see Table 7.3). Germany, the Netherlands, Spain and Turkey also forecast substantial increases in steam coal imports.

The Republic of South Africa has increased its exports from about 1 Mtoe before 1973 to a record level of 20 Mtoe in 1980. The development of the Richards Bay Harbour on the South African east coast made this possible. Development is under way to bring export levels to 31 Mtoe by the mid-1980's. Further expansion of coal exports was decided upon in September 1981 and could gradually reach a level of 56 Mtoe in the 1990's.

Colombia is developing a coalfield near the Atlantic coast with production expected to start in 1984 and reach 11 Mtoe in the early 1990's. Further development is projected toward the end of the century.

Poland's exports peaked in 1979 at 29 Mtoe but for the next five years Poland's exports to OECD Europe are unlikely to return to their 1979 level because Polish production is not expected to exceed the 1979 level and demand from within Poland and from other Eastern European countries (including the USSR) is likely to increase. If Poland continues to have problems in fulfilling its share as a traditional coal supplier, other major exporters could replace Poland's share in coal exports to the West. The USSR's exports are projected to remain mainly within the Eastern bloc region.

China's exports are likely to remain at a low level at least for the current decade, as the coal industry cannot make available for export more low sulphur coal.

Botswana has the potential to enter the international coal market toward the end of this century, depending on the solutions for the transportation issue. Coal exploitation in Botswana would need the construction of a 1,500 km railway and the upgrading of the port of Walvis Bay in Namibia.

The projected large potential increase in coal trade depends not only on potential growth in electricity and other industrial demand for coal but also on the ability of both exporting and importing countries to continue upgrading coal handling and inland transporation systems on time (see Chapter 7).

5. COAL PRICES

The distinct difference in price trends between coal and oil has been the main factor explaining the relative use of these fuels in the industrial sector and particularly for electricity generation. In addition to the incentive provided by relative price movements during the 1960's, environmental considerations and ease of handling also promoted the earlier shift away from coal, particularly given the less sophisticated technology available to control emissions from coal burning in the 1960's.

Following the considerable increase in oil prices after 1973, coal prices also tended to rise in all OECD countries, though at a slower rate than that of oil. In the United States real coal prices rose by about 70% from 1973 to 1980, while heavy fuel oil prices almost tripled over the same period. In other countries real coal prices rose by 4-8% per year between 1973 and 1980 compared with an 11-17% annual increase in real heavy fuel oil prices for industrial use. As a result, coal's competitive advantage increased dramatically after 1973 and this was further accentuated by the more recent increases in oil prices in 1979 and 1980. The price of internationally traded coal, however, increased rapidly during 1980 and 1981 due to the tight market caused by the sudden decrease of Polish exports and increased conversion from oil to coal. Since the beginning of 1982, however, world coal prices have declined in real terms due to lower demand under depressed economic conditions. Figure 5.1 compares steam coal and residual fuel oil prices for electricity generation, while Figure 5.2 compares coking coal and steam coal prices in the industrial sector in selected OECD countries.

Prices of coal vary significantly among OECD producing countries, due to differences in production costs and quality. Average coal export prices (FOB) for the major OECD coal exporters are shown in Table 5.1, while prices of delivered coking and steam coal to Japan and selected European countries are shown in Table 5.2. As can be seen from Table 5.2, prices of delivered coal also vary significantly depending on production and transportation costs, as well as terms of contract. Mine-mouth FOB costs account for about half of total delivered costs of imported coal. Though substantial differences exist, depending on specific trading patterns, roughly half of the non-production related costs are inland transportation costs to export terminals with the remaining half being maritime transportation and handling charges both at loading and unloading ports.

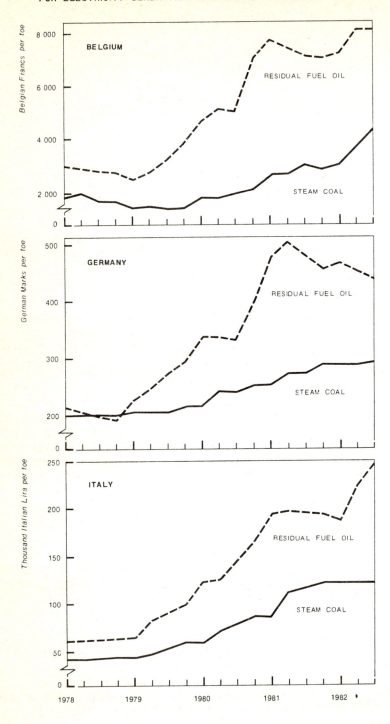

Figure 5.1 PRICES OF RESIDUAL FUEL OIL AND STEAM COAL FOR ELECTRICITY GENERATION IN SELECTED OECD COUNTRIES

Figure 5.1 Cont'd **PRICES OF RESIDUAL FUEL OIL AND STEAM COAL FOR ELECTRICITY GENERATION IN SELECTED OECD COUNTRIES**

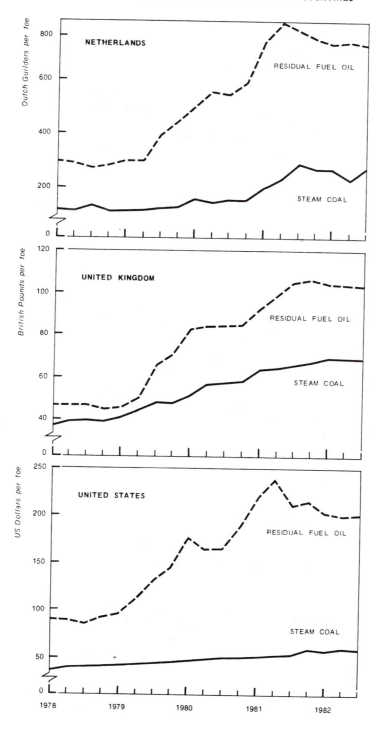

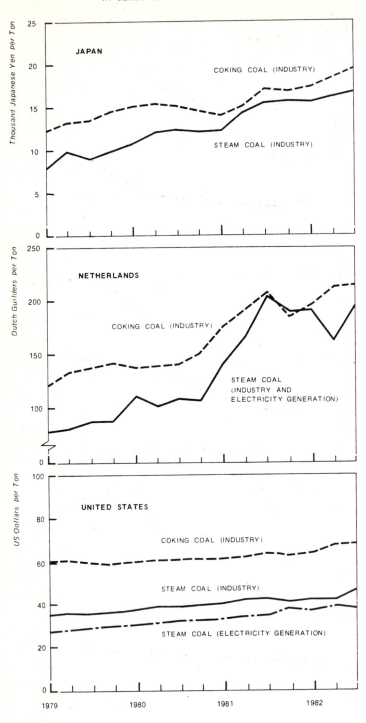

Figure 5.2 END-USE PRICES OF COKING COAL AND STEAM COAL IN SELECTED OECD COUNTRIES

JAPAN

COKING COAL (INDUSTRY)

STEAM COAL (INDUSTRY)

Thousand Japanese Yen per Ton

NETHERLANDS

COKING COAL (INDUSTRY)

STEAM COAL (INDUSTRY AND ELECTRICITY GENERATION)

Dutch Guilders per Ton

UNITED STATES

COKING COAL (INDUSTRY)

STEAM COAL (INDUSTRY)

STEAM COAL (ELECTRICITY GENERATION)

US Dollars per Ton

1979   1980   1981   1982

Table 5.1. **Selected Average Coal Export Prices (FOB)**[1]
in US$/metric ton

Countries/Coal Types	1980 Q1	1981 Q1	Q2	Q3	Q4	1982 Q1	Q2	Q3	Q4
Canada									
Hard Coal	44.26	48.16	53.22	53.80	57.87	57.11	64.63	61.76	62.69
United States									
Coking Coal	60.12	59.39	62.32	64.44	65.23	66.72	67.02	68.29	67.63
Steam Coal	44.50	47.86	49.58	51.45	51.74	54.46	54.44	54.17	52.97
Australia									
Coking Coal	40.43	42.54	47.88	50.72	50.40	56.33	55.25	57.78	58.89
Steam Coal	30.28	28.97	33.17	40.94	47.88	41.99	41.86	44.33	45.22

Sources: IEA/OECD Quarterly Coal Statistics.

1. Average unit values of coal export (FOB). The data have been computed from national foreign trade statistics. US$ exchange rates used are average monthly rates as compiled by the IMF.

Table 5.2. **Selected Average Coal Import Prices (CIF)** [1]
in US$/metric ton

Countries/Coal Types	1980 Q1	1981 Q1	Q2	Q3	Q4	1982 Q1	Q2	Q3	Q4
Japan									
Coking Coal	66.03	68.05	67.61	73.10	74.14	73.65	74.59	74.59	75.29
Steam Coal	54.17	59.09	64.64	66.31	68.73	66.36	65.84	64.22	63.71
Belgium									
Coking Coal	70.20	70.42	77.04	78.31	83.55	79.89	78.12	79.50	79.22
Steam Coal	44.61	53.34	52.83	54.12	54.17	56.53	58.00	57.37	56.95
Denmark									
Steam Coal	48.68	62.30	65.24	61.18	62.74	63.45	62.02	60.57	56.86
France									
Coking Coal	67.72	74.08	73.84	89.70	87.05	77.40	78.05	74.76	78.75
Steam Coal	51.82	59.47	60.92	60.09	58.21	55.88	57.98	53.30	51.75
Germany									
Coking Coal	68.10	65.16	61.06	69.49	78.98	78.86	79.05	75.36	73.97
Steam Coal	60.47	69.95	65.90	68.67	70.05	68.63	68.12	61.26	61.58
Italy									
Coking Coal	79.06	84.86	75.02	85.58	na	78.47	81.81	81.10	81.50
Steam Coal	49.26	54.91	51.19	74.06	62.70	64.20	65.90	67.47	65.48
Netherlands									
Coking Coal	70.89	76.65	74.75	78.11	77.78	76.11	80.75	78.02	79.99
Steam Coal	47.27	54.31	62.89	71.45	77.78	75.72	72.03	72.52	73.41

Sources: IEA/OECD Quarterly Coal Statistics.

1. Average unit values of coal import (CIF). The data have been computed from national foreign trade statistics. US$ exchange rates used are average monthly rates as compiled by the IMF.

The level of delivered prices resulting from deliveries performed under medium and long-term contracts was, for the first half of 1982, below the end-1981 level for both coking coal and steam coal. Some importers are covered by both medium and long-term supply and transportation contracts, while others have only supply coverage and use the voyage charter market for maritime transport, thus benefitting from the current low rates. Partly because of lower trade level, the loading conditions improved in both Australian and United States' ports which considerably lowered demurrage fees. At the same time, Poland was able to regain a prominent market share, offering substantial quantities at attractive prices. All these elements could more than offset some increases in the lowest FOB prices which occurred in most of the exporting countries for both qualities. Some FOB contract prices, negotiated in the course of the first boom for steam coal deliveries in 1980-81 and which were substantially higher, are currently being renegotiated towards lower levels.

However, the appreciation of the US dollar, particularly from the beginning of the second quarter 1982, has resulted in price increases for many importing countries, when expressed in national currencies, as can be seen from Table 5.3.

Table 5.3. **Index of Steam Coal Import Prices** [1]
Based on Prices in National Currencies (1980 = 100)

Countries	1981				1982			
	Q1	Q2	Q3	Q4	Q1	Q2	Q3	Q4
Japan	100.7	119.9	128.7	130.4	128.8	134.1	138.7	138.2
Belgium	137.2	149.5	164.1	155.6	178.6	198.7	207.6	211.2
Denmark	147.2	170.6	170.6	165.4	180.1	183.7	191.4	182.6
France	131.9	150.7	159.3	150.2	152.9	166.2	168.9	167.1
Italy	130.2	137.4	213.1	177.6	191.8	206.0	222.6	222.6
Netherlands	131.6	168.8	204.9	203.8	207.0	201.7	210.2	213.4

Source: IEA/OECD Quarterly Coal Statistics.

1.   Average unit value of imported steam coal (CIF). The data have been compiled from national foreign trade statistics.

Price will continue to be a determining factor of the pace at which coal use expands. Although current price relationships are such that coal enjoys a substantial competitive advantage, many potential purchasers are uncertain about how coal prices will evolve in the long run relative to oil prices and are therefore reluctant to undertake long-term commitments to switch to coal.

Although the price of coal relative to oil is critical to future fuel use decisions, it is not the only factor. Fuel use decisions must take into account not only the direct costs but also the indirect costs associated with the investment required to use coal, either directly in industry through installing new boilers or in the production of electricity through reconversion or the construction of new coal-fired plants.

6. COAL RESOURCES AND RESERVES*

The first comprehensive study of world coal resources was performed by the International Geological Congress in 1913. This exercise was not repeated until the World Power Conference issued its first Statistical Yearbook in 1929. These Yearbooks were superseded in 1960 by periodic Surveys of Energy Resources, now being produced by the successor organisation, the World Energy Conference (W.E.C.). The W.E.C. has developed the most widely used coal resource and reserve classifications. According to the W.E.C. standards, "proved reserves" represent the fraction of total resources that has been carefully measured and assessed as being exploitable in a particular country or region under present and expected future local economic conditions with current mining technology; "additional resources" embrace all coal-in-place, in addition to proved reserves, that at some time may acquire an economic value, while resources whose existence is entirely speculative are not included; "proved recoverable reserves" are the fraction of proved reserves estimated to be recovered (actually mined) from the coal fields in question under the above mentioned economic and technological limits. A summary of the latest W.E.C. findings (1980) is given in Table 6.1 and the more detailed principles and definitions are given in an annex to this chapter.

Table 6.1. **World Coal Resources and Reserves by Type of Coal**
(10^3 million metric tons)

	Bituminous Coal and Anthracite	Sub-Bituminous Coal	Brown Coal/ Lignite
Proved Reserves	774.6	221.6	323.5
of which: Proved Recoverable Reserves	487.7	143.0	251.1
Additional Resources	6161.4	3835.2	2159.6
Total Resources	6936.0	4056.8	2483.1

Source: World Energy Conference, Survey of Energy Resources 1980.

---

* This chapter was written with the assistance of the World Coal Resources and Reserves Data Bank Service of IEA Coal Research, London, who provided the factual information in it.

Total world coal resources, as indicated in Table 6.1, are so large that, if all were recoverable, they could provide thousands of years of supply at current consumption rates. These figures may be reassuring at first glance but they do not provide much information on possible constraints to coal development in this century.

The World Coal Resources and Reserves Data Bank Service (IEA Coal Research) has undertaken a detailed study of coal deposits throughout the world, identifying those considered capable of playing a significant part in world coal supply and trade within the next twenty years. For practical purposes, a period of 20 years is a reasonable planning horizon because:

1. Any coalfield which will make a significant impact on the world market in 20 years time must already be well known and reasonably accessible to existing facilities. Virgin fields remote from ports, railways and centres of population are unlikely to play a significant role within 20 years.

2. Any mining technology which will play a dominant part in 20 years time has already been invented and will take that long to spread across the industry.

For each of the "significant coalfields" (i.e. accessible to mining operations over the next 20 years) an estimate has been made of the amount of coal which is accessible for working by current technology. Therefore, these estimates refer to tonnages accessible within a twenty year period and do not relate directly to W.E.C. estimates of proved recoverable reserves for which there is no corresponding time limitation.

This survey of all the significant coalfields of the world is published in "Concise Guide to World Coalfields"* and a summary is given in Table 6.2, while the more detailed principles and definitions are given in an annex to this chapter.

A comparison of data in Tables 6.1 and 6.2 indicate that only 55 per cent of the proved recoverable reserves identified by the W.E.C. are considered by IEA Coal Research to be accessible over the next twenty years. This implies that actual reserve levels will not constrain development, though existing technology and economics do limit the availability of even proved recoverable reserves. According to Table 6.2, OECD reserves account for slightly over one third of total world accessible coal. The United States alone represents 21 per cent of total world accessible coal. However, one must keep in mind that the United States and the total OECD show large reserve estimates partly because these countries are well explored and resources are well identified. In future years, the non-OECD countries may greatly increase their share of total world accessible coal as the infrastructure is developed to mine and transport coal.

---

* *Concise Guide to World Coalfields* (IEA Coal Research, London 1983).

Table 6.2. Accessible Coal in Significant Coalfields by Regions/Countries and by Types of Coal
(Million metric tons)

	Bituminous Coal and Anthracite	Sub-Bituminous Coal	Brown Coal/ Lignite
TOTAL WORLD	357806	25205	100828
OECD	120379	23100	37073
Canada	1400	2000	1500
United States	73000	21000	8000
Australia	16000	-	10000
New Zealand	30	100	20
Japan	800	-	-
Austria	-	-	200
Belgium	800	-	-
France	350	-	40
Germany	13500	-	12000
Greece	-	-	1500
Spain	350	-	750
Turkey	100	-	3000
United Kingdom	14000	-	-
Other OECD	49	-	63
AFRICA	31592	105	20
Botswana	500	-	-
South Africa	30000	-	-
Zimbabwe	600	-	-
Other Africa	492	105	20
ASIA	89465	200	2160
China	72000	-	200
India	16000	-	650
Other Asia	1465	200	1310
USSR	81000	-	31000
EAST EUROPE	31220	-	30565
Czechoslovakia	1000	-	5000
Poland	30000	-	7000
Germany (East)	-	-	13000
Other East Europe	220	-	5565
CENTRAL & SOUTH AMERICA	4150	1800	10
Colombia	2000	-	-
Mexico	1000	-	-
Venezuela	750	-	-
Other C & S America	400	1800	10

Source: World Coal Resources and Reserves Data Bank Service (IEA Coal Research)

COAL RESOURCES AND RESERVES

PRINCIPLES AND DEFINITIONS*

A. EXPLORATION - PRINCIPLES AND METHODS

Many coal deposits have been easy to discover. In fact, coal was being taken from surface outcrops for use as a fuel before the dawn of history and there are still many places in the world where coal seams can be seen along river valleys and other steep hillsides. However before a large-scale modern mining undertaking can be established, it is necessary to know a great deal more about the available coal deposits.

Coal exploration is normally carried out in two distinct phases:

1. National Resource Assessment

Continuous long-term geological mapping and exploration is done by or for the government in order to maintain a national inventory for mineral exploitation policy and to control mineral leasing. The results of these investigations are normally published. In the case of coal, the results are summarised in periodic returns of national coal resources and reserves to the World Energy Conference.

The information collected will normally include:

(i) Seam thickness, depth, dip, folding, faulting and other features such as washouts or intrusions and other aspects of the geology of the seams.

(ii) Chemical analysis, including ash, moisture, sulphur content and other impurities; calorific value; coking properties, ash analysis and fusion temperature; petrographic analysis including maceral content and reflectivity.

---

* Source: World Coal Resources and Reserves Data Bank Service, IEA Coal Research, London

2. Project Reserves Estimation

A potential developer may, with the permission of the government authorities controlling the national minerals inventory, carry out detailed exploration within a clearly defined area with a view to devising a viable mining plan.

Information assembled for this purpose will of course be private to the organisation undertaking the survey, although it is normal for national authorities to demand copies of certain data which will be held by them on a confidential basis.

Reserve assessment calls for the combined judgement of experts in the fields of geology, mining engineering and marketing. Varying codes of practice have been developed, each designed specifically to serve the needs of a particular organisation. A fully comprehensive set of data will be required covering in greater detail the features outlined in (i) and (ii) in respect of national resource assessment with the addition of the following items:

(iii) Engineering properties of the rocks to be excavated; strength, fracture, hardness, abrasivity.

(iv) Hydrogeology i.e. permeability, water pressure and rate of flow at various horizons.

Recent developments in geophysics make it possible to deduce the overall structure at depth in a deposit with considerable accuracy from a limited number of boreholes combined with seismic surveys from surface stations.

From boreholes core samples are taken of the coal seams and host strata to be excavated for analysis and physical tests. If the coal is to be washed prior to disposal it may be necessary to obtain a larger sample by excavation or by sinking in a small shaft. Recent developments in the direct coupling of petrophysical logging to automated mapping have greatly reduced the need for core samples in many situations. Automated analytical methods together with computerised data handling and application of geostatistics now enable coal seam data to be evaluated with great speed.

B. RESOURCES AND RESERVES ASSESSMENT

1. Fundamental Principles

The term "resources" generally refers to quantities in the ground which are theoretically available almost irrespective of potential value. On the other hand, an estimate of "reserves" implies some economic evaluation, often very carefully defined in any particular instance. In respect of assessment of both resources and reserves there are certain basic common procedures:

(i) *Physical 'cut-off' limitations*

a) *Depth*

Coal basins have usually been identified first at points where seams occur at or near the surface. When circumstances are favourable to mining and as the industry develops a consensus of opinion within the country or region, they will conclude that coal below some specified depth is of no real interest and therefore should be ignored in any estimates of available tonnage. This depth limitation is normally applied not only to the reserves within reach of mining activity but also to the whole of the resources of the country or region.

In well developed mining areas, the depth limitation will reflect the ultimate capability of existing technology. New mining areas with large apparent resources relative to planned production will tend to be assessed in terms which ignore all coal below a much shallower cut-off point.

Furthermore, the depth to which local assessors are prepared to recognise the existence of resources (or reserves) depends on the value of the coal in question; cut-off being imposed at a shallower depth in respect of, say, lignite than in the case of bituminous coal.

b) *Thickness*

In the same way that within a country or region, current technology and local practice are both reflected in the choice of depth cut-off, so also these practical matters dictate the minimum thickness of seam which is to be recorded and assessed.

The economic value of the product may affect both depth and thickness cut-offs. In some countries the code includes thinner seams at greater depths for higher rank coals.

(ii) *Degree of certainty*

The precision with which resource or reserve assessments can be made is dependent on two main factors: the inherent structure and location of the coal basin and the extent to which exploration effort has been expended on it.

A flat shallow deposit with little apparent disturbance can be evaluated with confidence on the basis of a few relatively widely placed boreholes, whereas an evaluation of a complex disturbed basin concealed at depth beneath more recent strata must be recognised as bearing a considerable degree of uncertainty.

Assessments which carry a high degree of confidence are qualified by such descriptive labels as "proved" or "measured", whereas those which have not been so thoroughly investigated will be labelled, for example, as being "possible" or "inferred". There is as yet no uniform terminology in use throughout the world.

2. Assessment on a World Scale

At regular intervals the World Energy Conference (W.E.C.) publishes an authoritative review of energy resources, the most recent being published in 1980. All W.E.C. surveys have been based on replies received from national committees to questionnaires couched in carefully defined terms. The basic principles have been firmly established for many years with slight amendments from time to time in the precise terminology. The figures in Table 6.1 represent an aggregation of national figures (estimates being used by the W.E.C. to fill some gaps). The following comments emphasise important aspects of the definitions used.

(i)   *Proved Reserves*

This definition does not lay down precise procedures or criteria but recognises that each country must first judge in its own terms whether a coal deposit has been "carefully measured" and then assess whether it is "exploitable" under local economic conditions with existing available technology.

(ii) *Additional Resources*

This term refers to coal other than proved reserves; and when added to the proved reserves it makes up the total resources of the country. Any resources whose existence is "entirely speculative" are excluded. Furthermore, any resources which are included must be "of at least foreseeable economic interest".

(iii) *Proved Recoverable Reserves*

There are valid technical and economic reasons why the amount of coal actually mined from a certain location will represent only a (variable) fraction of the total amount in the ground. "Recovery factors", calculated in respect of particular coalfields, have been applied to the proved reserves, to produce estimates of proved *recoverable* reserves. There are no imposed standards, each such estimate being based on knowledge and experience within that particular country.

The national returns upon which the W.E.C. tables are based each record the limits of depth and thickness which have been applied by the compiler in respect of particular classes of coal reported. World totals thus represent an aggregation of diversely based national figures, each of which represents the amount of coal considered worthy of inclusion by those closest to it.

C.   ACCESSIBLE COAL IN SIGNIFICANT COALFIELDS

When, as is commonly the case, the national total of proved recoverable reserves is large in relation to the present and expected rate of recovery, then generally only part of the total area of coal involved will be exploited in the next 20 years (a reasonable period of time for energy forecasting purposes). To identify at least the coal in place in these areas, it is necessary to apply further "cut-offs" within proved recoverable reserves.

The World Coal Resources and Reserves Data Bank Service (IEA Coal Research) have therefore made a study of existing resource and reserve estimates in the light of detailed geological data and other related information in order to derive estimates of "Accessible Coal in Significant Coalfields", i.e. the amount of coal intrinsically capable of being mined by existing methods, lying in coalfields likely to play a significant part in coal trade during the next 20 years (see Table 6.2).

(i) *Selection of Significant Coalfields*

The relative significance of all of the coalfields in each nation was considered, both individually and in relation to one another, in terms of their present and expected rate of recovery. Their situation was considered both with respect to relative contribution to internal supply and where relevant, in terms of contribution to potential world coal trade in the nexts 20 years. This resulted in total deletion of some of the smallest coalfields whose contribution was minor relative to the collective contribution of other coalfields in that country. Partial or total cut-offs were also applied to large virgin coalfields clearly too remote in comparison with other already developed coalfields in the nation and to some coalfields in which the coal was very heavily disturbed and/or of very low quality relative to that more readily available elsewhere. The details of the procedure used in this study are given in 'Concise Guide to World Coalfields'.*

(ii) *Estimation of Accesible coal*

Estimation of the amount of accessible coal in each significant coalfield involved examination of available tonnage estimates in the light of knowledge of the code of terms used in each case and comparison with precise local geological data. Depth and thickness limiting criteria appropriate to deep mining were applied to the whole coalfield and where some or all of the area was suitable for surface mining, a separate figure was calculated to indicate that part considered to be accessible to surface mining methods.

The results of this study are summarised in Table 6.2, which lists only the national totals and the total tonnage for OECD countries and certain geographical groupings. Descriptions of all 'Significant Coalfields' together with tonnage estimates of 'Accessible Coal' in each of them are given in the "Concise Guide to World Coalfields".*

---

*   Concise Guide to World Coalfields (IEA Coal Research, London 1983)

7. COAL TRANSPORTATION INFRASTRUCTURE*

7.1. Historical Perspective

In the early part of this century, when coal was the predominant energy source in the industrialized world, there was an extensive coal transportation infrastructure. This used railways and waterways for inland movements and very small ships (by today's standards) for seaborne trade, which at its peak in 1913 was more than 100 million metric tons. However, during the era of cheap oil, coal generally became only a locally consumed fuel and much of the transportation infrastructure became redundant or was used for other commodities.

In the 1960's, a renewed demand for long distance transport came first for coking coal. Despite considerable research into other technologies like direct reduction, the most economic method of steel-making remained the traditional blast furnace using coke. In some areas, particularly in Japan, the steel industry's expanding demand for coking coal outstripped the local capability to supply it. As a result, long distance movements of coking coal developed, both from well-established coal-fields (Eastern United States) and from newly-exploited remote coalfields in Queensland (Australia) and Western Canada. Then, after the steep oil price rises in 1973/74, the demand for steam coal also began to exceed the local supply; this initiated further long distance movements of coal (maritime and inland), for example from South Africa to Western Europe and inter-regional trade in the United States.

Much new transportation infrastructure was needed for these new long distance movements of coking and steam coal. The railways developed the unit train concept and in some countries considerable lengths of new track were constructed, while elsewhere there was a need to substantially upgrade the existing track. On the Mississippi basin waterways, integrated tows of up to thirty barges became the normal method of moving coal. New port facilities were constructed in Australia, Canada, Poland, South Africa and elsewhere for exporting coal and the bulk carrier fleet was enlarged accordingly with Panamax vessels and large size coal carriers.

---

*   This chapter was written with the assistance of the Economic Assessment Service of IEA Coal Research, London, who provided the factual information in it.

No serious bottlenecks in the coal transportation system appeared before 1980. The expansion in the capacity of the infrastructure generally kept pace with, or exceeded, the need generated by the growing demand for coal as well as by the increasing distance between producers and consumers. However, in 1980 there was a sudden large increase in demand for coal from OECD exporters caused by oil price rises and by the sudden drop in exports from Poland due to unrest there, combined with stockbuilding in importing countries. Much of this extra demand was placed on the United States because it had adequate spare mining capacity but its coal port facilities were unprepared for a sudden, large increase in coal exports and large queues of ships waiting to load coal developed at Hampton Roads and other US ports. A similar queue also developed at Newcastle, New South Wales due to a number of strikes by transport workers as well as insufficient capacity of the physical infrastructure. These bottlenecks persisted for part of 1981 but disappeared completely in 1982 with port improvements, development of better procedures between traders and their transportation partners and because coal trade was slacker and better diversified with the come-back of Polish exports.

7.2. Current Situation

The coal transportation infrastructure of the world can be divided into three categories: inland, ports and ships.

The capacity of an inland transport system is usually extremely difficult to measure. In most countries coal is transported on a railway network whose coal capacity cannot reasonably be defined because of the availability of alternative routes and the use of the network for transporting other commodities and passengers. In addition substantial quantities of coal are transported on the Mississippi basin waterways in the United States and on an extensive network of waterways in Northern Europe. Only in a few specific cases is there a significant coal flow along a particular railway link with no reasonable alternative route or other transport facilities; for example, in South Africa all of the coal exported through Richards Bay must go along a rail track which is reported to have a capacity of 16 trains per day carrying 4890 metric tons of coal each and an additional 10 trains per day carrying other commodities. A similar situation exists in Western Canada and in some parts of Queensland. Overall, there were no serious bottlenecks in the world in 1982 concerning inland coal transport so it can be concluded that the capacity was sufficient for the current level of demand except possibly in China and the USSR.

The capacity of a port can be easier to define. The present nominal, or theoretical, capacity of coal importing and exporting ports in the world are summarised in Table 7.1. However, a distinction must be made between the nominal capacity of a port and the 'effective' capacity, which can be defined as the strain-free throughput which could be sustained over a whole year. The ratio of effective to nominal capacity depends on a number of factors including how much the port as a whole is used by other commodities, weather and tide conditions, the method of scheduling ship arrivals and the way in which the nominal capacity is defined.

An estimate of the effective port capacity in various countries and regions is given in Table 7.1. The Australian, Canadian and South African estimates are based on the experience of 1981 when these countries were working close to, or above, their effective capacities. For the United States, effective capacity is estimated at one third of nominal capacity which is the figure used by the US Maritime Administration for older multi-purpose terminals. In Europe and Japan, effective capacity is estimated at two thirds of nominal capacity: this is based on queuing theory which predicts that the average waiting time is twice the average berth time at this level of utilisation but rises very steeply for higher utilisations; this estimate also reflects the relatively high effective capacity one can expect from the considerable improvement of coal importing facilities in these industrial countries in recent years. For the rest of the world effective capacity is estimated at one half of the nominal capacity. Summary figures like those given in Table 7.1 can conceal the situation at individual ports: there can easily be inadequate capacity at one port in a country which in reality is not compensated for by excess capacity at another port despite the total figures appearing to be adequate. The current capacities of individual ports are recorded in detail in an annex to this chapter.

Table 7.1. **Summary of Current Coal Port Capacities**
(Million metric tons per annum)

	Seaborne[1] coal trade in 1981	Nominal port capacity in 1982	Trade as a percentage of nominal capacity	Estimated effective capacity[2]
Exporters				
Australia	51	90	57%	51
Canada	16	28	57%	17
China	3	14	21%	7
Poland	8	30	27%	15
South Africa	29	32	91%	29
United States[3]	86	262	33%	87
USSR	3	12	25%	6
Others	13	19	68%	10
World Total	209	487	43%	222
Importers				
EEC	85	205	41%	137
Other Europe	20	43	47%	29
Japan	78	119	66%	80
Other Asia	17	35	49%	18
South America	5	22	23%	11
Others	4	24	17%	12
World Total	209	448	47%	287

Source: Economic Assessment Service (IEA Coal Research).

1. An estimate of inland trade (including USA to Canada) has been subtracted from total world trade in 1981.
2. Work is continuing to quantify the relationship between actual throughput, nominal or rated capacity, and effective capacity. These figures for estimated effective capacity are only tentative indications.
3. Current nominal capacity includes 120 million metric tons per annum for the Great Lakes ports.

The number of ships available to carry coal in 1982 greatly exceeded the demand for seaborne trade. By the end of 1982 about 15 million dwt of bulk carriers were idle for more than two months. More capacity may be laid up if current rates, which in many instances hardly cover operating expenses, do not increase.

Specialised self-unloading ships are used to carry coal across the Great Lakes of North America. Some of these ships have also been used to take coal down the St. Lawrence Seaway for transshipping to ocean vessels. Freezing prevents ship movements for about three months per year; nevertheless the ships available have been sufficient to meet all the variations in demand.

7.3. Current Developments

Significant developments in the world coal transportation infrastructure are currently taking place in all three categories: inland, ports and ships.

In some countries the railway infrastructure is being improved. This is particularly marked in Canada, South Africa and to some extent in China. In the United States, a new waterway is being constructed between the Tennessee and Tombigbee rivers and some locks on existing waterways are being rebuilt due to previous congestion during high demand periods; these projects have frequently been delayed in the past and it is feared that they may not be ready in time for the next surge in demand. In Europe, very little has been reported on improvements in the inland coal transport infrastructure or the need for them, except the Rhine-Danube canal whose completion date is frequently being postponed.

Considerable coal port developments are taking place or are planned in many countries and if implemented according to schedule they could handle the trade currently anticipated by 1990 (see Table 4.6). These coal port developments are recorded in detail in an annex to this chapter and are summarised here in Table 7.2. If all the firm plans are realised by 1990, exporting ports will increase nominal capacity by 56% while importing ports will increase nominal capacity by 52%. In the 1990's, if all suggested plans materialize, capacity levels would be 130% and 83% over current capacities.

The uncertainty regarding plans for exporting ports is reflected in the United States, where the total nominal expansion in the period to 1990, from 262 (including 120 for the Great Lakes ports) to 396 million metric tons per annum, will account for 49% of the rise in world exporting capacity; an even bigger proportion accounts for the further world expansion indicated beyond 1990. One third of the United States capacity by the nineties would still be for deadweight sizes below 80,000 and 41 million metric tons per annum for vessels of 100,000 dwt and above, if deepening of access to some East Coast and Gulf ports can be accomplished in time. Australia and South Africa show an expansion of exporting capacity in line with expectations regarding their long-term coal production capacities, while Canadian prospects suggest a further stage of development at Roberts Bank beyond what is currently under construction. A notable feature is China's plan to almost treble its current nominal capacity (14 million metric tons per annum) by 1990, indicating that China is thinking of greater coal exports than anticipated in a number of recent studies on world coal trade. A substantial share of this additional capacity may, however, be aimed at coastal trade, though there are a number of foreign markets in the immediate area. However, these prospects should be considered carefully as plans are not yet adequately matured.

Table 7.2. **Summary of Current and Planned Nominal Port Capacities**
(Million Metric Tons per annum)

	Current	1985-1990	Beyond 1990[1]
Exporters			
Australia	90	126	180
Canada	28	50	69
China	14	37	115
Poland	30	30	40
South Africa	32	65	75
United States [2]	262	396	585
USSR	12	16	16
Others	19	41	68
World Total	487	761	1148
Importers			
EEC	205	312	325
Other Europe	43	88	88
Japan	119	174	174
Other Asia	35	48	142
South America	22	28	40
Others	24	32	52
World Total	448	682	821

Source: Economic Assessment Service (IEA Coal Research).

1. Figures for 1985-90 comprise expansion plans which are reasonably firm up to mid 1980s; figures for later period include all plans which have been suggested in recent years and which could go ahead if a sustained market growth develops with recovery from the current recession. These longer term projects are not included in the more detailed port tabulation [Tables 7.3 and 7.4 in the Annex].
2. Including the Great Lakes ports with current capacity of 120 million metric tons per annum and no projected expansions.

Table 7.2 shows that European importing port capacities are planned to rise from 248 to 400 million metric tons per annum by 1990 and to a possible 413 million metric tons in the 1990's which is in the aggregate more than adequate to meet currently anticipated import levels by that time (approximately 200 million metric tons by 1990). An uncertain feature of Northern Europe's plans is the extent of the transshipment capability. The Netherlands will get the greatest share of this (about 30% of the steam coal arriving at Rotterdam is now transshipped for other countries); but France, Denmark, Ireland, Norway and possibly the UK (using surplus deepwater ore-importing capacity) could all have transshipment ports. Firm plans for expansion in Southern Europe are gaining momentum but the longer-term trend which would raise the region's share of total European importing capacity from 29% currently to 45% is dependent on plans for Spain and Italy. Spain's import needs are now focussed on the definite development of Gijon-Musel, Carboneras and Algeciras; the latter two could also offer surplus capacity for transshipment to other Mediterranean countries. Italian plans for strategically located importing centres still have to be confirmed. Transshipment potential for central Europe could be developed in Trieste but no decision has yet been taken.

Overall importing port potential in Asia is indicated by a rise for Japan from 119 to 174 million metric tons per annum by 1990 and for the rest of Asia from 35 to 48 million metric tons. The Japanese figure is close to potential needs but reflects the absence of firm indications about the capacities of large steam coal importing centres.

There are a considerable number of new ships on order despite the current oversupply. Taking into account scrapping and losses, the world dry cargo fleet could grow at about 7% per year over the next few years. The demand for seaborne movements of all dry bulk commodities is unlikely to grow at this rate in the short term, though between 1970 and 1980 it grew at 11% per year (on a ton-mile basis). The Panamax is the most popular size of bulk carrier currently on order; but 20% of the deadweight tonnes on order are ships over 100,000 dwt.

7.4. Future Prospects

Inland transport infrastructure may give the greatest cause for concern in the longer term. As rail traffic volumes increase, the need for improvements in efficiency will become more acute in certain areas and very expensive capacity expansion projects will be needed for example in the Rockies of Canada. In the United States, some slurry pipelines are likely to be built by 1990 (one has been in operation for over ten years) and this will ease the demand on railroads and waterways.

Port expansion projects can easily be delayed either by uncertainty over the short term growth in demand or by indecision about how they are to be financed - this particularly applies to dredging in the United States. Very few of the port expansions described as 'beyond 1990' in Table 7.2 are yet firmly committed.

The supply of ships gives no cause for concern. The shipyards of the world have generally been under-utilised for a long time and governments are often willing to subsidise shipbuilding. Moreover, the lead times concerned are relatively short.

In summary, it is likely that in the short term the expansion of coal transportation infrastructure worldwide will keep ahead of the demand. However, continued expansion and renewing of outmoded infrastructure will be needed to meet the projected growth in coal production and use. The present situation could generate a dangerous complacency and there is always a need for an oversupply of infrastructure to allow for risks ranging from hurricane damage to political unrest like that in Poland.

CAPACITY AND CHARACTERISTICS OF COAL PORTS

PRINCIPLES AND DEFINITIONS

The survey of coal importing and exporting ports is based on a detailed questionnaire sent out in 1981 by the Economic Assessment Service of IEA Coal Research to all the ports tabulated. Coverage, from replies to the questionnaire, was supplemented in some instances by visits and further correspondence and an attempt has been made to bring the information up to date as of late 1982 in view of the uncertainties which have affected many development plans in the past 12 months. There are still gaps in the information which have been left as such in order to indicate the extent of the coverage achieved in the survey.

A critical conceptual difficulty in estimating current and future coal-handing capacities is that of defining port capacity in a realistic sense. The question of "effective" capacity, which might be defined as strain-free throughput which can be sustained over the course of a year, given demand, has to be answered in the first instance on a theoretical rather than an empirical basis. The US Maritime Administration (Marad) defines effective capacity as that "which for planning purposes a given port could be expected to sustain". To obtain effective capacity in the United States, Marad divides the design capacity of modern, specialised terminals by two and that of older, multi-purpose terminals by three. Another view is provided by Soros Associates, the engineering design firm. They base their view on an "Erlang berth occupancy curve" which has been derived from queuing theory and empirical analysis of port layout and patterns of ship arrivals and departures simulated by computer programmes. The curve relates the ratio of average waiting time/average berth time to the percentage of time that berths are occupied. This shows the former rising sharply from a ratio of 2, once berth occupancy rises above 66%. No account is taken of differences between old and new ports.

Both the Marad and Soros methods are proximate concepts attempting to measure site-specific port situations which in reality are influenced by the natural geography of harbours, port layout and the dedicated nature of port activity. Large ports specially designed for coal handling can probably operate at a significantly higher percentage level of occupancy but the capacity to berth ships without undue delay can be thrown out by the irregularity of arrivals due to weather and other effects on steaming times. Leaving aside these extraneous influences, however, it should be possible by detailed analysis of equipment capacities, hours operating and variable rates of discharge or load as operations proceed (e.g. slower discharge as holds are emptied) to achieve an empirically based estimate of effective capacity but this will require further improvements and extension of the port surveys carried out.

In the detailed tables the following information is included:

Country

Countries are listed alphabetically and ports within countries alphabetically except in a few instances like Australia, Canada and the United States which are first broken into regions. Some separate developments within large port complexes are also mentioned by name.

Year of Status

Current facilities are indicated as of end 1982. Where installations are under construction or plans are definite up to 1990 the year of completion is given.

Throughput

This is given in million metric tons per annum of likely achievable throughput as indicated by port authorities; it does not indicate the rated or effective capacity of the handling equipment working in optimum conditions with maximum realistic berth occupancy.

This question is being subjected to further investigation (see above for a discussion of nominal and effective port capacities). Where expansion is underway at the same port facility the total capacity resulting from expansion is indicated. Where expansion involves separate development in the same port which can be identified, the respective capacities of the individual developments are indicated.

Loading/Unloading Rates

Given in tonnes per hour for the combined facilities available. For simplicity the relevant loading rate only for exporting port and unloading rate for importing port is given. No information is tabulated for stacking and reclaiming from stock, belt capacities or for transshipment loading. Obviously any of these operations could be the constraining factor on effective handling rates.

Ground Storage

Where no figures are given this generally means that no significant storage space is available, though there may be some gaps in the coverage (see introductory paragraph above).

Maximum Draft and Deadweight

Figures relate generally to the maximum dimension of ship which can berth in normal operating conditions. Sometimes port entrance characteristics such as sealocks are constraining factors on breadth, length and draft.

Maximum Length and Beam

Figures relate generally to the maximum working length of the berth and outreach of handling equipment. For draft, deadweight, length and beam ranges of figures are given where several berths are available. If there are only two berths, two figures are given separated by a semi-colon.

Remarks

Anecdotal information is provided on developments which throw light on the port's role in coal handling or the stage of development with expansion plans but in a tabulation of this nature such remarks cannot be comprehensive.

Table 7.3 **CAPACITY AND CHARACTERISTICS OF COAL - IMPORTING PORTS**

Country/Port	Year of Status	Handling Capacity			Ship Characteristics				Remarks
		Through Put (Mt/a)	Unload Rate (t/h)	Ground Storage (kt)	Max. Draft (feet)	Max. DWT (kt)	Max. Length (feet)	Max. Beam (feet)	
Algeria									
Annaba	1982	2.5	2000		43	50			Coking coal for local steel plant, modernisation in progress.
Argentine									
Buenos-Aires	1981								Four berths for coal unloading, mostly for electric power plants.
Medanos	1985	1.5							Major bulk port development planned.
Belgium									
Antwerp	1982	21	11500	6000	20-47	100	985	UNL	4 berths; major expansion completed in 1981.
Ghent	1982	2.2	2400	3150	40	75	837	112	Major expansion completed end-1981.
Zeebrugge	1981 1985	1 8		500	35	45	788	109	New sea lock proposed that would allow capacity up to 150,000 dwt.

Source: Economic Assessment Service (IEA Coal Research)

Table 7.3 **CAPACITY AND CHARACTERISTICS OF COAL - IMPORTING PORTS**

Country/Port	Year of Status	Handling Capacity			Ship Characteristics				Remarks
		Through Put (Mt/a)	Unload Rate (t/h)	Ground Storage (kt)	Max. Draft (feet)	Max. DWT (kt)	Max. Length (feet)	Max. Beam (feet)	
Brazil									
Cubatao	1982		3000		39	55			Coking coal.
Santos	1982	0.5			33	40	985		Coking coal, three berths.
Sepetiba	1985	8.5	7000		79	300	1542		Currently bulk ore terminal, coal facilities being installed.
Vitoria (Esperito Santo State)									
1. Praia Mole	1982				56	80	890	144	Ore and coal imports.
2. Usiminas Quay	1982		500		36	55	240		Coking coal.
3. Capuaba	1982		1000		40	55			Coking coal.
Canada									
Ontario ports	1982	17			30				Coking and steam coal from U.S.

Source: Economic Assessment Service (IEA Coal Research)

Table 7.3 CAPACITY AND CHARACTERISTICS OF COAL - IMPORTING PORTS

Country/Port	Year of Status	Handling Capacity			Ship Characteristics				Remarks
		Through Put (Mt/a)	Unload Rate (t/h)	Ground Storage (kt)	Max. Draft (feet)	Max. DWT (kt)	Max. Length (feet)	Max. Beam (feet)	
Chile									
Huasco	1982		1600		43	100	589		Coking and steam coal.
Denmark									
Aabenraa	1982 1983	11	1500 3000	1000 2000	49 56	140 180	920 UNL	144 UNL	Ensted power station. Continuous unloader being installed. Barge transshipment.
Aarhus	1982	6	1500	600	43	130	880	UNL	
Asnaes	1982	4	3000	2500	40	70			Part-loads up to 150,000 dwt in combination with Stigsnaes or other ports.
Esbjerg	1982	1.2	600	650	31	20			For local power station and transshipment by train to Herning.
Stignaes	1982	4	800	1000	56	170	900	131	Expansion programme completed in 1982.
Studstrup	1982	1.5	600	1400	35				Private port of an electricity company, which is able to accept part loaded Panamax vessels.

Source: Economic Assessment Service (IEA Coal Research)

Table 7.3 CAPACITY AND CHARACTERISTICS OF COAL - IMPORTING PORTS

Country/Port	Year of Status	Handling Capacity			Ship Characteristics				Remarks
		Through Put (Mt/a)	Unload Rate (t/h)	Ground Storage (kt)	Max. Draft (feet)	Max. DWT (kt)	Max. Length (feet)	Max. Beam (feet)	
Finland									
Inkoo	1982 1983	2 5	2400	500	32 43	65			Dredging under way and new pier under construction.
Naantali	1982	0.8	370	950	33	40	393	59	Berth expansion 1981/82.
Dulu	1982	3		200	20	50	600	100	2 berths; new cranes by 1983.
Pori	1982	2	1200		33	35	476		Transshipment to other Finnish ports envisaged.
France									
Bordeaux	1982	1	900		32;41	30;70	724;1351		Expansion in advanced planning (Verdon)
Dunkirk	1982 1985	12 15	4400	1600	47 72	125 250	901	135	Can receive 2 ships simultaneously; coal and iron ore
Le Havre	1982 1983	8 12	2000 2000	120 450	50 55	150 170	965 1000	115;145 180	2 berths, serving power station. Industrial centre planned.

Source: Economic Assessment Service (IEA Coal Research)

Table 7.3 CAPACITY AND CHARACTERISTICS OF COAL - IMPORTING PORTS

Country/Port	Year of Status	Handling Capacity			Ship Characteristics				Remarks
		Through Put (Mt/a)	Unload Rate (t/h)	Ground Storage (kt)	Max. Draft (feet)	Max. DWT (kt)	Max. Length (feet)	Max. Beam (feet)	
Marseille-Fos	1982	2.5	1500	800	64	180			Coking and steam coal; transshipment to Italy.
	1983	3.5	2200						Under construction (continuous unloader).
	1986-88	6.5				250			Planned (in advanced stage). Planned to expand to 13 - 15 Mt/a in the early 1990's.
Nantes-St. Nazaire	1982	2	750	400	40	60	856	133	Power and cement works.
	1983	4.5	2500	600	51	120	919	141	New terminal planned for Montoir power station.
Rouen	1982	4			35	30-40	852		Turning basin restriction on length of vessel; transshipment for Paris Basin.
German Fed. Republic									
Bremerhaven	1982		500		32	40	606	79	Handles only iron ore at present.
Brunsbuttel	1982		1000		37	40	1640		Kiel Canal Lock limitations.
Emden	1982	1	1200	142	28	45	837	120	
Hamburg Hansaport	1982	10	5300		37-47	40-90	1050	73-130	Coal/ore terminal with 5 berths; local power station.

Source: Economic Assessment Service (IEA Coal Research)

Table 7.3 CAPACITY AND CHARACTERISTICS OF COAL - IMPORTING PORTS

Country/Port	Year of Status	Handling Capacity			Ship Characteristics				Remarks
		Through Put (Mt/a)	Unload Rate (t/h)	Ground Storage (kt)	Max. Draft (feet)	Max. DWT (kt)	Max. Length (feet)	Max. Beam (feet)	
Nordenham	1982	3.5	625	250	36	80	850		Privately run port serving local power station.
Wilhemshaven	1982	5	1000	350	47	80	869	UNL	2 berths for local power station, some transshipment.
	1985	10	2000		72	250			Combined coal and ore expansion in advanced planning stage (additional throughput of 10 Mt/a with access of 150,000 dwt bulk carriers).
Greece									
Milaki	1982	4		400		120			Primarily for local cement industry.
Piraeus	1982	1	60			33	820		
Hong Kong									
Castle Peak	1982	4			39	50			Local power stations: two berths initially developed to take 50,000 dwt and up to 120,000 dwt at a later stage.
	1990	11				120			
India									
Haldia	1982		3000	350	41	60	931	122	Local restrictions caused by traffic congestion. New berth for coal handling under construction.
	1983					100			

Source: Economic Assessment Service (IEA Coal Research)

Table 7.3 CAPACITY AND CHARACTERISTICS OF COAL - IMPORTING PORTS

Country/Port	Year of Status	Handling Capacity			Ship Characteristics				Remarks
		Through Put (Mt/a)	Unload Rate (t/h)	Ground Storage (kt)	Max. Draft (feet)	Max. DWT (kt)	Max. Length (feet)	Max. Beam (feet)	
Ireland									
Cork	1982	1	1000		29;42	15;60	500;800		Deepening of entrance and new berth completed.
Dublin	1982	1	1000		37	30	1117		Serving domestic and industrial markets.
Money Pt	1984	5	3500			160			To serve local power plant under construction. No rail connection with hinterland.
Israel									
Ashod	1990	8			39	150	820		Imports for cement industry; under consideration for local power station.
Hadera	1982				77	150	965		Temporary use was made of Haifa, pending completion.
Italy									
Ancona	1982	1.5	500	50	30-33	25	2130		Four berths.
Bagnoli	1982	1.5	500	600	40	75	984		Coking coal.

Source: Economic Assessment Service (IEA Coal Research)

Table 7.3 CAPACITY AND CHARACTERISTICS OF COAL - IMPORTING PORTS

Country/Port	Year of Status	Handling Capacity			Ship Characteristics				Remarks
		Through Put (Mt/a)	Unload Rate (t/h)	Ground Storage (kt)	Max. Draft (feet)	Max. DWT (kt)	Max. Length (feet)	Max. Beam (feet)	
Brindisi	1982	1	600	3000	25				Coal facilities planned for major expansion of local power station for which firm decision was taken
Civitavecchia	1982	1	800	40	28.5	12	180;220		
Genoa	1982	10	1500	1050	31-42	30	490	131	Coking coal; 5 berths; some transshipment planned.
La Spezia	1982	2	1440	450	31;42	50	590;830		2 berths; local power stations; received some coal from Marseille.
Livorno	1982	0.5	240	50	31	15	1000		Dredging in progress.
Piombino	1982	1.2	800		39	75	837	98	Coking coal.
Savona-Vado	1982	6	2000	1100	33-45	35-50	738-1412		3 berths. Revamping of terminal at Savona and new terminal at Vado planned which could give about 15 Mt/a throughput by the 1990's.
Taranto	1982 1983	6 1	800 1650	6000	39 50	50 120;170	1050;1148	148;164	2 berths for coking coal. New pier for the steel mill under construction.

Source: Economic Assessment Service (IEA Coal Research)

Table 7.3   CAPACITY AND CHARACTERISTICS OF COAL - IMPORTING PORTS

Country/Port	Year of Status	Handling Capacity			Ship Characteristics				Remarks
		Through Put (Mt/a)	Unload Rate (t/h)	Ground Storage (kt)	Max. Draft (feet)	Max. DWT (kt)	Max. Length (feet)	Max. Beam (feet)	
Trieste	1982	1.5	2600	400	45	60	1050	125	Coking coal; 2 berths. Major development (15 Mt/a throughput) planned for the future, allowing transshipment in 40,000 dwt vessels.
Venice	1982	3	700	50	45	25;30			To supply local power stations; other developments possible at Chiogga, St Leonardo.
Japan									
Chiba	1982	10	4600		36-56	60-250	820-1160	72	Coking coal.
Fukuyama	1982 1990	10 9	1500	950	16-39	70;200	820	49-115	Coking coal. Anticipated steam coal imports.
Higashi Harima	1982	10	700		53	165	984	110-125	Coking coal.
Himeji	1982	10	2900		43	110	869	66-85	Coking coal.
Ibaragi	1990	2				100			New steam coal port planned in central Honshu.
Kamaishi	1982	5	1190	120	43	120	886		Coking coal.

Source: Economic Assessment Service (IEA Coal Research)

Table 7.3 CAPACITY AND CHARACTERISTICS OF COAL - IMPORTING PORTS

Country/Port	Year of Status	Handling Capacity			Ship Characteristics				Remarks
		Through Put (Mt/a)	Unload Rate (t/h)	Ground Storage (kt)	Max. Draft (feet)	Max. DWT (kt)	Max. Length (feet)	Max. Beam (feet)	
Kashima	1982	5	4800	810	52;62	80;200	625;1240	115	Mostly coking coal, small transshipments of steam coal. Planned steam coal imports.
	1990	2							
Kisarazu	1982	10	2900		59	200-270	1148	59-112	Coking coal.
Kita Kyushu	1990	3				150			Steam coal imports planned for 1990s.
Kure	1982	10							Coking coal.
Kumamoto	1990	5				60			New steam coal port planned for Kyushu.
Matsushima	1982	2.5	1700	430	52	70	820	118	On-site power station. Can handle wide-bodied Panamax vessels.
Matsuura	1984	2				70;130			Possible onsite power developments requiring 9 Mt/a.
Muroran	1982	3	1800		52	160	984	115	Coking coal. Transshipment port, planned increases in steam coal handling by 1990.
	1982	1	750	350	40	30	1214	71	
Nagasaki	1990	5				60			New steam coal port planned.

Source: Economic Assessment Service (IEA Coal Research)

Table 7.3 CAPACITY AND CHARACTERISTICS OF COAL - IMPORTING PORTS

Country/Port	Year of Status	Handling Capacity			Ship Characteristics				Remarks
		Through Put (Mt/a)	Unload Rate (t/h)	Ground Storage (kt)	Max. Draft (feet)	Max. DWT (kt)	Max. Length (feet)	Max. Beam (feet)	
Nagoya	1982	4	4500	420	46	160	1148;2099	148	Coking coal.
	1982	2.5	600-900	100;135	33	34	636;659	89;98	Two separate steam coal handling berths.
Naoetsu	1990	1				55			Planned steam coal port in N. Honshu.
Noshiro	1990	2.5				60			Planned steam coal imports for N. Honshu.
Ogoshima	1982	12							Coking coal.
Oita	1982	10	2500		66-79	250-300	UNL	108;128	Coking coal.
	1990	9							Steam coal capacity planned.
Onahama	1990	2.5				60			Steam coal capacity planned for N. Honshu.
Sakai	1982	5	2700		38	80	820	32;67	Coking coal.
Shimane	1990	3							New steam coal port planned on Inland Sea.
Soma	1990	3				60			Steam coal capacity planned in N. Honshu.

Source: Economic Assessment Service (IEA Coal Research)

Table 7.3 **CAPACITY AND CHARACTERISTICS OF COAL - IMPORTING PORTS**

Country/Port	Year of Status	Handling Capacity			Ship Characteristics				Remarks
		Through Put (Mt/a)	Unload Rate (t/h)	Ground Storage (kt)	Max. Draft (feet)	Max. DWT (kt)	Max. Length (feet)	Max. Beam (feet)	
Tomakomai 1.	1982	3	400	1100	28	5	781		Handles Hokkaido produced coal for local cement industry and industrial use.
2.	1982	1	3500	350	46	80	918	115	Serving local power station.
	1986	3.5							Planned expansion of existing facilities as power station is expanded.
Ube	1990	3				50			Planned steam coal development for N. Honshu.
Yawata	1982	5	700		46	46-80	984	62-79	Coking coal, berth expansion underway.
Yokkaichi	1990	2				100			Steam coal handling planned for Central Honshu.
Korea, South									
Busan	1982	1.5	600			15	2156		Two berths' expansion in progress.
Masan	1982				26	20	1460		2 berths for anthracite.
Mugho	1982		600		26	25			Anthracite port.
Pohang	1982		1000		39	50			Coking coal.

Source: Economic Assessment Service (IEA Coal Research)

Table 7.3 CAPACITY AND CHARACTERISTICS OF COAL - IMPORTING PORTS

Country/Port	Year of Status	Handling Capacity			Ship Characteristics				Remarks
		Through Put (Mt/a)	Unload Rate (t/h)	Ground Storage (kt)	Max. Draft (feet)	Max. DWT (kt)	Max. Length (feet)	Max. Beam (feet)	
Malaysia									
Port Keelang	1984	0.5			40	60			Under study for local power stations; possibly 3 Mt/a., by 1990.
Mexico									
Lazaro Cardenas	1982				39	50	1969		Coking Coal.
Morocco									
Mohammedia	1990	2				80			Local power plant under discussion.
Netherlands									
Amsterdam	1982	2.5	4000	2000	45	100	1181	148	OBA (W port). de Raetlanden operation in E. Port (needs to move because of silting).
	1982	1.5			34	35			
Delfzyl	1982	1.5	375		29.5	30	656	98	Coking coal.
	1982	3	600	200					

Source: Economic Assessment Service (IEA Coal Research)

Table 7.3 **CAPACITY AND CHARACTERISTICS OF COAL - IMPORTING PORTS**

Country/Port	Year of Status	Handling Capacity			Ship Characteristics					Remarks
		Through Put (Mt/a)	Unload Rate (t/h)	Ground Storage (kt)	Max. Draft (feet)	Max. DWT (kt)	Max. Length (feet)	Max. Beam (feet)		
Ijmuiden	1982		1000		52	100	735		Possibly taking 180,000 DWT by 1990.
	1985	2	1500		65	150	1805		
Rotterdam									
1. Maasvlakte	1982	7	2000		68	250	UNL	144	EKOM.
	1984				72	300			
	1985	25							New MCT terminal proposed.
2. Botlek (Laurens- haven)	1982	4	1250		45	80	UNL	131	Frans Swarttouw.
	1982	3	900		45	80		131	Frans Swarttouw.
3. Vlaardingen	1982	3	900		39	60		106	Frans Swarttouw. Also Frans Swarttouw floating cranes division assisting in other coal-handling areas.
Terneuzen	1982		1200		40	55	1654		Bulk terminal used in part for coal transshipment and lightening for Belgian ports by offshore handling.
Norway									
Brevik	1983		700		47	100			
Rana	1982	2	400	300	36	40	427	90	Coking coal from Spitzbergen.

Source: Economic Assessment Service (IEA Coal Research)

Table 7.3 CAPACITY AND CHARACTERISTICS OF COAL - IMPORTING PORTS

| Country/Port | Year of Status | Handling Capacity | | | Ship Characteristics | | | | Remarks |
		Through Put (Mt/a)	Unload Rate (t/h)	Ground Storage (kt)	Max. Draft (feet)	Max. DWT (kt)	Max. Length (feet)	Max. Beam (feet)	
Phillipines									
Iligan City	1982	1	400			5			Serving local cement industry.
Poro (San Fernando)	1982	1	400			5			Serving local cement industry.
Semirara	1983		1050			10			Transshipment to cement industries in other centres.
Portugal									
Lisbon	1981	0.5	1000			55			Ore handling; potential for coal but lack of shore based facilities.
Sines	1984	1	2000	650	54		984		Development in two stages: 1) access to vessels of max. 30,000 dwt, 2) access to bulk carriers of 150,000 dwt.
Romania									
Constantza	1982		750		37	40			New facilities under construction.

Source: Economic Assessment Service (IEA Coal Research)

Table 7.3 **CAPACITY AND CHARACTERISTICS OF COAL - IMPORTING PORTS**

| Country/Port | Year of Status | Handling Capacity | | | Ship Characteristics | | | | Remarks |
		Through Put (Mt/a)	Unload Rate (t/h)	Ground Storage (kt)	Max. Draft (feet)	Max. DWT (kt)	Max. Length (feet)	Max. Beam (feet)	
Singapore									
Jurong	1982	8		89	40	60	902	98	General bulk facilities which can handle coal but do not at present; exploratory studies for coal terminal in progress. Handles coal for local steel industry.
Spain									
Algeciras	1985	5	8000			150			7 MT/a transshipment into 20-40,000 DWT vessels; further stages under study.
Bilbao	1982	2	1600	280	44	75	722-820		3 berths for local steel and cement industries which are being expanded.
	1985-90	4.5	4000			250			
Cadiz	1982	1	200		32	24	656		Proposed "Terminal" development.

Source: Economic Assessment Service (IEA Coal Research)

Table 7.3 CAPACITY AND CHARACTERISTICS OF COAL - IMPORTING PORTS

Country/Port	Year of Status	Handling Capacity			Ship Characteristics				Remarks
		Through Put (Mt/a)	Unload Rate (t/h)	Ground Storage (kt)	Max. Draft (feet)	Max. DWT (kt)	Max. Length (feet)	Max. Beam (feet)	
Carboneras	1984	3		600	55	150			Power plants and local cement works. Development to industrial capacity will depend on full implementation of local power generators.
Cartagena	1982	3	250		29	15	406		
Gijon-Musel	1982	3	2000	650	57	80	1805		Expansion at Musel; anticipated ultimate capacity 20 Mt/a.
	1984	12				UNL	2296		
Malaga	1982	3	1830	569	34.5	50	820	106	Extra cranes in 1982.
Pasajes	1982	1	250		33	17			
Sagunto	1982	1	300		49	81	886		3 berths for ore and coking coal; new integrated steel works proposed.
Tarragona	1982	2.5	430;1000	200	39-43	60	2208		3 berths available; major expansion underway at one berth.

Source: Economic Assessment Service (IEA Coal Research)

Table 7.3 CAPACITY AND CHARACTERISTICS OF COAL - IMPORTING PORTS

| Country/Port | Year of Status | Handling Capacity | | | Ship Characteristics | | | | Remarks |
		Through Put (Mt/a)	Unload Rate (t/h)	Ground Storage (kt)	Max. Draft (feet)	Max. DWT (kt)	Max. Length (feet)	Max. Beam (feet)	
Sweden									
Gothenburg	1982	0.2	250	20	20	60			Major expansion examined, but now seems unlikely.
Halmstad	1982	2	600	60-70	39	65			Multi-product terminal with some coal.
Hälsingborg	1982	1			26;34				Serving district heating plant.
Landskrona	1986		2000	1-2000	43	150	1148		Decision on development in 1983 to service proposed power plant.
Lulea	1982	1.5	1350	600	38	55	1887	UNL	Coking coal.
Norrköping	1982 1983	0.5 0.5	600 900		32 38				Serving local district heating; new terminal under construction.
Oxelösund	1982 1983	2 5	700	1400 1800	38 72	50 120	900	100 130	Mostly coking coal.

Source: Economic Assessment Service (IEA Coal Research)

Table 7.3 **CAPACITY AND CHARACTERISTICS OF COAL - IMPORTING PORTS**

Country/Port	Year of Status	Handling Capacity			Ship Characteristics					Remarks
		Through Put (Mt/a)	Unload Rate (t/h)	Ground Storage (kt)	Max. Draft (feet)	Max. DWT (kt)	Max. Length (feet)	Max. Beam (feet)		
Taiwan										
Hualion	1982	0.3	250		33	15	525			
Kaohsiung	1982	3.5	720		46	50	1050		Coking and steam coal.
Keelung	1982	1	500		38	39	787		
Su-Ao	1982	1.8	500		34	33	984		Proposed expansion to 5 Mt/a.
Taichung	1982	2.5	500		46	73	796		Proposed expansion to 5 Mt/a.
Turkey										
Eregli	1982		375		35	35			Coking coal.
Iskenderun	1982		1000		37	40			
United Kingdom										
Hunterston	1982	8	3000		115	250			
Port Talbot	1982	10	5000		47	110	995	140	About 0.6 Mt/a of coking coal.
Swansea	1982	3	1200	20	32.5	10-30	655	60-87	2 berths; exports only at present but import potential.

Source: Economic Assessment Service (IEA Coal Research)

Table 7.3 CAPACITY AND CHARACTERISTICS OF COAL - IMPORTING PORTS

| Country/Port | Year of Status | Handling Capacity | | | Ship Characteristics | | | | | Remarks |
		Through Put (Mt/a)	Unload Rate (t/h)	Ground Storage (kt)	Max. Draft (feet)	Max. DWT (kt)	Max. Length (feet)	Max. Beam (feet)	
Redcar	1982	8				140			Third ore unloader being installed; could be used for coal imports.
Thames	1982	4.5	2000	1650	27	13	454	64	CEGB imports largely from N.E. England; also imports at Meduray, Brighton, Ipswich.
USA-New England									
Providence (R.I.)	1981	0.5	500	100	37.5	45	650	100	Coastal importer but potential exporting port.
Somerset (Mass.)	1981		1200		34		645;1017		Coastal; 2 berths.
Searsport (Maine)	1981		1120		32		625		Coastal.
Yugoslavia									
Rijeka (Bakar)	1982 1985-90	6 15	3000 4500	300 500-1000	61 70	150 220	1260	150	Coal/ore port (one berth). Expansion will depend on transshipment markets in Danube Basin.

Source: Economic Assessment Service (IEA Coal Research)

Table 7.4 CAPACITY AND CHARACTERISTICS OF COAL - EXPORTING PORTS

Country/Port	Year of Status	Handling Capacity			Ship Characteristics					Remarks
		Through Put (Mt/a)	Loading Rate (t/h)	Ground Storage (kt)	Max. Draft (feet)	Max. DWT (kt)	Max. Length (feet)	Max. Beam (feet)		

Australia
- New South Wales

Newcastle

1. Basin	1982	25	2000	90	38	58	820	106		
2. PWCS	1982	25	7500	650	54	110	951	141		Throughput figures are for Basin and PWCS combined.
3. Kooragang Island	1984	15	6000	900		110				
Port Kembla	1982	15	5000	800	53	110	950	148		Being evaluated as offshore possibility.
	1985-90	25	10000	1400		250				

Sydney

1. Balmain	1982	3.5	1000	26	44	50	850	97		
	1984	5.5		65		55	1092	106		
2. Balls Hd	1982	1	2000		25		580	85		In progress.

Source: Economic Assessment Service (IEA Coal Research)

Table 7.4 **CAPACITY AND CHARACTERISTICS OF COAL - EXPORTING PORTS**

Country/Port	Year of Status	Handling Capacity			Ship Characteristics				Remarks
		Through Put (Mt/a)	Loading Rate (t/h)	Ground Storage (kt)	Max. Draft (feet)	Max. DWT (kt)	Max. Length (feet)	Max. Beam (feet)	
- Queensland									
Abbots Point	1984	10	2-4000	57-72	60-120	105-138	755-902	1984	
Bowen	1982	0.5	600		28	20	525	80	
Brisbane	1982	0.7	500		30	20	548		
Gladstone									
1. Auckland Point	1982	5	1600		37	50	700	105	
2. Barney Point	1982	8	2000		40	75	755	120	
3. Clinton Estate	1982	20	4000		54	120	112	138	
Hay Point	1982	20	10000	2500	55;56	150	1125;1200	95	2 berths; one with potential for doubling capacity.
Canada - East									
Sydney	1982		600		39	40	637		Coking coal for E. Canadian markets, plus some exports to Europe.
Quebec City	1982	5	2000	2000	48	150	UNL	UNL	Potential for quadrupling throughput.
Thunder Bay	1982	3	4000	1500					Lake Superior terminal for W. Canadian coal to Ontario.

Source: Economic Assessment Service (IEA Coal Research)

Table 7.4 CAPACITY AND CHARACTERISTICS OF COAL - EXPORTING PORTS

Country/Port	Year of Status	Handling Capacity			Ship Characteristics				Remarks
		Through Put (Mt/a)	Loading Rate (t/h)	Ground Storage (kt)	Max. Draft (feet)	Max. DWT (kt)	Max. Length (feet)	Max. Beam (feet)	
Canada - West									
Prince Rupert Vancouver	1984	12	8000	1000	65	250	1066	164	
1. Neptune	1982	2	4000		50	100	860		3 berths.
2. Pt. Moody	1982		3850		40	65	742	102	2 berths.
3. Roberts Banks	1982	17	8000		65	140	863		Preliminary investigations have begun on a further doubling of capacity.
	1983	27				225			
China									
Lianyungang	1982	2	1000		30	18	532	82	Some of the capacity is used for coastal trade.
Qinhuangdao	1982	12	500		31	10-35	597	82	
	1983-84	20	5500		41	50			
Shijiusuo	1990	15			49	25;100			2 berths planned.
Colombia									
Bahia Portete	1986	7	10000	1700	51	150			Increase to 15 Mt/a by 1989 according to proposed long-term development.

Source: Economic Assessment Service (IEA Coal Research)

Table 7.4 **CAPACITY AND CHARACTERISTICS OF COAL - EXPORTING PORTS**

Country/Port	Year of Status	Handling Capacity			Ship Characteristics				Remarks
		Through Put (Mt/a)	Loading Rate (t/h)	Ground Storage (kt)	Max. Draft (feet)	Max. DWT (kt)	Max. Length (feet)	Max. Beam (feet)	
Indonesia									
Lampung Bay	1983	3							Outlet to other parts of Indonesia for Bukit Assam mines. Some exports possible.
	1988	7							
Mozambique									
Maputo	1982	5	3000		30	25	656		Marginal capacity to take exports of S. African coal; dredging needed.
New Zealand									
Lyttleton	1982	1	1250	140	37	50	770	120	
Norway									
Spitzbergen	1982		400			20			Outlet of coking coal to mainland Norway.

Source: Economic Assessment Service (IEA Coal Research)

Table 7.4 CAPACITY AND CHARACTERISTICS OF COAL - EXPORTING PORTS

Country/Port	Year of Status	Handling Capacity			Ship Characteristics				Remarks
		Through Put (Mt/a)	Loading Rate (t/h)	Ground Storage (kt)	Max. Draft (feet)	Max. DWT (kt)	Max. Length (feet)	Max. Beam (feet)	
Poland									
Gdansk	1982	12	2660		31-49	32-100	708-837	129	3 berths.
Gdynia	1982	5	830		26-34	32	UNL	79	3 berths.
Szczecin	1982	7	750	199	27	20	579		New terminal being developed.
Swinoujscie	1982	6	2000	500	41	65	805		
South Africa									
Durban	1982	2	1200	112	36	38	781		Facilities being phased out.
Richards Bay	1982	30	13000		75	170			2 berths.
	1985	44				250			
	1990	65							
United Kingdom									
Cardiff	1982	0.5			34		650	85	Actual exports 0.5 Mt/a.
Garston	1982	1			29	20			0.25 Mt/a to Ireland.
Immingham	1982	6	4000		33.5	35	650		Actual exports about 5 Mt/a.
Leith	1982	1							Expanded exports indicated.
Newport	1982	1	500	100	33.5	35	800	96	Could be expanded to take 2-3 Mt/a.
Swansea	1982	2.5	1200	20	32.5	10:30	655	60:87	2 berths.

Source: Economic Assessment Service (IEA Coal Research)

Table 7.4 **CAPACITY AND CHARACTERISTICS OF COAL - EXPORTING PORTS**

Country/Port	Year of Status	Handling Capacity			Ship Characteristics				Remarks
		Through Put (Mt/a)	Loading Rate (t/h)	Ground Storage (kt)	Max. Draft (feet)	Max. DWT (kt)	Max. Length (feet)	Max. Beam (feet)	
United States									
Great Lakes									
Ashtabula	1982	24	7000	1500	27		1200	100	5 Mt/a steam coal to Canada.
Conneaut	1982	14	11000	6000	28				1.5 Mt exports in 1981.
Erie	1982	2		20	27		1100		Permanent export facility being studied.
Sandusky	1982	15	3500	950	25				8 Mt/a to Canada, mostly coking coal.
S. Chicago	1982	6	3000	UNL	27	25	730	76	Coking coal shipments to USA and Canada.
Superior	1982	24	10000	7000	27	67	1000	105	Western low sulpher coal for US Eastern markets.
Toledo	1982	35	6500	7000	28;30	25	730	76	2 berths.
East Coast									
Albany (NY)	1982	2	1000	500	32	30			Primarily for coastal shipments
Baltimore (Md)									
1. Curtis Bay	1982	27	6000	500	42	65			
2. Canton Marine Terminal	1983	10	7000		42	65	893		Major expansion completed in 1982.

Source: Economic Assessment Service (IEA Coal Research)

Table 7.4 CAPACITY AND CHARACTERISTICS OF COAL - EXPORTING PORTS

Country/Port	Year of Status	Handling Capacity			Ship Characteristics				Remarks
		Through Put (Mt/a)	Loading Rate (t/h)	Ground Storage (kt)	Max. Draft (feet)	Max. DWT (kt)	Max. Length (feet)	Max. Beam (feet)	
Hampton Roads (Virg.)									
1. Newport News									
Pier 14 & 15	1982	24	6000		38;45	65	687	155	Mostly coking coal.
Pier 9	1983	12		1500	55	65			Separate facilities from C&O pier.
Dominion Term.		12	6200						Development began in 1982.
2. Norfolk	1982	38	11500		36;46	80	1000	175	2 berths.
3. Portsmouth	1986	18			55	170	1000	175	Virginia Port Authority proposal which could involve dredging.
Jacksonville (Fla)	1984	4					997		
New York/ New Jersey	1985	10	6000	2000	60	200	1200	200	The preferred NY/NJ Port Authority development. Additionally, NY Port Authority has proposed a slurry system on Slater Island which is now moribund.

Source: Economic Assessment Service (IEA Coal Research)

Table 7.4 CAPACITY AND CHARACTERISTICS OF COAL - EXPORTING PORTS

Country/Port	Year of Status	Handling Capacity			Ship Characteristics				Remarks
		Through Put (Mt/a)	Loading Rate (t/h)	Ground Storage (kt)	Max. Draft (feet)	Max. DWT (kt)	Max. Length (feet)	Max. Beam (feet)	
Philadelphia (Pa.)									
1. Pier 124	1982	10	1500		80	60	1079		
2. Port Richmond									Unused at present.
Charleston (SC)	1983-84	4				60			Plans for major expansion in Savannah Port and Hutchinson Is.
Savannah (Ga)	1982	3			38	60			
Morehead (NC)	1982	3			30		650		Further expansion currently suspended.
Gulf									
Mobile(Alabama) (McDuffie Island Terminal)	1982	9	8000	1100	40	75	880	110	
	1983	20	12000	1900					
Lwr Mississipi									
1. Ryan Walsh Bulk Terminal	1982	3		750	36	40			
	1983	4.5							
2. Intl Marine Tmnl	1982	15		1100	45				

Source: Economic Assessment Service (IEA Coal Research)

Table 7.4 CAPACITY AND CHARACTERISTICS OF COAL - EXPORTING PORTS

Country/Port	Year of Status	Handling Capacity			Ship Characteristics				Remarks
		Through Put (Mt/a)	Loading Rate (t/h)	Ground Storage (kt)	Max. Draft (feet)	Max. DWT (kt)	Max. Length (feet)	Max. Beam (feet)	
3. Electro Coal Transfer	1982	4.5			40	55			Development started in 1982. Stevedoring companies have the capability of handling about 55 million tons a year by using mid-stream equipment. This system is currently under-utilized. There are several other shore-based proposals, currently suspended, which would increase capacity by 70 Mt/a.
4. Granada Export Tmnl	1983	25	4200	1250		55			
	1983-84	5							
West Coast									
Long Beach (Calif.)	1982	1.5	2400	80	34	70	800	100	Development at existing site. Cerritos Channel development under study. (15 Mt/a by 1986).
	1986	5	3000	200	45	100	900	130	
			5-8000	1500	60	150	1000	150	
Los Angeles (Calif.)	1982	2	800	200	35	70			Expansion of existing capacity at Berth 49-50 - 1 berth.
	1986	6							

Source: Economic Assessment Service (IEA Coal Research)

Table 7.4 CAPACITY AND CHARACTERISTICS OF COAL - EXPORTING PORTS

| Country/Port | Year of Status | Handling Capacity | | | Ship Characteristics | | | | | Remarks |
		Through Put (Mt/a)	Loading Rate (t/h)	Ground Storage (kt)	Max. Draft (feet)	Max. DWT (kt)	Max. Length (feet)	Max. Beam (feet)		
Portland (Or)	1984	11			40	50				Development started in 1982.
Alaska										
Anchorage	1982	0.5	300	50	65	60	800	100		Potential much larger; so far only test shipments of sub-bituminous coal.
USSR										
Ilichevsk (Black Sea)	1982		500		36	38	837			Being up-graded.
Nakhodka (Pacific)	1982		3000		29:49	60	586	73		
Vostochny (Pacific)	1982 1984	6 10	8000		50	120	1010	138		

Source: Economic Assessment Service (IEA Coal Research)

Part II

OECD COAL DATA
HISTORICAL AND PROJECTIONS

1. PRINCIPLES AND DEFINITIONS

SOURCES

A. Historical data.

The historical data in Part II of this report are all from IEA/OECD energy statistics based on annual submissions from Member countries.

1. IEA/OECD Coal Statistics.
   Annual statistics for all OECD countries covering all primary and derived solid fuels and manufactured gas with detailed supply/demand balances for each fuel, as well as information on coal trade by origin and destination. No separate publication.

2. IEA/OECD Electricity Statistics.
   Annual statistics for all OECD countries covering capacity and electricity production and heat production from combined heat and power plants, for public utilities and autoproducers, including information on electricity production by fuel type. The statistics also include detailed supply/demand balances for electricity and heat from combined heat and power plants. No separate publication.

3. IEA/OECD Oil and Gas Statistics
   Annual statistics for all OECD countries covering crude oil, NGL, refinery feedstock and natural gas, as well as derived petroleum products. Includes detailed supply/demand balances, trade by origin and destination and stock levels and changes. Annual publication "OECD Oil and Gas Statistics".

4. IEA/OECD Basic Energy Statistics.
   Annual statistics integrating data from the three IEA/OECD statistical systems listed above to provide a complete annual energy picture for each OECD country. It provides detailed statistics on production, trade and consumption for each source of energy, expressed in "original (basic) units" (e.g. metric tons, Kwh). Annual publication "Energy Statistics of OECD Countries".

5. IEA/OECD Energy Balances.
   Overall energy balances are constructed annually for all OECD countries from the basic energy statistics described above. The overall energy balance data are expressed in a common energy unit of tons of oil equivalent (toe) and presented in a standard matrix format. Annual publication "Energy Balances of OECD Countries".

B. Projections

All projections in Part II of this report are from submissions to the IEA from OECD countries. It should be noted that these national projections may reflect policy goals and are not necessarily "most likely case" forecasts. Elements of the projections are in some cases IEA Secretariat estimates. These are marked with an "e" in the country specific tables.

IEA Country Submissions

Annual submissions from all IEA countries to the IEA Standing Group of Long Term Co-operation of national projections and descriptions of energy policies.

The submissions give national projections of the main elements in an overall energy balance expressed in Mtoe for selected forecast years (e.g. 1985 and 1990 as presented in this report). Projections are also given for electricity generation capacity by fuel types and for production and trade of specific coal types.

Similar projections are collected on an ad hoc basis from OECD non-IEA Member countries.

UNITS

Historical coal data are generally expressed in metric tons in this report. Throughout this report 1 ton means 1 metric ton of 1,000 kg.

Historical data presenting interfuel relations and all projections are expressed in a common energy unit, tons of oil equivalent (toe). A ton of oil equivalent is defined as 10^7 Kcal (1 calorie = 4.1868 joules) a convenient measure although it is somewhat below (less than 1%) the average heat content of crude oil. This unit is used by the IEA/OECD in its energy balances (historical and projections).

Factors to convert from basic units to tons of oil equivalent can be found in the recent publication "OECD Energy Balances 1971 - 1981" IEA/OECD, Paris 1983.

In coal-related analysis tons of coal equivalent (tce) defined as 7 million kilocalories, is often used. The relation between tons of oil equivalent (toe) and tons of coal equivalent (tce) is then as follows:

   1 tce = 0.7 toe

Abbreviations used throughout this publication are as follows:

   t   - metric tons
   Mt   - million metric tons
   Mtoe - million tons of oil equivalent
   Mt/a - million metric tons per annum

NOTES REGARDING EACH SOURCE OF ENERGY

COAL

Coal is a family name for a variety of solid organic fuels and refers to a whole range of combustible sedimentary rock materials spanning a continuous quality scale. For convenience, this continuous series is divided into four main categories:
— Anthracite
— Bituminous Coal
— Sub-bituminous Coal
— Lignite

The origins of coal were plants, which accumulated in a bog and became a soggy mass of plant debris known as peat (a separate energy source of even lower quality than lignite). When peat was compressed and buried over 300 million years ago, it became lignite (the lowest quality of coal). Successive invasions of the sea and piling on of layer upon layer of material resulted in a rise in temperature and an expulsion of the moisture and thus lignite became sub-bituminous and bituminous coal. In some areas the layers of coal were subject to even larger compressive forces, resulting in anthracite, the highest quality coal.

The main constituents of coal are carbon and hydrogen, with small added quantities of sulphur, oxygen and nitrogen.

Classification of different types of coal into operational categories for use at an international level is difficult for two particular reasons:

a) Divisions between coal categories vary between classification systems, both national and international, based on calorific value, volatile matter content, fixed carbon content, caking and coking properties, or some combination of two or more of these criteria. Although the relative value of the coals within a particular category depends on the degree of dilution by moisture and ash and contamination by sulphur, chlorine, phosphorous and certain trace elements, these factors do not affect the divisions between categories.

b) Coal quality can vary and it is not always possible to ensure that available descriptive and analytical information is truly representative of the body of coal to which it refers.

The International Coal Classification (UN-ECE) recognises two broad categories of coal:

a) HARD COAL - defined as having a moist and ash-free basis with a calorific value above 5700 Kcal/Kg (gross calorific value), typed according to volatile matter, calorific value and coking properties.

b) BROWN COAL - defined as having a moist and ash-free basis with a calorific value below 5700 Kcal/Kg (gross calorific value), typed according to total moisture and low temperature yield.

In statistics relating to coal production, trade and use the IEA/OECD uses these two distinct categories of coal. To improve the information base for coal market analysis and projections these two main categories of coal have been further sub-divided in IEA/OECD Coal Statistics from 1978:

Hard Coal

— Coking coal, defined as hard coal with a quality that allows the production of coke suitable for blast furnace use (also called metallurgical coal).

— Other bituminous coal and anthracite, defined as all other hard coal not classified as coking coal. This category is known as steam coal or thermal coal.

Brown Coal (all low grade coals)

— Sub-bituminous coal, defined as non-agglomerating coals with a gross calorific value between 4165 and 5700 Kcal/Kg.

— Lignite, defined as non-agglomerating coal with a gross calorific value less than 4165 Kcal/Kg.

For time series consistency, a few exceptions to this breakdown of coal types were made. In the United States, Australia and New Zealand, sub-bituminous coal is included under Hard Coal because of its relatively high calorific value in these countries and the availability of data in the national statistical systems.

SOLID FUELS

Solid fuels include primary solid fuels (all types of coal described previously, plus other solid fuels) and derived solid fuels.

A. Primary Solid Fuels

Primary solid fuels are those solid fuels which have not yet been transformed, or manufactured, into a different substance. This category includes *Hard Coal* and *Brown Coal*, as described previously and other primary solid fuels.

Other Primary Solid Fuels

This category includes all primary solid fuels (not necessarily in solid form) which cannot be classified as coal. These fuels include peat, wood, wood-waste, bagasse, dung, vegetable waste, garbage and black liquor.

B. Derived Solid Fuels

Derived solid fuels are products resulting from the transformation or manufacturing of hard coal, brown coal on other primary solid fuels, sometimes with the addition of other materials.

Patent Fuel

Patent fuel is a composition fuel manufactured from coal fines by shaping with the addition of a binding agent such as pitch.

Coke Oven Coke

Coke oven coke is the solid product obtained after coal has been carbonized at a high temperature. Also included is semi-coke, a solid product obtained from the carbonization of coal at a low temperature.

Gas Coke

Gas coke is a by-product of coal used for production of gas works gas in gas works.

Brown Coal Briquettes (BKB)

BKB are composition fuels manufactured from brown coal, produced by briquetting under high pressure. Included are also peat briquettes, lignite coke, dried brown coal, fines and dust and brown coal breeze.

Coke Oven Gas

Coke oven gas is obtained as a by-product of solid fuel carbonization and gasification operations carried out by coke producers and iron and steel plants which are not connected with gas works and municipal gas plants.

Blast Furnace Gas

Blast furnace gas is obtained as a by-product in operating blast furnaces; recovered on leaving the furnace and used partly within the plant and partly in other steel industry processes or in power stations equipped to burn it.

OTHER FUELS

Oil

Includes crude oil, refinery feedstocks, natural gas liquids, hydrocarbons not of crude oil origin and all petroleum products: LPG, refinery gas, aviation gasoline, motor gasoline, jet fuel, kerosene, gas/diesel oil, residual fuel oil, naphtha, white spirit, lubricants, bitumen, paraffin waxes, petroleum coke and other petroleum products.

Gas

Includes natural gas (excluding natural gas liquids) and gas works gas (town gas).

Nuclear and Hydro/Geothermal

Shows the primary fossil fuel equivalent needed to produce the given output of hydro, nuclear and geothermal electricity. Hydro output includes output from pumped storage plants.

The primary equivalent of nuclear and hydro electricity is assumed to be identical to that of the electricity produced in conventional thermal power plants in each country.

2. OECD TOTAL

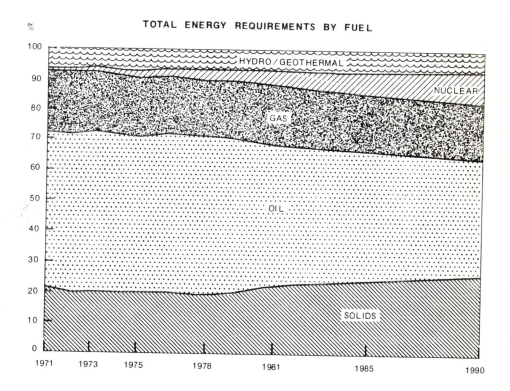

TOTAL ENERGY REQUIREMENTS BY FUEL

%

HYDRO / GEOTHERMAL

NUCLEAR

GAS

OIL

SOLIDS

OECD TOTAL

2/A1 TOTAL PRIMARY ENERGY REQUIREMENTS (TPER) BY FUEL

	1973	1978	1979	1980	1981	1985	1990
TPER (MTOE) [1]	3509.7	3782.7	3852.3	3738.3	3661.8	4068.0	4510.9
SOLID FUELS	702.2	730.1	777.2	809.0	825.8	976.1	1175.6
OIL	1850.8	1962.1	1943.0	1788.8	1667.9	1718.6	1690.4
GAS	697.2	708.6	741.5	739.7	735.1	762.4	827.6
NUCLEAR	45.3	135.4	137.4	147.4	176.0	300.0	454.9
HYDRO/GEOTHERMAL	214.0	246.5	253.0	253.2	256.9	290.0	317.4
OTHER [2]	-	-	-	-	-	21.8	47.2
TPER (PERCENTS) [1]	100.0	100.0	100.0	100.0	100.0	100.0	100.0
SOLID FUELS	20.0	19.3	20.2	21.7	22.6	24.0	26.1
OIL	52.8	51.9	50.4	47.9	45.6	42.2	37.5
GAS	19.9	18.7	19.2	19.8	20.1	18.7	18.3
NUCLEAR	1.3	3.6	3.6	3.9	4.8	7.4	10.1
HYDRO/GEOTHERMAL	6.0	6.5	6.6	6.7	7.0	7.1	7.0
OTHER [2]	-	-	-	-	-	.5	1.0

Source: IEA/OECD Energy Balances and IEA Country Submissions (1982).

1. Net imports of electricity are only included in totals.
2. Includes other electricity generation (solar, wind, etc.), traded synthetic fuels and direct use of solar, wind and other non-solid renewable energy sources.

2/A2 TOTAL FINAL ENERGY CONSUMPTION (TFC) BY FUEL

	1973	1978	1979	1980	1981	1985	1990
TFC (MTOE)	2628.8	2745.8	2826.2	2684.8	2602.1	2849.7	3085.0
SOLID FUELS	279.9	232.0	243.0	228.0	232.9	346.2	396.4
OIL	1543.3	1637.8	1671.4	1528.3	1434.7	1451.5	1473.6
GAS	493.3	506.0	529.1	535.5	537.4	580.4	635.2
ELECTRICITY	312.4	370.1	382.7	385.4	389.5	442.6	529.1
HEAT/OTHER [1]	na	na	na	7.6	7.6	29.0	50.7
TFC (PERCENTS)	100.0	100.0	100.0	100.0	100.0	100.0	100.0
SOLID FUELS	10.6	8.4	8.6	8.5	9.0	12.1	12.8
OIL	58.7	59.6	59.1	56.9	55.1	50.9	47.8
GAS	18.8	18.4	18.7	19.9	20.7	20.4	20.6
ELECTRICITY	11.9	13.5	13.5	14.4	15.0	15.5	17.1
HEAT/OTHER [1]	na	na	na	.3	.3	1.0	1.6

Source: IEA/OECD Energy Balances and IEA Country Submissions (1982).

1. Includes heat production from combined heat and power plants and direct use of solar, wind and other non-solid renewable energy sources.

OECD TOTAL

2/A3 TOTAL SOLID FUELS BALANCE
(MTOE)

	1973	1978	1979	1980	1981	1985	1990
PRODUCTION	662.3	688.7	757.0	799.4	808.2	958.9	1114.4
IMPORTS	107.3	115.1	134.0	147.8	154.6	177.3	243.3
EXPORTS	-82.3	-84.2	-100.9	-115.3	-134.4	-160.0	-182.1
MARINE BUNKERS	.0	-	-	-	-	-	-
STOCK CHANGE	15.0	10.5	-12.9	-22.9	-2.5	-.1	.1
PRIMARY REQUIREMENTS	702.2	730.1	777.2	809.0	825.8	976.1	1175.6
ELECTRICITY GENERATION	-382.7	-461.2	-500.8	-535.2	-555.4	-587.1	-718.8
GAS MANUFACTURE	-4.9	-2.0	-2.1	-1.7	-1.9	-2.3	-11.3
LIQUEFACTION	-	-	-	-	-	-8.0	-16.7
ENERGY SECTOR OWN USE + LOSSES [1]	-34.6	-34.8	-31.2	-43.9	-35.4	-31.5	-31.4
FINAL CONSUMPTION	279.9	232.0	243.0	228.0	232.9	346.2	396.4
INDUSTRY SECTOR	220.7	189.6	199.4	185.0	191.1	298.1	345.2
IRON AND STEEL	140.2	114.8	119.6	106.7	107.9	na	na
CHEMICAL/ PETROCHEMICAL	7.4	4.7	4.6	5.2	6.0	na	na
OTHER INDUSTRY	73.2	70.1	75.2	73.0	77.3	na	na
TRANSPORTATION SECTOR	2.2	.8	.6	.5	.6	.4	.7
OTHER SECTORS	57.0	41.6	42.9	42.5	41.2	46.3	49.0
AGRICULTURE	.1	.0	.0	.0	.1	.2	.2
COMMERCIAL AND PUBLIC SERVICE	2.3	1.9	1.9	1.6	1.5	3.3	3.8
RESIDENTIAL	51.9	38.7	39.8	39.8	38.6	25.4	25.9

Source: IEA/OECD Energy Balances and IEA Country Submissions (1982).

1. Includes statistical difference.

Note: See United Kingdom, Table 5/A3 for explanation of changes to IEA Country Submissions (1982) as published in "Energy Policies and Programmes of IEA Countries, 1982 Review".

OECD TOTAL

2/B1 SOLID FUELS INDIGENOUS PRODUCTION BY FUEL TYPE
(MTOE)

	1973	1978	1979	1980	1981	1985	1990
TOTAL SOLID FUELS	662.27	688.67	757.00	799.44	808.20	958.92	1114.37
HARD COAL	597.71	602.91	666.11	699.69	703.00	783.44	914.06
COKING COAL	na	153.81	173.22	170.23	173.57	196.58	202.34
STEAM COAL	na	449.10	492.88	529.46	529.43	586.86	711.72
BROWN COAL/LIGNITE	41.57	56.70	62.65	67.16	72.57	94.69	115.17
OTHER SOLID FUELS	22.99	29.06	28.24	32.59	32.63	80.79	85.14

Source: IEA/OECD Coal Statistics, IEA/OECD Energy Balances, IEA Country Submissions (1982).

Note: See Table 2/A3.

2/B2 COAL PRODUCTION BY TYPE
(THOUSAND METRIC TONS)

	1973	1978	1979	1980	1981
TOTAL COAL	1104425	1185916	1308336	1374707	1398410
HARD COAL	910451	940877	1040316	1093347	1097880
COKING COAL	na	234929	265652	260864	263606
STEAM COAL	na	705948	774664	832483	834274
BROWN COAL/LIGNITE	193974	245039	268020	281360	300530
DERIVED PRODUCTS:					
PATENT FUEL	8835	5314	5528	4734	4370
COKE OVEN COKE	216220	170721	182808	175386	169137
GAS COKE	7100	4239	4604	4458	3812
BROWN COAL BRIQUETTES	8083	6399	7940	8196	8113
COKE OVEN GAS(TCAL)	398742	330190	352920	338870	331275
BLAST FURNACE GAS(TCAL)	464451	373757	391324	343859	344682

Source: IEA/OECD Coal Statistics.

OECD TOTAL

2/C1 SOLID FUELS TRADE BY TYPE
(MTOE)

	1973	1978	1979	1980	1981	1985	1990
IMPORTS							
TOTAL SOLID FUELS	107.27	115.13	134.00	147.78	154.57	177.28	243.28
HARD COAL	94.55	103.23	120.70	135.86	143.54	172.19	238.72
COKING COAL	na	67.84	77.76	82.53	83.36	87.86	97.11
STEAM COAL	na	35.40	42.94	53.33	60.18	84.33	141.61
BROWN COAL/LIGNITE	.37	.42	.49	.62	.79	1.24	1.12
OTHER SOLID FUELS [1]	12.35	11.48	12.81	11.30	10.24	3.85	3.44
EXPORTS							
TOTAL SOLID FUELS	82.32	84.19	100.92	115.32	134.45	160.00	182.08
HARD COAL	70.37	73.16	86.53	103.24	123.68	158.98	181.06
COKING COAL	na	56.96	67.34	73.75	80.39	89.18	91.66
STEAM COAL	na	16.19	19.20	29.50	43.29	69.80	89.40
BROWN COAL/LIGNITE	.01	.03	.03	.01	.01	.01	.01
OTHER SOLID FUELS [1]	11.94	11.00	14.36	12.07	10.76	1.01	1.01

Source: IEA/OECD Coal Statistics, IEA/OECD Energy Balances, IEA Country Submissions (1982).

1. Includes derived coal products, e.g. coke.

Note: See Table 2/A3.

OECD TOTAL

2/C2 COAL IMPORTS BY ORIGIN
(THOUSAND METRIC TONS)

	1978	1979	1980	1981
HARD COAL IMPORTS	147471	173644	195428	205560
COKING COAL IMPORTS FROM:	95297	109873	116334	116829
AUSTRALIA	28771	31963	30243	34040
CANADA	11329	10299	10986	9676
GERMANY	10385	9352	8281	8398
UNITED KINGDOM	110	258	75	1678
UNITED STATES	23994	36312	46706	50803
OTHER OECD	405	326	810	698
CZECHOSLOVAKIA	832	970	1065	860
POLAND	10946	11160	9195	3942
USSR	4966	4676	4610	2142
CHINA	-	-	979	1148
SOUTH AFRICA	2571	3008	3348	3100
OTHER NON-OECD	988	1549	36	344
STEAM COAL IMPORTS FROM:	52174	63771	79094	88731
AUSTRALIA	4555	5975	8336	9031
CANADA	970	790	1080	2580
GERMANY	6642	6098	4134	3087
UNITED KINGDOM	2201	2367	4175	7889
UNITED STATES	8883	12049	23408	36064
OTHER OECD	1327	1532	1347	1469
CZECHOSLOVAKIA	243	203	255	217
POLAND	11665	13287	10932	2793
USSR	2353	2410	2016	1115
CHINA	-	70	1503	1809
SOUTH AFRICA	11983	17566	21102	22148
OTHER NON-OECD	1352	1414	806	529

Source: IEA/OECD Coal Statistics.

OECD TOTAL

2/D1 FUEL INPUT IN ELECTRICITY GENERATION

	1973	1978	1979	1980	1981	1985	1990
TOTAL (MTOE)	1007.1	1196.6	1242.0	1263.4	1288.7	1468.3	1760.3
SOLID FUELS	382.7	461.2	500.8	535.2	555.4	587.1	718.8
OIL	248.1	233.5	223.4	195.5	172.2	164.2	127.2
GAS	117.0	120.0	127.4	132.1	128.2	124.4	133.1
NUCLEAR	45.3	135.4	137.4	147.4	176.0	300.0	454.9
HYDRO/GEOTHERMAL	214.0	246.5	253.0	253.2	256.9	289.7	316.7
OTHER [1]	-	-	-	-	-	2.9	9.7
TOTAL (PERCENTS)	100.0	100.0	100.0	100.0	100.0	100.0	100.0
SOLID FUELS	38.1	38.5	40.3	42.4	43.2	40.0	40.8
OIL	24.7	19.5	18.0	15.5	13.4	11.2	7.2
GAS	11.6	10.0	10.3	10.5	10.0	8.5	7.6
NUCLEAR	4.5	11.3	11.1	11.6	13.5	20.4	25.8
HYDRO/GEOTHERMAL	21.1	20.6	20.4	20.0	19.9	19.7	18.0
OTHER [1]	-	-	-	-	-	.2	.6

Source: IEA/OECD Energy Balances and IEA Country Submissions (1982).

1. Solar, wind, etc.

2/D2 ELECTRICTY PRODUCTION BY FUEL TYPE

	1973	1978	1979	1980	1981	1985	1990
TOTAL (TWH)	4303.1	4907.8	5086.9	5150.0	5203.3	5737.9	6899.3
SOLID FUELS	1574.0	1735.4	1892.2	2032.4	2088.6	2208.0	2716.8
OIL	1081.9	1036.1	958.7	850.4	746.6	670.8	517.0
GAS	537.1	535.9	580.8	586.2	569.9	519.2	561.4
NUCLEAR	187.9	549.6	558.5	598.6	707.7	1173.6	1790.0
HYDRO/GEOTHERMAL	922.2	1050.8	1096.6	1082.4	1090.4	1157.4	1292.3
OTHER [1]	-	-	-	-	-	9.0	21.7
TOTAL (PERCENTS)	100.0	100.0	100.0	100.0	100.0	100.0	100.0
SOLID FUELS	36.6	35.4	37.2	39.5	40.1	38.5	39.4
OIL	25.1	21.1	18.8	16.5	14.3	11.7	7.5
GAS	12.5	10.9	11.4	11.4	11.0	9.0	8.1
NUCLEAR	4.4	11.2	11.0	11.6	13.6	20.5	25.9
HYDRO/GEOTHERMAL	21.4	21.4	21.6	21.0	21.0	20.2	18.7
OTHER [1]	-	-	-	-	-	.2	.3

Source: IEA/OECD Energy Balances and IEA Country Submissions (1982).

1. Solar, wind, etc.

OECD TOTAL

2/D4 ENERGY USE IN IRON AND STEEL INDUSTRY BY FUEL

	1973	1978	1979	1980	1981	1985	1990
TOTAL (MTOE)	231.9	186.7	194.4	179.5	174.0	na	na
SOLID FUELS	140.2	114.8	119.6	106.7	107.9	na	na
OIL	37.9	25.7	26.1	19.4	14.2	na	na
GAS	30.8	22.2	23.1	29.6	28.1	na	na
ELECTRICITY	23.0	24.0	25.5	23.8	23.9	na	na
HEAT [1]	-	-	-	-	-	-	-
TOTAL (PERCENTS)	100.0	100.0	100.0	100.0	100.0	na	na
SOLID FUELS	60.4	61.5	61.5	59.5	62.0	na	na
OIL	16.3	13.8	13.4	10.8	8.2	na	na
GAS	13.3	11.9	11.9	16.5	16.2	na	na
ELECTRICITY	9.9	12.8	13.1	13.2	13.7	na	na
HEAT [1]	-	-	-	-	-	-	-

Source: IEA/OECD Energy Balances and IEA Country Submissions (1982).

1. Includes only heat from combined heat and power plants.

2/D5 ENERGY USE IN OTHER INDUSTRIES BY FUEL

	1973	1978	1979	1980	1981	1985	1990
TOTAL (MTOE)	863.8	880.6	934.6	841.6	798.1	na	na
SOLID FUELS	80.5	74.8	79.8	78.3	83.3	na	na
OIL	431.3	458.0	494.9	400.4	357.1	na	na
GAS	220.8	202.7	209.8	212.9	208.2	na	na
ELECTRICITY	131.2	145.1	150.1	148.5	148.0	na	na
HEAT [1]	na	na	na	1.5	1.5	na	na
TOTAL (PERCENTS)	100.0	100.0	100.0	100.0	100.0	na	na
SOLID FUELS	9.3	8.5	8.5	9.3	10.4	na	na
OIL	49.9	52.0	53.0	47.6	44.7	na	na
GAS	25.6	23.0	22.5	25.3	26.1	na	na
ELECTRICITY	15.2	16.5	16.1	17.6	18.5	na	na
HEAT [1]	na	na	na	.2	.2	na	na

Source: IEA/OECD Energy Balances and IEA Country Submissions (1982).

1. Includes only heat from combined heat and power plants.

OECD TOTAL

2/D6 ENERGY USE IN INDUSTRY [1] BY FUEL

	1973	1978	1979	1980	1981	1985	1990
TOTAL (MTOE)	1095.7	1067.3	1128.9	1021.0	972.1	1247.2	1403.7
SOLID FUELS	220.7	189.6	199.4	185.0	191.1	299.6	346.6
OIL	469.2	483.7	521.0	419.8	371.3	447.6	484.2
GAS	251.6	225.0	233.0	242.5	236.3	290.8	323.5
ELECTRICITY	154.2	169.1	175.5	172.3	171.9	201.4	240.9
HEAT/OTHER [2]	na	na	na	1.5	1.5	7.7	8.5
TOTAL (PERCENTS)	100.0	100.0	100.0	100.0	100.0	100.0	100.0
SOLID FUELS	20.1	17.8	17.7	18.1	19.7	24.0	24.7
OIL	42.8	45.3	46.1	41.1	38.2	35.9	34.5
GAS	23.0	21.1	20.6	23.7	24.3	23.3	23.0
ELECTRICITY	14.1	15.8	15.5	16.9	17.7	16.2	17.2
HEAT/OTHER [2]	na	na	na	.2	.2	.6	.6

Source: IEA/OECD Energy Balances and IEA Country Submissions (1982).

1. Includes non-energy use of petroleum products.
2. Includes heat from combined heat and power plants and direct use of solar, wind and other non-solid renewable energy sources.

2/D7 ENERGY USE IN OTHER SECTORS [1] BY FUEL

	1973	1978	1979	1980	1981	1985	1990
TOTAL (MTOE)	861.5	896.3	914.0	905.8	889.9	846.1	917.9
SOLID FUELS	57.0	41.6	42.9	42.5	41.2	46.3	49.0
OIL	409.1	377.9	373.2	358.4	331.3	253.4	234.3
GAS	241.5	280.7	295.8	290.9	299.0	289.0	311.1
ELECTRICITY	153.9	196.1	202.1	208.0	212.4	236.5	282.9
HEAT/OTHER [2]	na	na	na	6.1	6.0	20.9	40.7
TOTAL (PERCENTS)	100.0	100.0	100.0	100.0	100.0	100.0	100.0
SOLID FUELS	6.6	4.6	4.7	4.7	4.6	5.5	5.3
OIL	47.5	42.2	40.8	39.6	37.2	29.9	25.5
GAS	28.0	31.3	32.4	32.1	33.6	34.2	33.9
ELECTRICITY	17.9	21.9	22.1	23.0	23.9	28.0	30.8
HEAT/OTHER [2]	na	na	na	.7	.7	2.5	4.4

Source: IEA/OECD Energy Balances and IEA Country Submissions (1982).

1. Includes use in agricultural, commercial, public services and residential sectors.
2. Includes heat from combined heat and power plants and direct use of solar, wind and other non-solid renewable energy sources.

3. OECD NORTH AMERICA

TOTAL ENERGY REQUIREMENTS BY FUEL

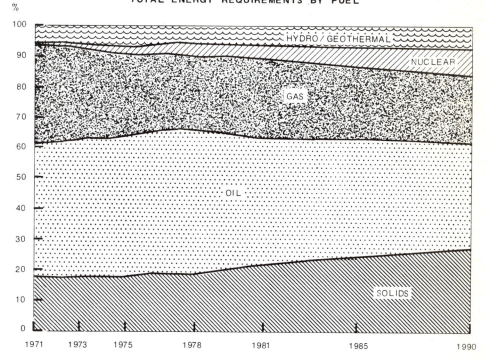

NORTH AMERICA

3/A1 TOTAL PRIMARY ENERGY REQUIREMENTS (TPER) BY FUEL

	1973	1978	1979	1980	1981	1985	1990
TPER (MTOE) [1]	1901.0	2098.8	2097.0	2034.3	1992.7	2183.3	2374.4
SOLID FUELS	337.4	391.0	408.9	425.9	435.8	535.4	640.4
OIL	860.2	986.0	953.7	872.1	816.0	837.9	818.4
GAS	565.5	513.6	532.3	530.7	527.6	506.8	527.9
NUCLEAR	24.3	80.4	75.8	75.8	82.7	141.4	209.5
HYDRO/GEOTHERMAL	113.5	128.0	126.3	130.0	130.7	155.0	164.2
OTHER [2]	-	-	-	-	-	9.0	16.0
TPER (PERCENTS) [1]	100.0	100.0	100.0	100.0	100.0	100.0	100.0
SOLID FUELS	17.8	18.6	19.5	20.9	21.9	24.5	27.0
OIL	45.3	47.0	45.5	42.9	41.0	38.4	34.5
GAS	29.8	24.5	25.4	26.1	26.5	23.2	22.2
NUCLEAR	1.3	3.8	3.6	3.7	4.2	6.5	8.8
HYDRO/GEOTHERMAL	5.9	6.1	6.0	6.4	6.6	7.1	6.9
OTHER [2]	-	-	-	-	-	.4	.7

Source: IEA/OECD Energy Balances and IEA Country Submissions (1982).

1. Net imports of electricity are only included in totals.
2. Includes other electricity generation (solar, wind, etc.), traded synthetic fuels and direct use of solar, wind and other non-solid renewable energy sources.

3/A2 TOTAL FINAL ENERGY CONSUMPTION (TFC) BY FUEL

	1973	1978	1979	1980	1981	1985	1990
TFC (MTOE)	1426.4	1509.6	1529.4	1463.6	1421.8	1521.1	1608.0
SOLID FUELS	97.9	92.7	96.1	80.3	82.5	158.7	178.1
OIL	774.4	861.0	862.9	803.2	754.8	744.1	744.4
GAS	384.4	356.1	366.7	373.6	374.7	378.9	398.8
ELECTRICITY	169.6	199.6	203.6	206.0	209.0	231.3	274.4
HEAT/OTHER [1]	na	na	na	.5	.7	8.1	12.3
TFC (PERCENTS)	100.0	100.0	100.0	100.0	100.0	100.0	100.0
SOLID FUELS	6.9	6.1	6.3	5.5	5.8	10.4	11.1
OIL	54.3	57.0	56.4	54.9	53.1	48.9	46.3
GAS	27.0	23.6	24.0	25.5	26.4	24.9	24.8
ELECTRICITY	11.9	13.2	13.3	14.1	14.7	15.2	17.1
HEAT/OTHER [1]	na	na	na	.0	.1	.5	.8

Source: IEA/OECD Energy Balances and IEA Country Submissions (1982).

1. Includes heat production from combined heat and power plants and direct use of solar, wind and other non-solid renewable energy sources.

3/A3 TOTAL SOLID FUELS BALANCE
(MTOE)

	1973	1978	1979	1980	1981	1985	1990
PRODUCTION [1]	354.4	400.6	463.2	494.2	488.8	615.2	735.3
IMPORTS	11.1	15.0	15.0	11.4	10.5	11.8	11.8
EXPORTS	-39.8	-33.6	-48.0	-64.4	-76.0	-91.6	-106.7
MARINE BUNKERS	-	-	-	-	-	-	-
STOCK CHANGE	11.8	9.0	-21.2	-15.3	12.5	-	-
PRIMARY REQUIREMENTS	337.4	391.0	408.9	425.9	435.8	535.4	640.4
ELECTRICITY GENERATION	-229.8	-282.1	-306.7	-331.8	-347.9	-371.7	-445.3
GAS MANUFACTURE	-.2	-	-	-	-	-	-6.0
LIQUEFACTION	-	-	-	-	-	-3.0	-9.0
ENERGY SECTOR							
OWN USE + LOSSES [2]	-9.4	-16.1	-6.1	-13.8	-5.4	-1.0	-1.0
FINAL CONSUMPTION	97.9	92.7	96.1	80.3	82.5	158.7	178.1
INDUSTRY SECTOR	91.9	87.1	91.2	76.2	78.6	152.8	171.7
IRON AND STEEL	50.0	42.7	43.9	33.8	36.4	52.1	52.8
CHEMICAL/							
PETROCHEMICAL	.2	-	.0	.3	.3	-	-
OTHER INDUSTRY	41.7	44.4	47.2	42.1	41.9	100.7	118.9
TRANSPORTATION SECTOR	.2	.0	.0	.0	.0	-	-
OTHER SECTORS	5.8	5.6	4.9	4.1	3.9	5.9	6.4
AGRICULTURE	-	-	-	-	-	-	-
COMMERCIAL AND							
PUBLIC SERVICE	.0	.0	.1	.2	.2	3.2	3.4
RESIDENTIAL	5.8	5.6	4.8	3.9	3.8	2.7	3.0

Source: IEA/OECD Energy balances and IEA Country Submissions (1982).

1. Wood and wood waste production and use for the United States is only included in 1985 and 1990 projections. Historical data were not available for inclusion in this table-set. See detailed note on Table 5/B1 for the United States.
2. Includes statistical difference.

3/B1 SOLID FUELS INDIGENOUS PRODUCTION BY FUEL TYPE
(MTOE)

	1973	1978	1979	1980	1981	1985	1990
TOTAL SOLID FUELS [1]	354.41	400.61	463.18	494.22	488.78	615.20	735.30
HARD COAL	346.97	380.34	440.57	467.09	460.21	526.30	637.90
COKING COAL	na	65.33	83.90	84.26	82.85	102.10	104.00
STEAM COAL	na	315.01	356.67	382.83	377.37	424.20	533.90
BROWN COAL/LIGNITE	7.44	15.13	17.27	19.32	21.28	31.60	36.40
OTHER SOLID FUELS [1]	na	na	na	na	na	57.30	61.00

Source: IEA/OECD Coal Statistics, IEA/OECD Energy Balances, IEA Country Submissions (1982).

1. Wood and wood waste production for the United States is only included in 1985 and 1990 projections. Historical data were not available for inclusion in this table-set. See detailed note on Table 5/B1 for the United States.

3/B2 COAL PRODUCTION BY TYPE
(THOUSAND METRIC TONS)

	1973	1978	1979	1980	1981
TOTAL COAL	563484	638439	741646	789356	784560
HARD COAL	541973	593934	689052	730536	719801
COKING COAL	na	101508	131294	131859	129654
STEAM COAL	na	492426	557758	598677	590147
BROWN COAL/LIGNITE	21511	44505	52594	58820	64759
DERIVED PRODUCTS:					
PATENT FUEL	-	-	-	-	-
COKE OVEN COKE	68173	49428	53804	47100	43474
GAS COKE	150	-	-	-	-
BROWN COAL BRIQUETTES	-	-	-	-	-
COKE OVEN GAS(TCAL)	140674	119556	127332	112290	104956
BLAST FURNACE GAS(TCAL)	134954	119206	116380	90118	95470

Source: IEA/OECD Coal Statistics.

NORTH AMERICA

3/C1 SOLID FUELS TRADE BY TYPE
(MTOE)

	1973	1978	1979	1980	1981	1985	1990
IMPORTS							
TOTAL SOLID FUELS	11.09	14.97	14.99	11.41	10.55	11.80	11.80
HARD COAL	10.19	11.17	12.32	10.74	10.03	11.40	11.40
COKING COAL	na	3.57	4.45	4.06	3.40	4.70	4.70
STEAM COAL	na	7.61	7.87	6.69	6.62	6.70	6.70
BROWN COAL/LIGNITE	-	-	-	-	-	-	-
OTHER SOLID FUELS [1]	.90	3.80	2.67	.67	.52	.40	.40
EXPORTS							
TOTAL SOLID FUELS	39.83	33.59	48.02	64.39	76.02	91.60	106.70
HARD COAL	38.67	32.98	46.98	62.92	75.26	91.60	106.70
COKING COAL	na	26.04	37.51	45.58	46.61	51.10	53.60
STEAM COAL	na	6.94	9.47	17.34	28.66	40.50	53.10
BROWN COAL/LIGNITE	-	.03	.02	-	-	-	-
OTHER SOLID FUELS [1]	1.16	.58	1.02	1.47	.76	-	-

Source: IEA/OECD Coal Statistics, IEA/OECD Energy Balances, IEA Country Submissions (1982).

1.   Includes derived coal products, e.g. coke.

NORTH AMERICA

3/C2 COAL IMPORTS BY ORIGIN
(THOUSAND METRIC TONS)

	1978	1979	1980	1981
HARD COAL IMPORTS	16798	19392	16912	15781
COKING COAL IMPORTS FROM:	5323	7011	6389	5355
AUSTRALIA	-	-	-	-
CANADA	-	-	-	-
GERMANY	-	-	-	-
UNITED KINGDOM	-	-	-	-
UNITED STATES	5323	7011	6389	5355
OTHER OECD	-	-	-	-
CZECHOSLOVAKIA	-	-	-	-
POLAND	-	-	-	-
USSR	-	-	-	-
CHINA	-	-	-	-
SOUTH AFRICA	-	-	-	-
OTHER NON-OECD	-	-	-	-
STEAM COAL IMPORTS FROM:	11475	12381	10523	10426
AUSTRALIA	933	156	-	-
CANADA	49	53	37	145
GERMANY	-	-	-	-
UNITED KINGDOM	-	-	-	3
UNITED STATES	8796	10513	9440	9481
OTHER OECD	-	-	-	-
CZECHOSLOVAKIA	-	-	-	-
POLAND	645	635	287	28
USSR	-	-	-	-
CHINA	-	-	-	-
SOUTH AFRICA	996	1024	699	738
OTHER NON-OECD	56	-	60	31

Source: IEA/OECD Coal Statistics.

NORTH AMERICA

3/C3 HARD COAL EXPORTS BY DESTINATION
(THOUSAND METRIC TONS)

	1978	1979	1980	1981
TOTAL EXPORTS	50903	73555	98496	117801
EXPORTS TO:				
AUSTRALIA	-	3	-	-
AUSTRIA	-	-	-	-
BELGIUM	174	2962	4223	3859
CANADA	14192	17637	15830	16525
DENMARK	309	281	1749	3838
FINLAND	-	-	253	1140
FRANCE	1468	3579	7066	8840
GERMANY	1052	2960	2898	4514
GREECE	-	55	368	124
ICELAND	-	-	6	-
IRELAND	-	257	357	459
ITALY	3069	4654	6518	9571
JAPAN	20464	24710	32051	33941
LUXEMBOURG	-	-	-	-
NETHERLANDS	982	1824	4224	6262
NEW ZEALAND	-	-	-	-
NORWAY	68	210	252	439
PORTUGAL	265	280	329	256
SPAIN	838	1379	3141	5832
SWEDEN	453	834	1112	1335
SWITZERLAND	-	-	99	1213
TURKEY	409	758	796	516
UNITED KINGDOM	360	1225	3756	2123
UNITED STATES	4	12	1	67
TOTAL NON-OECD	6796	9935	13467	16947

Source: IEA/OECD Coal Statistics.

NORTH AMERICA

3/D1 FUEL INPUT IN ELECTRICITY GENERATION

	1973	1978	1979	1980	1981	1985	1990
TOTAL (MTOE)	547.6	662.2	676.6	693.4	706.9	794.8	936.1
SOLID FUELS	229.8	282.1	306.7	331.8	347.9	371.7	445.3
OIL	89.9	94.3	83.2	66.7	57.7	56.6	43.8
GAS	89.9	77.3	84.5	89.1	87.8	69.1	69.3
NUCLEAR	24.3	80.4	75.8	75.8	82.7	141.4	209.5
HYDRO/GEOTHERMAL	113.5	128.0	126.3	130.0	130.7	155.0	164.2
OTHER [1]	-	-	-	-	-	1.0	4.0
TOTAL (PERCENTS)	100.0	100.0	100.0	100.0	100.0	100.0	100.0
SOLID FUELS	42.1	42.6	45.3	47.9	49.2	46.8	47.6
OIL	16.5	14.2	12.3	9.6	8.2	7.1	4.7
GAS	16.5	11.7	12.5	12.9	12.4	8.7	7.4
NUCLEAR	4.4	12.1	11.2	10.9	11.7	17.8	22.4
HYDRO/GEOTHERMAL	20.6	19.3	18.7	18.7	18.5	19.5	17.5
OTHER [1]	-	-	-	-	-	.1	.4

Source: IEA/OECD Energy Balances and IEA Country Submissions (1982).

1. Solar, wind, etc.

3/D2 ELECTRICTY PRODUCTION BY FUEL TYPE

	1973	1978	1979	1980	1981	1985	1990
TOTAL (TWH) [1]	2357.1	2661.2	2719.7	2768.5	2790.1	2977.3	3561.5
SOLID FUELS	1019.6	1080.9	1189.8	1284.2	1330.9	1389.8	1668.6
OIL	343.6	400.5	334.9	272.8	227.7	214.9	172.3
GAS	414.1	332.2	358.6	374.1	372.8	265.1	269.7
NUCLEAR	104.4	322.4	303.7	302.1	325.9	529.8	796.1
HYDRO/GEOTHERMAL	475.3	525.2	532.7	535.4	532.7	573.7	639.8
OTHER [2]	-	-	-	-	-	4.0	15.0
TOTAL (PERCENTS)	100.0	100.0	100.0	100.0	100.0	100.0	100.0
SOLID FUELS	43.3	40.6	43.7	46.4	47.7	46.7	46.9
OIL	14.6	15.0	12.3	9.9	8.2	7.2	4.8
GAS	17.6	12.5	13.2	13.5	13.4	8.9	7.6
NUCLEAR	4.4	12.1	11.2	10.9	11.7	17.8	22.4
HYDRO/GEOTHERMAL	20.2	19.7	19.6	19.3	19.1	19.3	18.0
OTHER [2]	-	-	-	-	-	.1	.4

Source: IEA/OECD Energy Balances and IEA Country Submissions (1982).

1. Electricity generation by autoproducers is not included for the United States.
2. Solar, wind, etc.

3/D4 ENERGY USE IN IRON AND STEEL INDUSTRY BY FUEL

	1973	1978	1979	1980	1981	1985	1990
TOTAL (MTOE)	83.9	69.5	71.6	66.1	67.5	na	na
SOLID FUELS	50.0	42.7	43.9	33.8	36.4	52.1	52.8
OIL	6.3	6.0	6.2	5.0	3.8	na	na
GAS	19.3	12.9	13.3	20.1	19.4	na	na
ELECTRICITY	8.2	7.8	8.2	7.2	7.8	na	na
HEAT/OTHER [1]	-	-	-	-	-	na	na
TOTAL (PERCENTS)	100.0	100.0	100.0	100.0	100.0	na	na
SOLID FUELS	59.7	61.5	61.3	51.1	54.0	na	na
OIL	7.5	8.6	8.7	7.5	5.6	na	na
GAS	23.1	18.6	18.5	30.4	28.8	na	na
ELECTRICITY	9.8	11.3	11.4	10.9	11.6	na	na
HEAT/OTHER [1]	-	-	-	-	-	-	-

Source: IEA/OECD Energy Balances and IEA Country Submissions (1982).

1. Includes only heat from combined heat and power plants.

3/D5 ENERGY USE IN OTHER INDUSTRIES BY FUEL

	1973	1978	1979	1980	1981	1985	1990
TOTAL (MTOE)	424.8	440.4	458.9	415.9	394.0	na	na
SOLID FUELS [1]	41.9	44.4	47.3	42.4	42.2	100.7	118.9
OIL	143.9	183.3	194.0	153.9	135.7	na	na
GAS	175.1	141.4	145.2	148.6	144.8	na	na
ELECTRICITY	64.0	71.3	72.5	71.0	71.3	na	na
HEAT/OTHER [2]	-	-	-	-	-	na	na
TOTAL (PERCENTS)	100.0	100.0	100.0	100.0	100.0	na	na
SOLID FUELS	9.9	10.1	10.3	10.2	10.7	na	na
OIL	33.9	41.6	42.3	37.0	34.5	na	na
GAS	41.2	32.1	31.6	35.7	36.8	na	na
ELECTRICITY	15.1	16.2	15.8	17.1	18.1	na	na
HEAT/OTHER [2]	-	-	-	-	-	na	na

Source: IEA/OECD Energy Balances and IEA Country Submissions (1982).

1. Wood and wood waste production and use for the United States is only included in 1985 and 1990 projections. Historical data was not available for inclusion in this table-set. See detailed note on Table 5/B1 for the United States.
2. Includes only heat from combined heat and power plants.

3/D6 ENERGY USE IN INDUSTRY [1] BY FUEL

	1973	1978	1979	1980	1981	1985	1990
TOTAL (MTOE)	508.7	509.9	530.5	482.0	461.4	646.4	725.9
SOLID FUELS	91.9	87.1	91.2	76.2	78.6	152.8	171.7
OIL	150.2	189.3	200.2	158.9	139.5	200.0	227.8
GAS	194.4	154.3	158.5	168.7	164.2	190.7	203.7
ELECTRICITY	72.2	79.2	80.7	78.2	79.1	97.8	119.5
HEAT/OTHER [2]	-	-	-	-	-	5.1	3.2
TOTAL (PERCENTS)	100.0	100.0	100.0	100.0	100.0	100.0	100.0
SOLID FUELS	18.1	17.1	17.2	15.8	17.0	23.6	23.7
OIL	29.5	37.1	37.7	33.0	30.2	30.9	31.4
GAS	38.2	30.3	29.9	35.0	35.6	29.5	28.1
ELECTRICITY	14.2	15.5	15.2	16.2	17.1	15.1	16.5
HEAT/OTHER [2]	-	-	-	-	-	.8	.4

Source: IEA/OECD Energy Balances and IEA Country Submissions (1982).

1. Includes non-energy use of petroleum products.
2. Includes heat from combined heat and power plants and direct use of solar, wind and other non-solid renewable energy sources.

3/D7 ENERGY USE IN OTHER SECTORS [1] BY FUEL

	1973	1978	1979	1980	1981	1985	1990
TOTAL (MTOE)	473.2	486.2	490.3	493.1	486.6	414.3	440.3
SOLID FUELS	5.8	5.6	4.9	4.1	3.9	5.9	6.4
OIL	180.4	158.6	154.6	157.9	143.7	83.7	74.8
GAS	190.0	201.8	208.3	203.2	208.7	188.2	195.1
ELECTRICITY	97.0	120.1	122.5	127.4	129.5	133.5	154.9
HEAT/OTHER [2]	na	na	na	.5	.7	3.0	9.1
TOTAL (PERCENTS)	100.0	100.0	100.0	100.0	100.0	100.0	100.0
SOLID FUELS	1.2	1.2	1.0	.8	.8	1.4	1.5
OIL	38.1	32.6	31.5	32.0	29.5	20.2	17.0
GAS	40.2	41.5	42.5	41.2	42.9	45.4	44.3
ELECTRICITY	20.5	24.7	25.0	25.8	26.6	32.2	35.2
HEAT/OTHER [2]	na	na	na	.1	.1	.7	2.1

Source: IEA/OECD Energy Balances and IEA Country Submissions (1982).

1. Includes use in agricultural, commercial, public services and residential sectors.
2. Includes heat from combined heat and power plants and direct use of solar, wind and other non-solid renewable energy sources.

CANADA

TOTAL ENERGY REQUIREMENTS BY FUEL

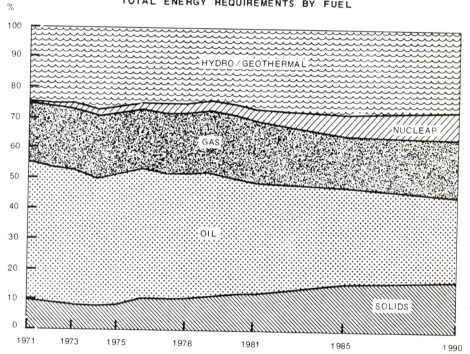

CANADA

3/A1 TOTAL PRIMARY ENERGY REQUIREMENTS (TPER) BY FUEL

	1973	1978	1979	1980	1981	1985	1990
TPER (MTOE) [1]	190.0	216.5	217.3	227.6	224.5	254.3	281.4
SOLID FUELS	15.0	23.1	23.5	27.3	27.0	40.4	46.4
OIL	84.5	89.5	90.4	87.4	82.6	81.9	80.4
GAS	38.7	43.5	45.4	49.0	47.5	43.8	54.9
NUCLEAR	3.8	6.9	7.3	8.3	8.6	18.4	25.5
HYDRO/GEOTHERMAL	49.1	55.3	53.2	58.0	61.7	74.0	78.2
OTHER [2]	-	-	-	-	-	-	-
TPER (PERCENTS) [1]	100.0	100.0	100.0	100.0	100.0	100.0	100.0
SOLID FUELS	8.0	10.7	10.8	12.0	12.0	15.9	16.5
OIL	44.7	41.3	41.6	38.4	36.8	32.2	28.6
GAS	20.5	20.1	20.9	21.5	21.1	17.2	19.5
NUCLEAR	2.0	3.2	3.3	3.6	3.8	7.2	9.1
HYDRO/GEOTHERMAL	25.5	25.5	24.5	25.5	27.5	29.1	27.8
OTHER [2]	-	-	-	-	-	-	-

Source: IEA/OECD Energy Balances and IEA Country Submissions (1982).

1. Net imports of electricity are only included in totals.
2. Includes other electricity generation (solar, wind, etc.) and traded synthetic fuels.

3/A2 TOTAL FINAL ENERGY CONSUMPTION (TFC) BY FUEL

	1973	1978	1979	1980	1981	1985	1990
TFC (MTOE)	126.9	146.2	150.9	160.7	155.4	160.1	181.0
SOLID FUELS	6.3	10.3	11.0	13.2	12.3	17.7	22.1
OIL	78.5	81.1	84.4	83.6	78.7	73.1	72.4
GAS	23.4	30.9	31.0	37.8	37.4	38.9	49.8
ELECTRICITY	18.7	23.9	24.5	25.7	26.3	30.3	36.4
HEAT [1]	na	na	na	.5	.7	.1	.3
TFC (PERCENTS)	100.0	100.0	100.0	100.0	100.0	100.0	100.0
SOLID FUELS	5.0	7.1	7.3	8.2	7.9	11.1	12.2
OIL	61.9	55.5	55.9	52.0	50.7	45.7	40.0
GAS	18.4	21.1	20.5	23.5	24.1	24.3	27.5
ELECTRICITY	14.7	16.3	16.3	16.0	16.9	18.9	20.1
HEAT [1]	na	na	na	.3	.5	.1	.2

Source: IEA/OECD Energy Balances and IEA Country Submissions (1982).

1. Includes only heat production from combined heat and power plants.

CANADA

3/A3 TOTAL SOLID FUELS BALANCE
(MTOE)

	1973	1978	1979	1980	1981	1985	1990
PRODUCTION	11.5	22.0	22.4	26.7	27.9	48.2	64.3
IMPORTS	10.4	9.8	11.4	10.3	9.6	10.8	10.8
EXPORTS	-7.6	-9.5	-8.8	-9.9	-10.0	-18.6	-28.7
MARINE BUNKERS	-	-	-	-	-	-	-
STOCK CHANGE	.7	.8	-1.4	.2	-.5	-	-
PRIMARY REQUIREMENTS	15.0	23.1	23.5	27.3	27.0	40.4	46.4
ELECTRICITY GENERATION	-8.2	-12.3	-12.2	-13.9	-15.0	-22.7	-24.3
GAS MANUFACTURE	-	-	-	-	-	-	-
LIQUEFACTION	-	-	-	-	-	-	-
ENERGY SECTOR							
OWN USE + LOSSES [1]	-.5	-.5	-.4	-.2	.3	-	-
FINAL CONSUMPTION	6.3	10.3	11.0	13.2	12.3	17.7	22.1
INDUSTRY SECTOR	5.8	10.1	10.8	13.0	12.2	16.8	20.7
IRON AND STEEL	3.7	3.7	4.2	4.1	3.8	6.1	6.8
CHEMICAL/							
PETROCHEMICAL	.0	-	.0	.3	.3	-	-
OTHER INDUSTRY	2.1	6.4	6.6	8.7	8.1	10.7	13.9
TRANSPORTATION SECTOR	.1	.0	.0	.0	.0	-	-
OTHER SECTORS	.4	.2	.1	.1	.1	.9	1.4
AGRICULTURE	-	-	-	-	-	-	-
COMMERCIAL AND							
PUBLIC SERVICE	.0	-	.0	.1	.0	.2	.4
RESIDENTIAL	.4	.2	.1	.1	.1	.7	1.0

Source: IEA/OECD Energy Balances and IEA Country Submissions (1982).

1. Includes statistical difference.

CANADA

3/B1 SOLID FUELS INDIGENOUS PRODUCTION BY FUEL TYPE
(MTOE)

	1973	1978	1979	1980	1981	1985	1990
TOTAL SOLID FUELS	11.53	22.01	22.43	26.70	27.86	48.20	64.30
HARD COAL	8.27	11.48	11.70	12.80	13.80	25.30	39.90
COKING COAL	na	9.23	8.89	8.98	9.62	18.10	20.00
STEAM COAL	na	2.25	2.81	3.82	4.18	7.20	19.90
BROWN COAL/LIGNITE	3.26	5.39	5.38	6.09	6.77	13.60	16.40
OTHER SOLID FUELS	-	5.14	5.35	7.81	7.29	9.30	8.00

Source: IEA/OECD Coal Statistics, IEA/OECD Energy Balances, IEA Country Submissions (1982).

3/B2 COAL PRODUCTION BY TYPE
(THOUSAND METRIC TONS)

	1973	1978	1979	1980	1981
TOTAL COAL	20472	30477	33013	36664	40088
HARD COAL	12337	17141	18432	20151	21739
COKING COAL	na	13780	13999	14140	15149
STEAM COAL	na	3361	4433	6011	6590
BROWN COAL/LIGNITE	8135	13336	14581	16513	18349
DERIVED PRODUCTS:					
PATENT FUEL	-	-	-	-	-
COKE OVEN COKE	5370	4968	5775	5250	4659
GAS COKE	-	-	-	-	-
BROWN COAL BRIQUETTES	-	-	-	-	-
COKE OVEN GAS(TCAL)	11083	11171	12987	11805	10456
BLAST FURNACE GAS(TCAL)	14589	13200	14300	14300	14500

Source: IEA/OECD Coal Statistics.

CANADA

3/C1 SOLID FUELS TRADE BY TYPE
(MTOE)

	1973	1978	1979	1980	1981	1985	1990
IMPORTS							
TOTAL SOLID FUELS	10.35	9.78	11.38	10.32	9.63	10.80	10.80
HARD COAL	10.11	9.46	11.13	10.05	9.42	10.80	10.80
COKING COAL	na	3.57	4.45	4.06	3.40	4.70	4.70
STEAM COAL	na	5.89	6.68	5.99	6.02	6.10	6.10
BROWN COAL/LIGNITE	-	-	-	-	-	-	-
OTHER SOLID FUELS [1]	.24	.32	.25	.27	.21	-	-
EXPORTS							
TOTAL SOLID FUELS	7.55	9.54	8.84	9.91	10.02	18.60	28.70
HARD COAL	7.31	9.37	8.68	9.70	9.97	18.60	28.70
COKING COAL	na	8.72	8.10	8.97	8.76	15.10	17.60
STEAM COAL	na	.65	.58	.73	1.21	3.50	11.10
BROWN COAL/LIGNITE	-	-	.01	-	-	-	-
OTHER SOLID FUELS [1]	.24	.17	.15	.21	.05	-	-

Source: IEA/OECD Coal Statistics, IEA/OECD Energy Balances, IEA Country Submissions (1982).

1. Includes derived coal products, e.g. coke.

CANADA

3/C2 COAL IMPORTS BY ORIGIN
(THOUSAND METRIC TONS)

	1978	1979	1980	1981
HARD COAL IMPORTS	14119	17524	15829	14836
COKING COAL	5323	7011	6389	5355
IMPORTS FROM:				
AUSTRALIA	-	-	-	-
CANADA	-	-	-	-
GERMANY	-	-	-	-
UNITED KINGDOM	-	-	-	-
UNITED STATES	5323	7011	6389	5355
OTHER OECD	-	-	-	
CZECHOSLOVAKIA	-	-	-	-
POLAND	-	-	-	-
USSR	-	-	-	-
CHINA	-	-	-	-
SOUTH AFRICA	-	-	-	-
OTHER NON-OECD	-	-	-	-
STEAM COAL	8796	10513	9440	9481
IMPORTS FROM:				
AUSTRALIA	-	-	-	-
CANADA	-	-	-	-
GERMANY	-	-	-	
UNITED KINGDOM	-	-	-	-
UNITED STATES	8796	10513	9440	9481
OTHER OECD	-	-	-	
CZECHOSLOVAKIA	-	-	-	-
POLAND	-	-	-	-
USSR	-	-	-	-
CHINA	-	-	-	-
SOUTH AFRICA	-	-	-	-
OTHER NON-OECD	-	-	-	-

Source: IEA/OECD Coal Statistics.

CANADA

3/C3 HARD COAL EXPORTS BY DESTINATION
(THOUSAND METRIC TONS)

	1978	1979	1980	1981
TOTAL EXPORTS	13988	13669	15269	15705
EXPORTS TO:				
AUSTRALIA	-	-	-	-
AUSTRIA	-	-	-	-
BELGIUM	174	56	21	56
CANADA	-	-	-	-
DENMARK	309	133	252	319
FINLAND	-	-	-	-
FRANCE	-	72	-	-
GERMANY	492	641	603	607
GREECE	-	55	316	-
ICELAND	-	-	-	-
IRELAND	-	-	-	-
ITALY	164	132	48	71
JAPAN	11017	10485	11123	10486
LUXEMBOURG	-	-	-	-
NETHERLANDS	-	-	-	75
NEW ZEALAND	-	-	-	-
NORWAY	-	-	-	-
PORTUGAL	-	-	-	-
SPAIN	150	153	48	54
SWEDEN	154	163	192	261
SWITZERLAND	-	-	-	-
TURKEY	-	-	-	-
UNITED KINGDOM	-	-	-	-
UNITED STATES	4	12	1	67
TOTAL NON-OECD	1524	1767	2665	3709

Source: IEA/OECD Statistics.

CANADA

3/C4 COKING COAL EXPORTS BY DESTINATION
(THOUSAND METRIC TONS)

	1978	1979	1980	1981
TOTAL EXPORTS	13017	12752	14127	13796
EXPORTS TO:				
AUSTRALIA	-	-	-	-
AUSTRIA	-	-	-	-
BELGIUM	147	56	21	56
CANADA	-	-	-	-
DENMARK	-	-	-	-
FINLAND	-	-	-	-
FRANCE	-	50	-	-
GERMANY	-	-	161	152
GREECE	-	55	316	-
ICELAND	-	-	-	-
IRELAND	-	-	-	-
ITALY	164	132	48	71
JAPAN	10934	10376	10711	9434
LUXEMBOURG	-	-	-	-
NETHERLANDS	-	-	-	75
NEW ZEALAND	-	-	-	-
NORWAY	-	-	-	-
PORTUGAL	-	-	-	-
SPAIN	150	153	48	54
SWEDEN	154	163	157	261
SWITZERLAND	-	-	-	-
TURKEY	-	-	-	-
UNITED KINGDOM	-	-	-	-
UNITED STATES	-	-	-	22
TOTAL NON-OECD	1468	1767	2665	3671

Source: IEA/OECD Statistics.

CANADA

3/C5 STEAM COAL EXPORTS BY DESTINATION
(THOUSAND METRIC TONS)

	1978	1979	1980	1981
TOTAL EXPORTS	971	917	1142	1909
EXPORTS TO:				
AUSTRALIA	-	-	-	-
AUSTRIA	-	-	-	-
BELGIUM	27	-	-	-
CANADA	-	-	-	-
DENMARK	309	133	252	319
FINLAND	-	-	-	-
FRANCE	-	22	-	-
GERMANY	492	641	442	455
GREECE	-	-	-	-
ICELAND	-	-	-	-
IRELAND	-	-	-	-
ITALY	-	-	-	-
JAPAN	83	109	412	1052
LUXEMBOURG	-	-	-	-
NETHERLANDS	-	-	-	-
NEW ZEALAND	-	-	-	-
NORWAY	-	-	-	-
PORTUGAL	-	-	-	-
SPAIN	-	-	-	-
SWEDEN	-	-	35	-
SWITZERLAND	-	-	-	-
TURKEY	-	-	-	-
UNITED KINGDOM	-	-	-	-
UNITED STATES	4	12	1	45
TOTAL NON-OECD	56	-	-	38

Source: IEA/OECD Statistics.

CANADA

3/D1 FUEL INPUT IN ELECTRICITY GENERATION

	1973	1978	1979	1980	1981	1985	1990
TOTAL (MTOE)	68.1	80.2	78.1	84.8	88.8	118.8	131.1
SOLID FUELS	8.2	12.3	12.2	13.9	15.0	22.7	24.3
OIL	2.6	3.5	3.2	2.8	2.1	1.6	.8
GAS	4.3	2.2	2.1	1.8	1.5	2.1	2.3
NUCLEAR	3.8	6.9	7.3	8.3	8.6	18.4	25.5
HYDRO/GEOTHERMAL	49.1	55.3	53.2	58.0	61.7	74.0	78.2
OTHER [1]	-	-	-	-	-	-	-
TOTAL (PERCENTS)	100.0	100.0	100.0	100.0	100.0	100.0	100.0
SOLID FUELS	12.2	15.3	15.6	16.4	16.8	19.1	18.5
OIL	3.9	4.4	4.2	3.3	2.3	1.3	.6
GAS	6.4	2.8	2.7	2.2	1.7	1.8	1.8
NUCLEAR	5.6	8.6	9.3	9.8	9.7	15.5	19.5
HYDRO/GEOTHERMAL	71.8	69.0	68.2	68.4	69.4	62.3	59.6
OTHER [1]	-	-	-	-	-	-	-

Source: IEA/OECD Energy Balances and IEA Country Submissions (1982).

1. Solar, wind, etc.

3/D2 ELECTRICTY PRODUCTION BY FUEL TYPE

	1973	1978	1979	1980	1981	1985	1990
TOTAL (TWH)	270.2	341.3	358.2	367.5	379.4	431.3	529.5
SOLID FUELS	34.9	51.1	55.1	58.3	61.1	74.8	82.6
OIL	9.1	15.4	14.7	13.3	10.0	7.9	10.3
GAS	16.2	10.0	10.9	8.8	8.0	13.1	17.7
NUCLEAR	15.3	29.4	33.3	35.9	36.9	66.8	103.1
HYDRO/GEOTHERMAL	194.8	235.4	244.2	251.2	263.4	268.7	315.8
OTHER [1]	-	-	-	-	-	-	-
TOTAL (PERCENTS)	100.0	100.0	100.0	100.0	100.0	100.0	100.0
SOLID FUELS	12.9	15.0	15.4	15.9	16.1	17.3	15.6
OIL	3.4	4.5	4.1	3.6	2.6	1.8	1.9
GAS	6.0	2.9	3.1	2.4	2.1	3.0	3.3
NUCLEAR	5.6	8.6	9.3	9.8	9.7	15.5	19.5
HYDRO/GEOTHERMAL	72.1	69.0	68.2	68.4	69.4	62.3	59.6
OTHER [1]	-	-	-	-	-	-	-

Source: IEA/OECD Energy Balances and IEA Country Submissions (1982).

1. Solar, wind, etc.

CANADA

3/D3 PROJECTIONS OF ELECTRICITY GENERATION CAPACITY BY FUEL
(GW)

	Nuclear	Hydro/ Geoth.	Solid Fuels	Oil	Gas	Multi Firing	Total
I 1981							
Operating	5.6	49.7	16.1	7.7	4.8	-	83.9
II 1981-1985							
Capacity:							
Under Construction	4.9	9.0	1.8	0.2	0.1	-	16.0
Authorised	-	-	-	-	-	-	-
Other Planned	-	-	-	-	-	-	-
Conversion	-	-	-	-	-	-	-
Decommissioning	-	-	-	-	-	-	-
Total Operating 1985	10.5	58.7	17.9	7.9	4.9	-	99.9
III 1986-1990							
Capacity:							
Under Construction	5.1	8.0	2.7	-	2.0	-	17.8
Authorised	-	-	-	-	-	-	-
Other Planned	-	-	-	-	0.7	-	0.7
Conversion	-	-	-	1.2^e	-1.2	-	-
Decommissioning	-	-	-	-	-	-	-
Total Operating 1990	15.6	66.7	21.8	7.4	6.9	-	118.4

Source: IEA/OECD Electricity Statistics and IEA Country Submissions (1982).

CANADA

3/D4 ENERGY USE IN IRON AND STEEL INDUSTRY BY FUEL

	1973	1978	1979	1980	1981	1985	1990
TOTAL (MTOE)	5.8	6.7	7.3	6.8	6.4	9.1	10.2
SOLID FUELS	3.7	3.7	4.2	4.1	3.8	6.1	6.8
OIL	.5	.9	.9	.4	.4	.6	.4
GAS	1.1	1.4	1.4	1.5	1.4	1.4	1.8
ELECTRICITY	.5	.7	.8	.8	.8	1.0	1.2
HEAT [1]	-	-	-	-	-	-	-
TOTAL (PERCENTS)	100.0	100.0	100.0	100.0	100.0	100.0	100.0
SOLID FUELS	63.2	55.1	57.8	59.7	60.0	67.0	66.7
OIL	9.1	13.7	12.3	6.0	5.7	6.6	3.9
GAS	18.8	20.2	19.1	22.2	22.4	15.4	17.6
ELECTRICITY	8.9	10.9	10.7	12.0	11.9	11.0	11.8
HEAT [1]	-	-	-	-	-	-	-

Source: IEA/OECD Energy Balances and IEA Country Submissions (1982).

1. Includes only heat from combined heat and power plants.

3/D5 ENERGY USE IN OTHER INDUSTRIES BY FUEL

	1973	1978	1979	1980	1981	1985	1990
TOTAL (MTOE)	42.5	48.5	51.3	56.7	56.0	62.3	74.7
SOLID FUELS	2.1	6.4	6.6	9.0	8.3	10.7	13.9
OIL	21.6	19.8	20.9	20.8	20.0	20.4	21.4
GAS	10.1	12.6	14.0	16.7	17.0	18.3	23.9
ELECTRICITY	8.6	9.7	9.7	10.3	10.7	12.8	15.3
HEAT [1]	-	-	-	-	-	.1	.2
TOTAL (PERCENTS)	100.0	100.0	100.0	100.0	100.0	100.0	100.0
SOLID FUELS	5.0	13.3	13.0	15.8	14.9	17.2	18.6
OIL	50.8	40.7	40.8	36.7	35.7	32.7	28.6
GAS	23.8	26.1	27.2	29.4	30.3	29.4	32.0
ELECTRICITY	20.3	19.9	19.0	18.1	19.2	20.5	20.5
HEAT [1]	-	-	-	-	-	.2	.3

Source: IEA/OECD Energy Balances and IEA Country Submissions (1982).

1. Includes only heat from combined heat and power plants.

CANADA

3/D6 ENERGY USE IN INDUSTRY [1] BY FUEL

	1973	1978	1979	1980	1981	1985	1990
TOTAL (MTOE)	48.3	55.2	58.5	63.6	62.4	71.4	84.9
SOLID FUELS	5.8	10.1	10.8	13.0	12.2	16.8	20.7
OIL	22.1	20.7	21.8	21.2	20.3	21.0	21.8
GAS	11.2	14.0	15.4	18.2	18.4	19.7	25.7
ELECTRICITY	9.1	10.4	10.5	11.1	11.5	13.8	16.5
HEAT [2]	-	-	-	-	-	.1	.2
TOTAL (PERCENTS)	100.0	100.0	100.0	100.0	100.0	100.0	100.0
SOLID FUELS	12.0	18.4	18.5	20.5	19.5	23.5	24.4
OIL	45.8	37.5	37.3	33.4	32.6	29.4	25.7
GAS	23.2	25.3	26.2	28.6	29.5	27.6	30.3
ELECTRICITY	18.9	18.8	18.0	17.5	18.4	19.3	19.4
HEAT [2]	-	-	-	-	-	.1	.2

Source: IEA/OECD Energy Balances and IEA Country Submissions (1982).

1. Includes non-energy use of petroleum products.
2. Includes only heat from combined heat and power plants.

3/D7 ENERGY USE IN OTHER SECTORS [1] BY FUEL

	1973	1978	1979	1980	1981	1985	1990
TOTAL (MTOE)	43.4	49.6	49.2	52.7	51.8	49.3	54.3
SOLID FUELS	.4	.2	.1	.1	.1	.9	1.4
OIL	21.3	19.2	19.7	19.8	19.2	12.7	8.8
GAS	12.2	16.9	15.6	17.9	17.2	19.2	24.1
ELECTRICITY	9.5	13.3	13.8	14.4	14.6	16.5	19.9
HEAT [2]	na	na	na	.5	.7	-	.1
TOTAL (PERCENTS)	100.0	100.0	100.0	100.0	100.0	100.0	100.0
SOLID FUELS	.9	.3	.2	.2	.2	1.8	2.6
OIL	49.0	38.8	40.0	37.6	37.0	25.8	16.2
GAS	28.1	34.1	31.8	34.0	33.3	38.9	44.4
ELECTRICITY	22.0	26.8	28.1	27.3	28.2	33.5	36.6
HEAT [2]	na	na	na	.9	1.4	-	.2

Source: IEA/OECD Energy Balances and IEA Country Submissions (1982).

1. Includes use in agricultural, commercial, public services and residential sectors.
2. Includes only heat from combined heat and power plants.

UNITED STATES

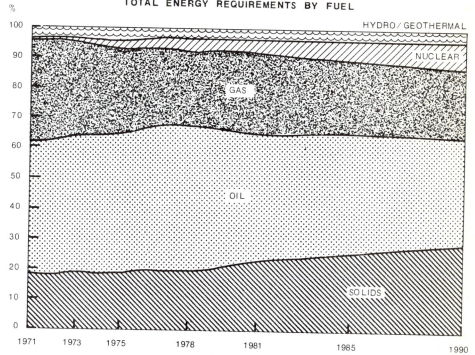

TOTAL ENERGY REQUIREMENTS BY FUEL

%

HYDRO/GEOTHERMAL

NUCLEAR

GAS

OIL

SOLIDS

100
90
80
70
60
50
40
30
20
10
0

1971   1973   1975   1978   1981   1985   1990

UNITED STATES

3/A1 TOTAL PRIMARY ENERGY REQUIREMENTS (TPER) BY FUEL

	1973	1978	1979	1980	1981	1985	1990
TPER (MTOE) [1]	1711.0	1882.3	1879.8	1806.7	1768.2	1929.0	2093.0
SOLID FUELS [2]	322.4	367.9	385.4	398.6	408.9	495.0	594.0
OIL	775.7	896.5	863.3	784.7	733.4	756.0	738.0
GAS	526.7	470.1	487.0	481.7	480.1	463.0	473.0
NUCLEAR	20.5	73.5	68.5	67.5	74.1	123.0	184.0
HYDRO/GEOTHERMAL	64.5	72.7	73.1	72.0	69.0	81.0	86.0
OTHER [3]	-	-	-	-	-	9.0	16.0
TPER (PERCENTS) [1]	100.0	100.0	100.0	100.0	100.0	100.0	100.0
SOLID FUELS [2]	18.8	19.5	20.5	22.1	23.1	25.7	28.4
OIL	45.3	47.6	45.9	43.4	41.5	39.2	35.3
GAS	30.8	25.0	25.9	26.7	27.2	24.0	22.6
NUCLEAR	1.2	3.9	3.6	3.7	4.2	6.4	8.8
HYDRO/GEOTHERMAL	3.8	3.9	3.9	4.0	3.9	4.2	4.1
OTHER [3]	-	-	-	-	-	.5	.8

Source: IEA/OECD Energy Balances and IEA Country Submissions (1982).

1. Net imports of electricity are only included in totals.
2. Projections for 1985 and 1990 include about 50 Mtoe wood and wood-waste not reported for historical years. See note to Table 5/B1.
3. Includes other electricity generation (solar, wind, etc.), traded and direct use of solar, wind and other non-solid renewable energy sources.

3/A2 TOTAL FINAL ENERGY CONSUMPTION (TFC) BY FUEL

	1973	1978	1979	1980	1981	1985	1990
TFC (MTOE)	1299.4	1363.4	1378.5	1302.9	1266.4	1361.0	1427.0
SOLID FUELS	91.6	82.4	85.1	67.1	70.3	141.0	156.0
OIL	695.9	779.9	778.5	719.6	676.1	671.0	672.0
GAS	361.0	325.2	335.7	335.8	337.3	340.0	349.0
ELECTRICITY	150.9	175.8	179.1	180.4	182.8	201.0	238.0
OTHER [1]	-	-	-	-	-	8.0	12.0
TFC (PERCENTS)	100.0	100.0	100.0	100.0	100.0	100.0	100.0
SOLID FUELS	7.0	6.0	6.2	5.1	5.5	10.4	10.9
OIL	53.6	57.2	56.5	55.2	53.4	49.3	47.1
GAS	27.8	23.9	24.4	25.8	26.6	25.0	24.5
ELECTRICITY	11.6	12.9	13.0	13.8	14.4	14.8	16.7
OTHER [1]	-	-	-	-	-	.6	.8

Source: IEA/OECD Energy Balances and IEA Country Submissions (1982).

1. Includes heat production from combined heat and power plants and direct use of solar, wind and other non-solid renewable energy sources.

UNITED STATES

3/A3 TOTAL SOLID FUELS BALANCE
(MTOE)

	1973	1978	1979	1980	1981	1985	1990
PRODUCTION [1]	342.9	378.6	440.7	467.5	460.9	567.0	671.0
IMPORTS	.7	5.2	3.6	1.1	.9	1.0	1.0
EXPORTS	-32.3	-24.1	-39.2	-54.5	-66.0	-73.0	-78.0
MARINE BUNKERS	-	-	-	-	-	-	-
STOCK CHANGE	11.0	8.2	-19.8	-15.5	13.0	-	-
PRIMARY REQUIREMENTS	322.4	367.9	385.4	398.6	408.9	495.0	594.0
ELECTRICITY GENERATION	-221.6	-269.9	-294.5	-317.9	-332.9	-349.0	-421.0
GAS MANUFACTURE	-.2	-	-	-	-	-	-6.0
LIQUEFACTION	-	-	-	-	-	-3.0	-9.0
ENERGY SECTOR							
OWN USE + LOSSES [2]	-8.9	-15.6	-5.7	-13.6	-5.6	-1.0	-1.0
FINAL CONSUMPTION	91.6	82.4	85.1	67.1	70.3	141.0	156.0
INDUSTRY SECTOR [1]	86.1	76.9	80.3	63.1	66.4	136.0	151.0
IRON AND STEEL	46.4	39.0	39.7	29.7	32.6	46.0	46.0
CHEMICAL/							
PETROCHEMICAL	.2	-	-	-	-	-	-
OTHER INDUSTRY [1]	39.6	37.9	40.6	33.4	33.9	90.0	105.0
TRANSPORTATION SECTOR	.1	-	-	-	-	-	-
OTHER SECTORS	5.4	5.5	4.8	4.0	3.8	5.0	5.0
AGRICULTURE	-	-	-	-	-	-	-
COMMERCIAL AND							
PUBLIC SERVICE	-	.0	.0	.2	.1	3.0	3.0
RESIDENTIAL	5.4	5.5	4.8	3.8	3.7	2.0	2.0

Source: IEA/OECD Energy Balances and IEA Country Submissions (1982).

1. Projections for 1985 and 1990 include about 50 Mtoe wood and wood-waste not reported for historical years. See note to Table 5/B1.
2. Includes statistical difference.

UNITED STATES

3/B1 SOLID FUELS INDIGENOUS PRODUCTION BY FUEL TYPE
(MTOE)

	1973	1978	1979	1980	1981	1985	1990
TOTAL SOLID FUELS [1]	342.88	378.61	440.75	467.52	460.92	567.00	671.00
HARD COAL	338.70	368.86	428.86	454.29	446.41	501.00	598.00
COKING COAL	na	56.10	75.01	75.28	73.23	84.00	84.00
STEAM COAL	na	312.76	353.85	379.01	373.18	417.00e	514.00e
BROWN COAL/LIGNITE	4.18	9.75	11.89	13.23	14.51	18.00e	20.00e
OTHER SOLID FUELS [1]	na	na	na	na	na	48.00	53.00

Source: IEA/OECD Coal Statistics, IEA/OECD Energy Balances, IEA Country Submissions (1982).

1. Projections for 1985 and 1990 include wood and wood-waste mostly for use in the pulp and paper industries. Historical data was not available for inclusion in this table-set. More recent information gives the following historical data in Mtoe: 1973-38.5; 1978-51.3; 1979-54.1; 1980-57.0; 1981-58.0. This information indicates that the projections for 1985 and 1990 listed above should be revised to about 60 Mtoe.

3/B2 COAL PRODUCTION BY TYPE
(THOUSAND METRIC TONS)

	1973	1978	1979	1980	1981
TOTAL COAL	543012	607962	708633	752692	744472
HARD COAL	529636	576793	670620	710385	698062
COKING COAL	na	87728	117295	117719	114505
STEAM COAL	na	489065	553325	592666	583557
BROWN COAL/LIGNITE	13376	31169	38013	42307	46410
DERIVED PRODUCTS:					
PATENT FUEL	-	-	-	-	
COKE OVEN COKE	62803	44460	48029	41850	38815
GAS COKE	150	-	-	-	-
BROWN COAL BRIQUETTES	-	-	-	-	
COKE OVEN GAS(TCAL)	129591	108385	114345	100485	94500
BLAST FURNACE GAS(TCAL)	120365	106006	102080	75818	80970

Source: IEA/OECD Coal Statistics.

UNITED STATES

3/C1 SOLID FUELS TRADE BY TYPE
(MTOE)

	1973	1978	1979	1980	1981	1985	1990
IMPORTS							
TOTAL SOLID FUELS	.73	5.19	3.61	1.09	.92	1.00	1.00
HARD COAL	.07	1.71	1.19	.69	.60	.60e	.60e
COKING COAL	na	-	-	-	-	-	-
STEAM COAL	na	1.71	1.19	.69	.60	.60e	.60e
BROWN COAL/LIGNITE	-	-	-	-	-	-	-
OTHER SOLID FUELS [1]	.66	3.48	2.42	.40	.32	.40	.40
EXPORTS							
TOTAL SOLID FUELS	32.27	24.05	39.18	54.48	66.00	73.00	78.00
HARD COAL	31.36	23.61	38.30	53.22	65.29	73.00	78.00
COKING COAL	na	17.32	29.41	36.61	37.84	36.00e	36.00e
STEAM COAL	na	6.29	8.89	16.61	27.45	37.00e	42.00e
BROWN COAL/LIGNITE	-	.02	.01	-	-	-	-
OTHER SOLID FUELS [1]	.91	.42	.87	1.26	.71	-	-

Source: IEA/OECD Coal Statistics, IEA/OECD Energy Balances, IEA Country Submissions (1982).

1. Includes derived coal products, e.g. coke.

UNITED STATES

3/C2 COAL IMPORTS BY ORIGIN
(THOUSAND METRIC TONS)

	1978	1979	1980	1981
HARD COAL IMPORTS	2679	1868	1083	945
COKING COAL	-	-	-	-
STEAM COAL	2679	1868	1083	945
IMPORTS FROM:				
AUSTRALIA	933	156	-	-
CANADA	49	53	37	145
GERMANY	-	-	-	-
UNITED KINGDOM	-	-	-	3
UNITED STATES	-	-	-	-
OTHER OECD	-	-	-	-
CZECHOSLOVAKIA	-	-	-	-
POLAND	645	635	287	28
USSR	-	-	-	-
CHINA	-	-	-	-
SOUTH AFRICA	996	1024	699	738
OTHER NON-OECD	56	-	60	31

Source: IEA/OECD Coal Statistics.

UNITED STATES

3/C3 HARD COAL EXPORTS BY DESTINATION
(THOUSAND METRIC TONS)

	1978	1979	1980	1981
TOTAL EXPORTS	36915	59886	83227	102096
EXPORTS TO:				
AUSTRALIA	-	3	-	-
AUSTRIA	-	-	-	-
BELGIUM	-	2906	4202	3803
CANADA	14192	17637	15830	16525
DENMARK	-	148	1497	3519
FINLAND	-	-	253	1140
FRANCE	1468	3507	7066	8840
GERMANY	560	2319	2295	3907
GREECE	-	-	52	124
ICELAND	-	-	6	-
IRELAND	-	257	357	459
ITALY	2905	4522	6470	9500
JAPAN	9447	14225	20928	23455
LUXEMBOURG	-	-	-	-
NETHERLANDS	982	1824	4224	6187
NEW ZEALAND	-	-	-	-
NORWAY	68	210	252	439
PORTUGAL	265	280	329	256
SPAIN	688	1226	3093	5778
SWEDEN	299	671	920	1074
SWITZERLAND	-	-	99	1213
TURKEY	409	758	796	516
UNITED KINGDOM	360	1225	3756	2123
UNITED STATES	-	-	-	-
TOTAL NON-OECD	5272	8168	10802	13238

Source: IEA/OECD Statistics.

3/C4 COKING COAL EXPORTS BY DESTINATION
(THOUSAND METRIC TONS)

	1978	1979	1980	1981
TOTAL EXPORTS	27078	45992	57246	59179
EXPORTS TO:				
AUSTRALIA	-	-	-	-
AUSTRIA	-	-	-	-
BELGIUM	-	2724	2819	2650
CANADA	5410	6852	5694	5273
DENMARK	-	7	141	496
FINLAND	-	-	20	41
FRANCE	1468	3193	4082	4592
GERMANY	560	1678	1549	1861
GREECE	-	-	48	48
ICELAND	-	-	6	-
IRELAND	-	147	198	283
ITALY	2905	4505	5546	6421
JAPAN	8991	13839	19914	19869
LUXEMBOURG	-	-	-	-
NETHERLANDS	929	1688	2390	2775
NEW ZEALAND	-	-	-	-
NORWAY	68	210	252	244
PORTUGAL	265	280	305	196
SPAIN	688	1007	2123	2723
SWEDEN	299	671	916	982
SWITZERLAND	-	-	99	376
TURKEY	409	758	796	516
UNITED KINGDOM	360	1148	1881	1539
UNITED STATES	-	-	-	-
TOTAL NON-OECD	4726	7285	8467	8294

Source: IEA/OECD Statistics.

UNITED STATES

3/C5 STEAM COAL EXPORTS BY DESTINATION
(THOUSAND METRIC TONS)

	1978	1979	1980	1981
TOTAL EXPORTS	9837	13894	25981	42917
EXPORTS TO:				
AUSTRALIA	-	3	-	-
AUSTRIA	-	-	-	-
BELGIUM	-	182	1383	1153
CANADA	8782	10785	10136	11252
DENMARK	-	141	1356	3023
FINLAND	-	-	233	1099
FRANCE	-	314	2984	4248
GERMANY	-	641	746	2046
GREECE	-	-	4	76
ICELAND	-	-	-	-
IRELAND	-	110	159	176
ITALY	-	17	924	3079
JAPAN	456	386	1014	3586
LUXEMBOURG	-	-	-	-
NETHERLANDS	53	136	1834	3412
NEW ZEALAND	-	-	-	-
NORWAY	-	-	-	195
PORTUGAL	-	-	24	60
SPAIN	-	219	970	3055
SWEDEN	-	-	4	92
SWITZERLAND	-	-	-	837
TURKEY	-	-	-	-
UNITED KINGDOM	-	77	1875	584
UNITED STATES	-	-	-	-
TOTAL NON-OECD	546	883	2335	4944

Source: IEA/OECD Statistics.

UNITED STATES

3/D1 FUEL INPUT IN ELECTRICITY GENERATION

	1973	1978	1979	1980	1981	1985	1990
TOTAL (MTOE)	479.5	582.0	598.5	608.7	618.1	676.0	805.0
SOLID FUELS	221.6	269.9	294.5	317.9	332.9	349.0	421.0
OIL	87.3	90.9	79.9	64.0	55.7	55.0	43.0
GAS	85.6	75.1	82.3	87.3	86.3	67.0	67.0
NUCLEAR	20.5	73.5	68.5	67.5	74.1	123.0	184.0
HYDRO/GEOTHERMAL	64.5	72.7	73.1	72.0	69.0	81.0	86.0
OTHER [1]	-	-	-	-	-	1.0	4.0
TOTAL (PERCENTS)	100.0	100.0	100.0	100.0	100.0	100.0	100.0
SOLID FUELS	46.2	46.4	49.2	52.2	53.9	51.6	52.3
OIL	18.2	15.6	13.4	10.5	9.0	8.1	5.3
GAS	17.9	12.9	13.8	14.3	14.0	9.9	8.3
NUCLEAR	4.3	12.6	11.5	11.1	12.0	18.2	22.9
HYDRO/GEOTHERMAL	13.4	12.5	12.2	11.8	11.2	12.0	10.7
OTHER [1]	-	-	-	-	-	.1	.5

Source: IEA/OECD Energy Balances and IEA Country Submissions (1982).

1. Solar, wind, etc.

3/D2 ELECTRICTY PRODUCTION BY FUEL TYPE [1]

	1973	1978	1979	1980	1981	1985	1990
TOTAL (TWH)	2086.9	2319.9	2361.5	2401.0	2410.7	2546.0	3032.0
SOLID FUELS	984.7	1029.8	1134.7	1225.9	1269.8	1315.0	1586.0
OIL	334.6	385.1	320.2	259.5	217.8	207.0	162.0
GAS	397.9	322.2	347.6	365.3	364.8	252.0	252.0
NUCLEAR	89.2	293.0	270.5	266.2	289.0	463.0	693.0
HYDRO/GEOTHERMAL	280.5	289.8	288.5	284.1	269.3	305.0	324.0
OTHER [2]	-	-	-	-	-	4.0	15.0
TOTAL (PERCENTS)	100.0	100.0	100.0	100.0	100.0	100.0	100.0
SOLID FUELS	47.2	44.4	48.1	51.1	52.7	51.6	52.3
OIL	16.0	16.6	13.6	10.8	9.0	8.1	5.3
GAS	19.1	13.9	14.7	15.2	15.1	9.9	8.3
NUCLEAR	4.3	12.6	11.5	11.1	12.0	18.2	22.9
HYDRO/GEOTHERMAL	13.4	12.5	12.2	11.8	11.2	12.0	10.7
OTHER [2]	-	-	-	-	-	.2	.5

Source: IEA/OECD Energy Balances and IEA Country Submissions (1982).

1. Includes only electricity generated by public utilities.
2. Solar, wind, etc.

UNITED STATES

3/D3 PROJECTIONS OF ELECTRICITY GENERATION CAPACITY BY FUEL
(GW)

	Nuclear	Hydro/ Geoth.	Solid Fuels	Oil	Gas	Multi Firing [1]	Total
I 1980							
Operating	60.1	77.1	253.0	148.1	76.5	21.0	635.8
II 1980-1985							
Capacity:							
Under Construction[2]	29.9	11.9	26.0	-		-	67.8
Authorised	-	-	-	-	-	-	-
Other Planned	-	-	-	-	-	-	-
Conversion	-	-	10.0	-10.0		-	-
Decommissioning	-	-	-	-		-	-
Total Operating 1985	90.0	89.0	289.0	214.6		21.0	683.6
III 1986-1990							
Capacity:							
Under Construction [2]	31.0	13.0	39.0	-		-	83.0
Authorised	-	-	-	-	-	-	-
Other Planned	-	-	-	-	-	-	-
Conversion	-	-	5.0	-5.0		-	-
Decommissioning	-	-	-	-		-	-
Total Operating 1990	121.0	102.0	333.0	209.6		21.0	766.6

Source: IEA/OECD Electricity Statistics and IEA Country Submissions (1982).
1. Includes "Other" capacity.
2. Includes authorised.

UNITED STATES

3/D4 ENERGY USE IN IRON AND STEEL INDUSTRY BY FUEL

	1973	1978	1979	1980	1981	1985	1990
TOTAL (MTOE)	78.0	62.8	64.3	59.3	61.1	na	na
SOLID FUELS	46.4	39.0	39.7	29.7	32.6	46.0	46.0
OIL	5.7	5.1	5.3	4.6	3.4	na	na
GAS	18.3	11.6	11.9	18.6	18.0	na	na
ELECTRICITY	7.7	7.1	7.4	6.4	7.1	na	na
OTHER [1]	-	-	-	-	-	na	na
TOTAL (PERCENTS)	100.0	100.0	100.0	100.0	100.0	100.0	100.0
SOLID FUELS	59.4	62.2	61.7	50.1	53.3	na	na
OIL	7.4	8.1	8.3	7.7	5.6	na	na
GAS	23.4	18.4	18.4	31.3	29.5	na	na
ELECTRICITY	9.8	11.3	11.5	10.8	11.6	na	na
OTHER [1]	-	-	-	-	-	-	-

Source: IEA/OECD Energy Balances and IEA Country Submissions (1982).

1. Includes heat from combined heat and power plants and direct use of solar, wind and other non-solid renewable energy sources.

3/D5 ENERGY USE IN OTHER INDUSTRIES BY FUEL

	1973	1978	1979	1980	1981	1985	1990
TOTAL (MTOE)	382.4	391.9	407.7	359.1	338.0	na	na
SOLID FUELS	39.7	37.9	40.6	33.4	33.9	90.0	105.0
OIL	122.3	163.6	173.1	133.1	115.8	na	na
GAS	165.0	128.7	131.2	131.9	127.8	na	na
ELECTRICITY	55.4	61.7	62.8	60.7	60.5	na	na
OTHER [1]	-	-	-	-	-	na	na
TOTAL (PERCENTS)	100.0	100.0	100.0	100.0	100.0	100.0	100.0
SOLID FUELS	10.4	9.7	10.0	9.3	10.0	na	na
OIL	32.0	41.7	42.5	37.1	34.3	na	na
GAS	43.1	32.8	32.2	36.7	37.8	na	na
ELECTRICITY	14.5	15.7	15.4	16.9	17.9	na	na
OTHER [1]	-	-	-	-	-	na	na

Source: IEA/OECD Energy Balances and IEA Country Submissions (1982).

1. Includes heat from combined heat and power plants and direct use of solar, wind and other non-solid renewable energy sources.

UNITED STATES

3/D6 ENERGY USE IN INDUSTRY [1] BY FUEL

	1973	1978	1979	1980	1981	1985	1990
TOTAL (MTOE)	460.4	454.7	472.0	418.4	399.0	575.0	641.0
SOLID FUELS	86.1	76.9	80.3	63.1	66.4	136.0	151.0
OIL	128.1	168.6	178.4	137.6	119.2	179.0	206.0
GAS	183.2	140.3	143.1	150.5	145.8	171.0	178.0
ELECTRICITY	63.0	68.8	70.2	67.1	67.6	84.0	103.0
OTHER [2]	-	-	-	-	-	5.0	3.0
TOTAL (PERCENTS)	100.0	100.0	100.0	100.0	100.0	100.0	100.0
SOLID FUELS	18.7	16.9	17.0	15.1	16.6	23.7	23.6
OIL	27.8	37.1	37.8	32.9	29.9	31.1	32.1
GAS	39.8	30.9	30.3	36.0	36.5	29.7	27.8
ELECTRICITY	13.7	15.1	14.9	16.0	16.9	14.6	16.1
OTHER [2]	-	-	-	-	-	.9	.5

Source: IEA/OECD Energy Balances and IEA Country Submissions (1982).

1. Includes non-energy use of petroleum products.
2. Includes heat from combined heat and power plants and direct use of solar, wind and other non-solid renewable energy sources.

3/D7 ENERGY USE IN OTHER SECTORS [1] BY FUEL

	1973	1978	1979	1980	1981	1985	1990
TOTAL (MTOE)	429.8	436.6	441.1	440.4	434.7	365.0	386.0
SOLID FUELS	5.4	5.5	4.8	4.0	3.8	5.0	5.0
OIL	159.1	139.4	135.0	138.1	124.6	71.0	66.0
GAS	177.8	184.9	192.6	185.3	191.4	169.0	171.0
ELECTRICITY	87.5	106.8	108.7	113.0	114.9	117.0	135.0
OTHER [2]	-	-	-	-	-	3.0	9.0
TOTAL (PERCENTS)	100.0	100.0	100.0	100.0	100.0	100.0	100.0
SOLID FUELS	1.3	1.3	1.1	.9	.9	1.4	1.3
OIL	37.0	31.9	30.6	31.4	28.7	19.5	17.1
GAS	41.4	42.4	43.7	42.1	44.0	46.3	44.3
ELECTRICITY	20.4	24.5	24.6	25.7	26.4	32.1	35.0
OTHER [2]	-	-	-	-	-	.8	2.3

Source: IEA/OECD Energy Balances and IEA Country Submissions (1982).

1. Includes use in agricultural, commercial, public services and residential sectors.
2. Includes heat from combined heat and power plants and direct use of solar, wind and other non-solid renewable energy sources.

4. OECD PACIFIC

TOTAL ENERGY REQUIREMENTS BY FUEL

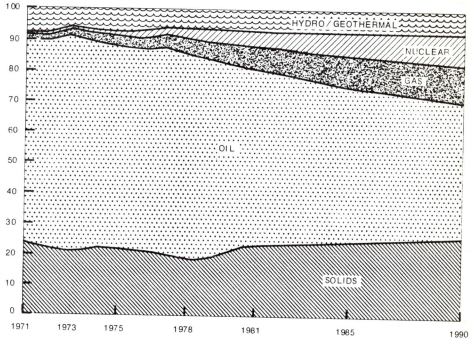

PACIFIC

4/A1 TOTAL PRIMARY ENERGY REQUIREMENTS (TPER) BY FUEL

	1973	1978	1979	1980	1981	1985	1990
TPER (MTOE) [1]	408.2	446.1	465.1	453.0	451.1	535.2	637.4
SOLID FUELS	86.0	82.7	89.1	97.8	104.3	130.0	161.0
OIL	286.1	298.4	301.0	272.9	261.3	268.1	280.0
GAS	9.1	23.6	27.1	30.4	32.3	55.2	78.1
NUCLEAR	2.4	14.5	17.2	20.2	21.5	39.4	62.5
HYDRO/GEOTHERMAL	24.6	26.8	30.7	31.7	31.7	37.8	45.1
OTHER [2]	-	-	-	-	-	4.7	10.6
TPER (PERCENTS) [1]	100.0	100.0	100.0	100.0	100.0	100.0	100.0
SOLID FUELS	21.1	18.5	19.2	21.6	23.1	24.3	25.3
OIL	70.1	66.9	64.7	60.2	57.9	50.1	43.9
GAS	2.2	5.3	5.8	6.7	7.2	10.3	12.3
NUCLEAR	.6	3.3	3.7	4.5	4.8	7.4	9.8
HYDRO/GEOTHERMAL	6.0	6.0	6.6	7.0	7.0	7.1	7.1
OTHER [2]	-	-	-	-	-	.9	1.7

Source: IEA/OECD Energy Balances and IEA Country Submissions (1982).

1. Net imports of electricity are only included in totals.
2. Includes other electricity generation (solar, wind, etc.), traded synthetic fuels and direct use of solar, wind and other non-solid renewable energy sources.

4/A2 TOTAL FINAL ENERGY CONSUMPTION (TFC) BY FUEL

	1973	1978	1979	1980	1981	1985	1990
TFC (MTOE)	296.1	327.3	348.1	317.3	311.3	365.4	428.4
SOLID FUELS	47.0	40.4	42.7	47.7	49.2	62.8	73.6
OIL	197.8	222.5	237.3	201.9	192.6	206.0	228.9
GAS	9.7	13.7	15.0	14.9	16.1	26.6	36.3
ELECTRICITY	41.6	50.7	53.2	52.7	53.3	66.6	81.9
HEAT/OTHER [1]	na	na	na	.1	.1	3.4	7.7
TFC (PERCENTS)	100.0	100.0	100.0	100.0	100.0	100.0	100.0
SOLID FUELS	15.9	12.3	12.3	15.0	15.8	17.2	17.2
OIL	66.8	68.0	68.2	63.6	61.9	56.4	53.4
GAS	3.3	4.2	4.3	4.7	5.2	7.3	8.5
ELECTRICITY	14.0	15.5	15.3	16.6	17.1	18.2	19.1
HEAT/OTHER [1]	na	na	na	.0	.0	.9	1.8

Source: IEA/OECD Energy Balances and IEA Country Submissions (1982).

1. Includes heat production from combined heat and power plants and direct use of solar, wind and other non-solid renewable energy sources.

PACIFIC

4/A3 TOTAL SOLID FUELS BALANCE
(MTOE)

	1973	1978	1979	1980	1981	1985	1990
PRODUCTION	63.3	71.2	72.5	74.1	84.1	109.2	126.4
IMPORTS	42.4	38.6	43.4	50.1	57.0	75.0	97.9
EXPORTS	-18.7	-25.9	-27.8	-29.3	-34.6	-54.3	-63.3
MARINE BUNKERS	-	-	-	-	-	-	-
STOCK CHANGE	-1.0	-1.2	.9	2.9	-2.3	-	-
TPE REQUIREMENTS	86.0	82.7	89.1	97.8	104.3	130.0	161.0
ELECTRICITY GENERATION	-26.5	-32.8	-34.6	-34.4	-38.1	-52.9	-73.3
GAS MANUFACTURE	-2.4	-1.7	-1.7	-1.4	-1.8	-1.2	-1.2
LIQUEFACTION	-	-	-	-	-	-	-
ENERGY SECTOR OWN USE + LOSSES [1]	-10.1	-7.8	-10.1	-14.4	-15.3	-13.0	-12.9
FINAL CONSUMPTION	47.0	40.4	42.7	47.7	49.2	62.8	73.6
INDUSTRY SECTOR	44.4	39.1	41.2	45.6	47.1	59.5	70.6
IRON AND STEEL	37.0	30.6	32.5	33.9	32.1	na	na
CHEMICAL/ PETROCHEMICAL	.3	.5	.6	.6	.7	na	na
OTHER INDUSTRY	7.1	8.0	8.2	11.1	14.2	na	na
TRANSPORTATION SECTOR	.2	.0	.0	.0	.0	.1	.1
OTHER SECTORS	2.3	1.3	1.4	2.1	2.1	3.2	2.9
AGRICULTURE	.0	-	-	.0	.0	.0	.0
COMMERCIAL AND PUBLIC SERVICE	.1	.4	.3	.3	.3	.1	.1
RESIDENTIAL	2.2	.9	1.1	1.8	1.8	2.0	2.0

Source: IEA/OECD Energy Balances and IEA Country Submissions (1982).

1. Includes statistical difference.

PACIFIC

4/B1 SOLID FUELS INDIGENOUS PRODUCTION BY FUEL TYPE
(MTOE)

	1973	1978	1979	1980	1981	1985	1990
TOTAL SOLID FUELS	63.27	71.15	72.54	74.11	84.09	109.24	126.38
HARD COAL	55.94	61.55	62.67	64.27	74.15	97.45	111.25
COKING COAL	na	34.61	35.04	32.57	37.52	41.85	45.60
STEAM COAL	na	26.94	27.63	31.70	36.62	55.60	65.65
BROWN COAL/LIGNITE	5.42	6.81	7.29	7.33	7.36	8.86	13.00
OTHER SOLID FUELS	1.91	2.79	2.58	2.51	2.58	2.93	2.13

Source: IEA/OECD Coal Statistics, IEA/OECD Energy Balances, IEA Country submissions (1982).

4/B2 COAL PRODUCTION BY TYPE
(THOUSAND METRIC TONS)

	1973	1978	1979	1980	1981
TOTAL COAL	107262	123085	127094	129692	145122
HARD COAL	82898	92412	94261	96663	111941
COKING COAL	na	52263	53052	49327	57055
STEAM COAL	na	40149	41209	47336	54886
BROWN COAL/LIGNITE	24364	30673	32833	33029	33181
DERIVED PRODUCTS:					
PATENT FUEL	1215	430	530	441	338
COKE OVEN COKE	57303	49001	53108	52940	52321
GAS COKE	4895	3347	3388	3546	3487
BROWN COAL BRIQUETTES	1221	1101	1213	1280	1059
COKE OVEN GAS(TCAL)	84834	75920	82627	85344	83441
BLAST FURNACE GAS(TCAL)	125893	99300	103570	105569	99898

Source: IEA/OECD Coal Statistics.

PACIFIC

4/C1 SOLID FUELS TRADE BY TYPE
(MTOE)

	1973	1978	1979	1980	1981	1985	1990
IMPORTS							
TOTAL SOLID FUELS	42.40	38.65	43.44	50.14	57.01	75.00	97.90
HARD COAL	42.40	38.65	43.42	50.13	56.99	75.00	97.90
COKING COAL	na	37.19	41.42	45.49	47.76	53.00	59.00
STEAM COAL	na	1.46	2.00	4.64	9.23	22.00	38.90
BROWN COAL/LIGNITE	-	-	-	-	-	-	-
OTHER SOLID FUELS [1]	-	-	.02	.01	.02	-	-
EXPORTS							
TOTAL SOLID FUELS	18.65	25.95	27.76	29.30	34.56	54.28	63.26
HARD COAL	18.25	25.12	26.22	27.85	33.24	54.28	63.26
COKING COAL	na	21.59	22.54	22.04	26.59	28.28	30.36
STEAM COAL	na	3.53	3.69	5.81	6.65	26.00	32.90
BROWN COAL/LIGNITE	-	-	-	-	-	-	-
OTHER SOLID FUELS [1]	.40	.83	1.54	1.45	1.32	-	-

Source: IEA/OECD Coal Statistics, IEA/OECD Energy Balances, IEA Country Submissions (1982).

1.   Includes derived coal products, e.g. coke.

PACIFIC

4/C2 COAL IMPORTS BY ORIGIN
(THOUSAND METRIC TONS)

	1978	1979	1980	1981
HARD COAL IMPORTS	52872	59403	68572	77960
COKING COAL IMPORTS FROM:	**50876**	**56660**	**62227**	**65333**
AUSTRALIA	24149	27013	25917	29148
CANADA	10895	9777	10759	9252
GERMANY	398	223	-	-
UNITED KINGDOM	-	-	-	-
UNITED STATES	9956	13987	19414	21761
OTHER OECD	-	-	70	124
CZECHOSLOVAKIA	-	-	-	-
POLAND	429	489	387	65
USSR	2244	1932	1898	1112
CHINA	-	-	979	1148
SOUTH AFRICA	2360	2389	2803	2723
OTHER NON-OECD	445	850	-	-
STEAM COAL	**1996**	**2743**	**6345**	**12627**
IMPORTS FROM:				
AUSTRALIA	669	1287	3462	5427
CANADA	105	72	352	1199
GERMANY	-	-	-	-
UNITED KINGDOM	-	-	-	-
UNITED STATES	1	10	385	2080
OTHER OECD	-	-	1	19
CZECHOSLOVAKIA	-	-	-	-
POLAND	-	-	-	-
USSR	149	147	233	286
CHINA	-	-	1086	1614
SOUTH AFRICA	157	143	386	1626
OTHER NON-OECD	915	1084	440	376

Source: IEA/OECD Coal Statistics.

PACIFIC

4/C3 HARD COAL EXPORTS BY DESTINATION
(THOUSAND METRIC TONS)

	1978	1979	1980	1981
TOTAL EXPORTS	38739	40445	42950	51245
EXPORTS TO:				
AUSTRALIA	-	-	-	-
AUSTRIA	-	-	-	-
BELGIUM	276	228	175	330
CANADA	-	-	-	-
DENMARK	43	-	-	-
FINLAND	-	-	-	-
FRANCE	3655	1114	903	1133
GERMANY	213	2936	-	39
GREECE	296	253	48	-
ICELAND	-	-	-	-
IRELAND	-	-	-	-
ITALY	1030	867	1163	1785
JAPAN	25174	26941	30173	35165
LUXEMBOURG	-	-	-	-
NETHERLANDS	631	414	2835	3088
NEW ZEALAND	-	-	-	-
NORWAY	-	-	-	-
PORTUGAL	-	-	-	-
SPAIN	559	515	608	636
SWEDEN	-	-	-	53
SWITZERLAND	-	-	-	-
TURKEY	-	-	-	-
UNITED KINGDOM	1735	1857	1865	2109
UNITED STATES	419	156	61	15
TOTAL NON-OECD	4708	5164	5119	6892

Source: IEA/OECD Coal Statistics.

PACIFIC

4/D1 FUEL INPUT IN ELECTRICITY GENERATION

	1973	1978	1979	1980	1981	1985	1990
TOTAL (MTOE)	129.3	147.7	157.7	156.2	159.4	204.9	259.8
SOLID FUELS	26.5	32.8	34.6	34.4	38.1	52.9	73.3
OIL	72.8	61.5	60.3	51.5	48.8	44.8	38.1
GAS	2.9	12.1	14.9	18.4	19.4	30.2	41.4
NUCLEAR	2.4	14.5	17.2	20.2	21.5	39.4	62.5
HYDRO/GEOTHERMAL	24.6	26.8	30.7	31.7	31.7	37.5	44.4
OTHER [1]	-	-	-	-	-	-	-
TOTAL (PERCENTS)	100.0	100.0	100.0	100.0	100.0	100.0	100.0
SOLID FUELS	20.5	22.2	21.9	22.0	23.9	25.8	28.2
OIL	56.3	41.6	38.2	33.0	30.6	21.9	14.7
GAS	2.3	8.2	9.5	11.8	12.2	14.8	15.9
NUCLEAR	1.8	9.8	10.9	13.0	13.5	19.2	24.1
HYDRO/GEOTHERMAL	19.1	18.2	19.4	20.3	19.9	18.3	17.1
OTHER [1]	-	-	-	-	-	-	-

Source: IEA/OECD Energy Balances and IEA Country Submissions (1982).

1. Solar, wind, etc.

4/D2 ELECTRICTY PRODUCTION BY FUEL TYPE

	1973	1978	1979	1980	1981	1985	1990
TOTAL (TWH) [1]	553.6	609.3	641.7	638.8	654.0	821.8	1024.3
SOLID FUELS	87.4	95.1	99.9	109.4	119.3	178.0	247.2
OIL	343.9	284.6	268.8	230.2	226.5	202.5	163.2
GAS	13.3	63.1	80.7	89.7	93.4	139.7	193.3
NUCLEAR	9.7	59.3	70.4	82.6	87.8	159.7	253.0
HYDRO/GEOTHERMAL	99.3	107.2	121.9	126.9	127.0	141.8	167.7
OTHER [2]	-	-	-	-	-	-	-
TOTAL (PERCENTS)	100.0	100.0	100.0	100.0	100.0	100.0	100.0
SOLID FUELS	15.8	15.6	15.6	17.1	18.2	21.7	24.1
OIL	62.1	46.7	41.9	36.0	34.6	24.6	15.9
GAS	2.4	10.4	12.6	14.0	14.3	17.0	18.9
NUCLEAR	1.8	9.7	11.0	12.9	13.4	19.4	24.7
HYDRO/GEOTHERMAL	17.9	17.6	19.0	19.9	19.4	17.3	16.4
OTHER [2]	-	-	-	-	-	-	-

Source: IEA/OECD Energy Balances and IEA Country Submissions (1982).

1. Electricity generation by autoproducers is not included for Japan.
2. Solar, wind, etc.

PACIFIC

4/D4 ENERGY USE IN IRON AND STEEL INDUSTRY BY FUEL

	1973	1978	1979	1980	1981	1985	1990
TOTAL (MTOE)	56.1	45.5	48.3	47.0	42.6	na	na
SOLID FUELS	37.0	30.6	32.5	33.9	32.1	na	na
OIL	12.8	8.0	8.4	6.0	3.4	na	na
GAS	.0	.1	.1	.1	.3	na	na
ELECTRICITY	na	6.8	7.3	7.0	6.8	na	na
HEAT [1]	-	-	-	-	-	-	-
TOTAL (PERCENTS)	na	100.0	100.0	100.0	100.0	na	na
SOLID FUELS	na	67.3	67.3	72.1	75.4	na	na
OIL	na	17.7	17.3	12.7	8.0	na	na
GAS	na	.2	.3	.3	.8	na	na
ELECTRICITY	na	14.9	15.0	14.9	15.9	na	na
HEAT [1]	-	-	-	-	-	-	-

Source: IEA/OECD Energy Balances and IEA Country Submissions (1982).

1. Includes only heat from combined heat and power plants.

4/D5 ENERGY USE IN OTHER INDUSTRIES BY FUEL

	1973	1978	1979	1980	1981	1985	1990
TOTAL (MTOE)	119.2	138.8	154.6	126.3	124.0	na	na
SOLID FUELS	7.4	8.5	8.8	11.7	15.0	na	na
OIL	86.8	100.5	114.6	83.7	78.9	na	na
GAS	3.6	6.1	6.4	6.5	6.1	na	na
ELECTRICITY	na	23.7	24.8	24.4	24.1	na	na
HEAT [1]	-	-	-	-	-	-	-
TOTAL (PERCENTS)	na	100.0	100.0	100.0	100.0	na	na
SOLID FUELS	na	6.1	5.7	9.3	12.1	na	na
OIL	na	72.4	74.2	66.3	63.6	na	na
GAS	na	4.4	4.1	5.2	4.9	na	na
ELECTRICITY	na	17.1	16.1	19.3	19.5	na	na
HEAT [1]	-	-	-	-	-	-	-

Source: IEA/OECD Energy Balances and IEA Country Submissions (1982).

1. Includes only heat from combined heat and power plants.

4/D6 ENERGY USE IN INDUSTRY [1] BY FUEL

	1973	1978	1979	1980	1981	1985	1990
TOTAL (MTOE)	175.3	184.3	202.8	173.3	166.7	193.5	224.4
SOLID FUELS	44.4	39.1	41.2	45.6	47.1	59.5	70.6
OIL	99.6	108.6	123.0	89.7	82.3	85.2	93.9
GAS	3.7	6.1	6.5	6.7	6.4	14.1	20.4
ELECTRICITY	27.5	30.5	32.1	31.4	30.9	34.7	39.5
HEAT [2]	-	-	-	-	-	-	-
TOTAL (PERCENTS)	100.0	100.0	100.0	100.0	100.0	100.0	100.0
SOLID FUELS	25.4	21.2	20.3	26.3	28.2	30.7	31.5
OIL	56.8	58.9	60.6	51.8	49.4	44.0	41.9
GAS	2.1	3.3	3.2	3.8	3.9	7.3	9.1
ELECTRICITY	15.7	16.6	15.8	18.1	18.5	17.9	17.6
HEAT [2]	-	-	-	-	-	-	-

Source: IEA/OECD Energy Balances and IEA Country Submissions (1982).

1. Includes non-energy use of petroleum products.
2. Includes only heat from combined heat and power plants.

4/D7 ENERGY USE IN OTHER SECTORS [1] BY FUEL

	1973	1978	1979	1980	1981	1985	1990
TOTAL (MTOE)	60.9	68.2	70.7	75.6	76.2	86.2	108.5
SOLID FUELS	2.3	1.3	1.4	2.1	2.1	3.2	2.9
OIL	39.7	40.4	41.1	45.2	43.4	36.8	41.4
GAS	6.1	7.6	8.5	8.2	9.7	12.4	15.7
ELECTRICITY	12.8	18.9	19.7	20.0	21.0	30.4	40.9
HEAT/OTHER [2]	na	na	na	.1	.1	3.4	7.7
TOTAL (PERCENTS)	100.0	100.0	100.0	100.0	100.0	100.0	100.0
SOLID FUELS	3.8	1.9	2.0	2.8	2.7	3.7	2.7
OIL	65.2	59.3	58.1	59.8	56.9	42.7	38.1
GAS	10.0	11.1	12.0	10.9	12.7	14.4	14.5
ELECTRICITY	21.1	27.7	27.9	26.4	27.6	35.3	37.6
HEAT/OTHER [2]	na	na	na	.2	.1	3.9	7.1

Source: IEA/OECD Energy Balances and IEA Country Submissions (1982).

1. Includes use in agricultural, commercial, public services and residential sectors.
2. Includes heat from combined heat and power plants and direct use of solar, wind and other non-solid renewable energy sources.

AUSTRALIA

TOTAL ENERGY REQUIREMENTS BY FUEL

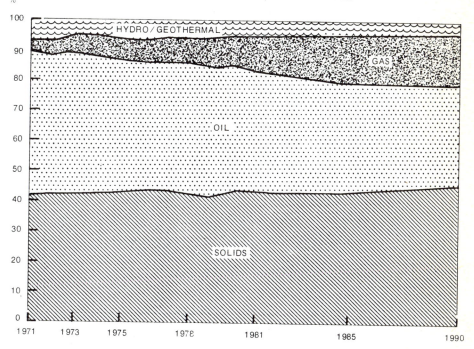

AUSTRALIA

4/A1 TOTAL PRIMARY ENERGY REQUIREMENTS (TPER) BY FUEL

	1973	1978	1979	1980	1981	1985	1990
TPER (MTOE) [1]	58.2	72.4	76.9	76.2	77.2	89.1	99.5
SOLID FUELS	25.0	30.2	31.6	33.0	33.2	39.0	45.3
OIL	26.5	31.6	33.3	31.4	30.6	31.5	33.0
GAS	3.4	6.2	7.1	7.6	8.9	14.3	16.7
NUCLEAR	-	-	-	-	-	-	-
HYDRO/GEOTHERMAL	3.4	4.3	5.0	4.2	4.4	4.3	4.5
OTHER [2]	-	-	-	-	-	-	-
TPER (PERCENTS) [1]	100.0	100.0	100.0	100.0	100.0	100.0	100.0
SOLID FUELS	42.9	41.7	41.1	43.3	43.0	43.8	45.5
OIL	45.5	43.7	43.2	41.2	39.7	35.4	33.2
GAS	5.9	8.6	9.2	10.0	11.5	16.0	16.8
NUCLEAR	-	-	-	-	-	-	-
HYDRO/GEOTHERMAL	5.8	6.0	6.4	5.5	5.8	4.8	4.5
OTHER [2]	-	-	-	-	-	-	-

Source: IEA/OECD Energy Balances and IEA Country Submissions (1982).

1. Net imports of electricity are only included in totals.
2. Includes other electricity generation (solar, wind, etc.) and traded synthetic fuels.

4/A2 TOTAL FINAL ENERGY CONSUMPTION (TFC) BY FUEL

	1973	1978	1979	1980	1981	1985	1990
TFC (MTOE)	39.9	50.5	48.0	48.3	49.1	58.8	64.0
SOLID FUELS	8.1	8.3	8.4	8.4	8.0	12.1	12.8
OIL	24.9	31.4	27.8	27.6	27.7	28.1	29.5
GAS	2.4	4.5	5.1	5.4	6.1	9.3	10.8
ELECTRICITY	4.5	6.3	6.7	6.9	7.3	9.3	10.9
HEAT [1]	na	na	na	.1	.1	-	-
TFC (PERCENTS)	100.0	100.0	100.0	100.0	100.0	100.0	100.0
SOLID FUELS	20.3	16.5	17.4	17.3	16.2	20.6	20.0
OIL	62.4	62.1	58.0	57.0	56.3	47.8	46.1
GAS	6.1	8.8	10.6	11.1	12.4	15.8	16.9
ELECTRICITY	11.3	12.6	14.0	14.3	14.8	15.8	17.0
HEAT [1]	na	na	na	.3	.2	-	-

Source: IEA/OECD Energy Balances and IEA Country Submissions (1982).

1. Includes only heat production from combined heat and power plants.

AUSTRALIA

4/A3 TOTAL SOLID FUELS BALANCE
(MTOE)

	1973	1978	1979	1980	1981	1985	1990
PRODUCTION	43.2	55.6	57.8	59.0	69.1	93.0	108.2
IMPORTS	-	.0	.0	.0	.0	-	-
EXPORTS	-18.2	-25.2	-26.3	-27.9	-33.1	-54.0	-62.9
MARINE BUNKERS	-	-	-	-	-	-	-
STOCK CHANGE	.0	-.2	.1	1.9	-2.8	-	-
PRIMARY REQUIREMENTS	25.0	30.2	31.6	33.0	33.2	39.0	45.3
ELECTRICITY GENERATION	-13.9	-19.6	-20.8	-22.4	-22.6	-26.2	-31.5
GAS MANUFACTURE	.0	.0	.0	.0	.0	-	-
LIQUEFACTION	-	-	-	-	-	-	-
ENERGY SECTOR							
OWN USE + LOSSES [1]	-3.0	-2.3	-2.5	-2.2	-2.6	-.7	-1.0
FINAL CONSUMPTION	8.1	8.3	8.4	8.4	8.0	12.1	12.8
INDUSTRY SECTOR	7.8	8.1	8.2	8.2	7.8	10.3	11.0
IRON AND STEEL	3.1	3.1	3.4	3.3	3.0	4.5	4.6
CHEMICAL/							
PETROCHEMICAL	.3	.2	.2	.1	.1	.2	.3
OTHER INDUSTRY	4.4	4.8	4.7	4.7	4.7	5.6	6.1
TRANSPORTATION SECTOR	.0	.0	.0	.0	.0	.1	.1
OTHER SECTORS	.3	.2	.2	.2	.2	1.7	1.7
AGRICULTURE	-	-	-	-	-	-	-
COMMERCIAL AND							
PUBLIC SERVICE	.1	.1	.1	.1	.1	-	-
RESIDENTIAL	.2	.1	.1	.1	.1	1.6	1.6

Source: IEA/OECD Energy Balances and IEA Country Submissions (1982).

1. Includes statistical difference.

AUSTRALIA

4/B1 SOLID FUELS INDIGENOUS PRODUCTION BY FUEL TYPE
(MTOE)

	1973	1978	1979	1980	1981	1985	1990
TOTAL SOLID FUELS	43.23	55.63	57.76	58.96	69.08	93.00	108.20
HARD COAL	35.97	43.97	45.76	46.56	56.64	81.80e	93.70e
COKING COAL	na	28.13	29.27	27.32	32.74	37.00e	40.00e
STEAM COAL	na	18.44	19.20	22.31	26.97	44.80e	53.70e
BROWN COAL/LIGNITE	5.35	6.77	7.24	7.28	7.32	8.80e	12.90e
OTHER SOLID FUELS	1.91	2.29	2.06	2.06	2.06	2.40	2.50

Source: IEA/OECD Coal Statistics, IEA/OECD Energy Balances, IEA Country Submissions (1982).

4/B2 COAL PRODUCTION BY TYPE
(THOUSAND METRIC TONS)

	1973	1978	1979	1980	1981
TOTAL COAL	79604	102316	107357	109347	125065
HARD COAL	55483	71831	74762	76553	92104
COKING COAL	na	43386	45153	42141	50496
STEAM COAL	na	28445	29609	34412	41608
BROWN COAL/LIGNITE	24121	30485	32595	32794	32961
DERIVED PRODUCTS:					
PATENT FUEL	-	-	-	-	-
COKE OVEN COKE	4983	5003	5475	5474	5077
GAS COKE	64	60	70	13	10
BROWN COAL BRIQUETTES	1221	1101	1213	1280	1059
COKE OVEN GAS(TCAL)	10504	11500	12897	10277	9799
BLAST FURNACE GAS(TCAL)	11403	10750	11700	11711	10277

Source: IEA/OECD Coal Statistics.

AUSTRALIA

4/C1 SOLID FUELS TRADE BY TYPE
(MTOE)

	1973	1978	1979	1980	1981	1985	1990
TOTAL SOLID FUELS	-	.01	.03	.01	.02	-	-
HARD COAL	-	.01	.01	-	-	-	-
COKING COAL	-	-	-	-	-	-	-
STEAM COAL	-	.01	.01	-	-	-	-
BROWN COAL/LIGNITE	-	-	-	-	-	-	-
OTHER SOLID FUELS [1]	-	-	.02	.01	.02	-	-
EXPORTS							
TOTAL SOLID FUELS	18.25	25.24	26.26	27.87	33.09	54.00	62.90
HARD COAL	18.25	25.08	26.18	27.76	33.04	54.00e	62.90e
COKING COAL	na	21.56	22.52	21.97	26.43	28.00e	30.00e
STEAM COAL	na	3.52	3.66	5.79	6.61	26.00e	32.90e
BROWN COAL/LIGNITE	-	-	-	-	-	-	-
OTHER SOLID FUELS [1]	-	.16	.08	.11	.05	-	-

Source: IEA/OECD Coal Statistics, IEA/OECD Energy Balances, IEA Country Submissions (1982).

1. Includes derived coal products, e.g. coke.

AUSTRALIA

4/C3 HARD COAL EXPORTS BY DESTINATION
(THOUSAND METRIC TONS)

	1978	1979	1980	1981
TOTAL EXPORTS	38683	40390	42819	50967
EXPORTS TO:				
AUSTRALIA	-	-	-	-
AUSTRIA	-	-	-	-
BELGIUM	276	228	175	330
CANADA	-	-	-	-
DENMARK	43	-	-	-
FINLAND	-	-	-	-
FRANCE	3655	1114	903	1133
GERMANY	213	2936	-	39
GREECE	296	253	48	-
ICELAND	-	-	-	-
IRELAND	-	-	-	-
ITALY	1030	867	1163	1785
JAPAN	25174	26931	30103	34994
LUXEMBOURG	-	-	-	-
NETHERLANDS	631	414	2835	3088
NEW ZEALAND	-	-	-	-
NORWAY	-	-	-	-
PORTUGAL	-	-	-	-
SPAIN	559	515	608	636
SWEDEN	-	-	-	53
SWITZERLAND	-	-	-	-
TURKEY	-	-	-	-
UNITED KINGDOM	1735	1857	1865	2109
UNITED STATES	419	156	61	15
TOTAL NON-OECD	4652	5119	5058	6785

Source: IEA/OECD Statistics.

AUSTRALIA

4/C4 COKING COAL EXPORTS BY DESTINATION
(THOUSAND METRIC TONS)

	1978	1979	1980	1981
TOTAL EXPORTS	33257	34738	33882	40774
EXPORTS TO:				
AUSTRALIA	-	-	-	-
AUSTRIA	-	-	-	-
BELGIUM	276	228	175	330
CANADA	-	-	-	-
DENMARK	-	-	-	-
FINLAND	-	-	-	-
FRANCE	1073	1114	903	1133
GERMANY	198	98	-	39
GREECE	296	253	48	-
ICELAND	-	-	-	-
IRELAND	-	-	-	-
ITALY	1030	867	1163	1785
JAPAN	24354	25833	26212	29261
LUXEMBOURG	-	-	-	-
NETHERLANDS	631	414	274	736
NEW ZEALAND	-	-	-	-
NORWAY	-	-	-	-
PORTUGAL	-	-	-	-
SPAIN	559	515	608	636
SWEDEN	-	-	-	53
SWITZERLAND	-	-	-	-
TURKEY	-	-	-	-
UNITED KINGDOM	750	876	527	935
UNITED STATES	105	104	-	-
TOTAL NON-OECD	3985	4436	3972	5866

Source: IEA/OECD Statistics.

AUSTRALIA

4/C5 STEAM COAL EXPORTS BY DESTINATION
(THOUSAND METRIC TONS)

	1978	1979	1980	1981
TOTAL EXPORTS	5426	5652	8937	10193
EXPORTS TO:				
AUSTRALIA	-	-	-	-
AUSTRIA	-	-	-	-
BELGIUM	-	-	-	-
CANADA	-	-	-	-
DENMARK	43	-	-	-
FINLAND	-	-	-	-
FRANCE	2582	-	-	-
GERMANY	15	2838	-	-
GREECE	-	-	-	-
ICELAND	-	-	-	-
IRELAND	-	-	-	-
ITALY	-	-	-	-
JAPAN	820	1098	3891	5733
LUXEMBOURG	-	-	-	-
NETHERLANDS	-	-	2561	2352
NEW ZEALAND	-	-	-	
NORWAY	-	-	-	-
PORTUGAL	-	-	-	-
SPAIN	-	-	-	-
SWEDEN	-	-	-	-
SWITZERLAND	-	-	-	-
TURKEY	-	-	-	-
UNITED KINGDOM	985	981	1338	1174
UNITED STATES	314	52	61	15
TOTAL NON-OECD	667	683	1086	919

Source: IEA/OECD Statistics.

AUSTRALIA

4/D1 FUEL INPUT IN ELECTRICITY GENERATION

	1973	1978	1979	1980	1981	1985	1990
TOTAL (MTOE)	18.4	25.5	27.8	29.3	30.3	34.7	39.7
SOLID FUELS	13.9	19.6	20.8	22.4	22.6	26.2	31.5
OIL	.4	.5	.8	1.1	1.2	.9	.7
GAS	.8	1.1	1.3	1.5	2.1	3.3	3.0
NUCLEAR	-	-	-	-	-	-	-
HYDRO/GEOTHERMAL	3.4	4.3	5.0	4.2	4.4	4.3	4.5
OTHER [1]	-	-	-	-	-	-	-
TOTAL (PERCENTS)	100.0	100.0	100.0	100.0	100.0	100.0	100.0
SOLID FUELS	75.1	76.7	74.7	76.6	74.6	75.5	79.3
OIL	2.3	2.1	2.7	3.8	3.8	2.6	1.8
GAS	4.4	4.3	4.7	5.3	6.9	9.5	7.6
NUCLEAR	-	-	-	-	-	-	-
HYDRO/GEOTHERMAL	18.2	16.9	17.8	14.3	14.7	12.4	11.3
OTHER [1]	-	-	-	-	-	-	-

Source: IEA/OECD Energy Balances and IEA Country Submissions (1982).

1.   Solar, wind, etc.

4/D2 ELECTRICTY PRODUCTION BY FUEL TYPE

	1973	1978	1979	1980	1981	1985	1990
TOTAL (TWH)	64.8	86.0	90.9	96.2	101.6	127.9	150.0
SOLID FUELS	48.6	63.5	66.3	70.0	71.9	96.8	119.4
OIL	1.7	2.7	2.2	5.4	5.3	3.3	2.6
GAS	2.8	5.2	6.2	6.9	9.5	12.2	11.4
NUCLEAR	-	-	-	-	-	-	-
HYDRO/GEOTHERMAL	11.8	14.5	16.2	13.8	14.9	15.6	16.6
OTHER [1]	-	-	-	-	-	-	-
TOTAL (PERCENTS)	100.0	100.0	100.0	100.0	100.0	100.0	100.0
SOLID FUELS	75.0	73.9	73.0	72.8	70.7	75.7	79.6
OIL	2.6	3.1	2.4	5.6	5.2	2.6	1.8
GAS	4.2	6.1	6.8	7.2	9.3	9.5	7.6
NUCLEAR	-	-	-	-	-	-	-
HYDRO/GEOTHERMAL	18.2	16.9	17.8	14.3	14.7	12.2	11.1
OTHER [1]	-	-	-	-	-	-	-

Source: IEA/OECD Energy Balances and IEA Country Submissions (1982).

1.   Solar, wind, etc.

AUSTRALIA

4/D3 PROJECTIONS OF ELECTRICITY GENERATION CAPACITY BY FUEL
(GW)

	Nuclear	Hydro/ Geoth.	Solid Fuels	Oil	Gas	Multi Firing [1]	Total
I 1981							
Operating [2]	-	6.2	15.2	1.5	2.4	0.6	25.9
II 1981-1985							
Capacity:							
Under Construction	-	0.7	7.1	-	-	-	7.8
Authorised	-	-	-	-	-	-	-
Other Planned	-	-	-	-	-	-	-
Conversion	-	-	0.2 e	-0.2	-	-	-
Decommissioning	-	-	-	-	-	-	-
Total Operating 1985	-	6.9	22.5	1.3	2.4	0.6	33.7
III 1986-1990							
Capacity:							
Under Construction	-	0.4	6.7	-	-	-	7.1
Authorised	-	-	1.1	-	-	-	1.1
Other Planned	-	-	1.1	-	-	-	1.1
Conversion	-	-	-	-	-	-	-
Decommissioning	-	-	-	-	-	-	-
Total Operating 1990	-	7.3	31.4	1.3	2.4	0.6	43.0

Source: IEA/OECD Electricity Statistics and IEA Country Submissions (1982)

1. Includes "Other" capacity.
2. 1981 data are for the fiscal year ending 30 June.

4/D4 ENERGY USE IN IRON AND STEEL INDUSTRY BY FUEL

	1973	1978	1979	1980	1981	1985	1990
TOTAL (MTOE)	4.2	4.8	5.0	4.4	4.0	5.5	5.6
SOLID FUELS	3.1	3.1	3.4	3.3	3.0	4.5	4.6
OIL	1.1	1.1	1.1	.5	.3	.1	.1
GAS	.0	.1	.1	.1	.3	.4	.4
ELECTRICITY	na	.4	.4	.4	.4	.5	.5
HEAT [1]	-	-	-	-	-	-	-
TOTAL (PERCENTS)	100.0	100.0	100.0	100.0	100.0	100.0	100.0
SOLID FUELS	na	66.2	67.3	76.1	74.4	81.8	82.1
OIL	na	24.2	21.7	11.1	6.5	1.8	1.8
GAS	na	1.4	2.6	2.9	8.2	7.3	7.1
ELECTRICITY	na	8.2	8.4	9.8	11.0	9.1	8.9
HEAT [1]	-	-	-	-	-	-	-

Source: IEA/OECD Energy Balances and IEA Country Submissions (1982).

1. Includes only heat from combined heat and power plants.

4/D5 ENERGY USE IN OTHER INDUSTRIES BY FUEL

	1973	1978	1979	1980	1981	1985	1990
TOTAL (MTOE)	15.8	18.6	16.9	17.9	18.6	21.6	24.1
SOLID FUELS	4.7	5.0	4.8	4.9	4.8	5.8	6.4
OIL	7.7	8.1	6.1	6.2	7.3	5.6	5.9
GAS	1.5	3.2	3.5	4.2	3.7	6.4	7.4
ELECTRICITY	na	2.4	2.5	2.5	2.7	3.8	4.4
HEAT [1]	-	-	-	-	-	-	-
TOTAL (PERCENTS)	100.0	100.0	100.0	100.0	100.0	100.0	100.0
SOLID FUELS	na	26.8	28.5	27.3	26.0	26.9	26.6
OIL	na	43.3	36.2	34.9	39.5	25.9	24.5
GAS	na	17.1	20.5	23.7	20.1	29.6	30.7
ELECTRICITY	na	12.9	14.8	14.2	14.4	17.6	18.3
HEAT [1]	-	-	-	-	-	-	-

Source: IEA/OECD Energy Balances and IEA Country Submissions (1982).

1. Includes only heat from combined heat and power plants.

AUSTRALIA

4/D6 ENERGY USE IN INDUSTRY [1] BY FUEL

	1973	1978	1979	1980	1981	1985	1990
TOTAL (MTOE)	20.0	23.4	21.9	22.2	22.6	27.1	29.7
SOLID FUELS	7.8	8.1	8.2	8.2	7.8	10.3	11.0
OIL	8.8	9.2	7.2	6.7	7.6	5.7	6.0
GAS	1.5	3.2	3.6	4.4	4.1	6.8	7.8
ELECTRICITY	2.0	2.8	2.9	3.0	3.1	4.3	4.9
HEAT [2]	-	-	-	-	-	-	-
TOTAL (PERCENTS)	100.0	100.0	100.0	100.0	100.0	100.0	100.0
SOLID FUELS	38.8	34.8	37.3	36.8	34.6	38.0	37.0
OIL	43.7	39.4	32.9	30.2	33.6	21.0	20.2
GAS	7.5	13.9	16.5	19.6	18.0	25.1	26.3
ELECTRICITY	9.9	11.9	13.3	13.4	13.8	15.9	16.5
HEAT [2]	-	-	-	-	-	-	-

Source: IEA/OECD Energy Balances and IEA Country Submissions (1982).

1. Includes non-energy use of petroleum products.
2. Includes only heat from combined heat and power plants.

4/D7 ENERGY USE IN OTHER SECTORS [1] BY FUEL

	1973	1978	1979	1980	1981	1985	1990
TOTAL (MTOE)	7.4	8.2	9.1	9.0	8.8	11.7	13.5
SOLID FUELS	.3	.2	.2	.2	.2	1.7	1.7
OIL	3.8	3.3	3.7	3.8	2.4	2.6	2.9
GAS	.9	1.2	1.5	1.0	2.0	2.5	3.0
ELECTRICITY	2.5	3.5	3.7	3.9	4.1	4.9	5.9
HEAT [2]	na	na	na	.1	.1	-	-
TOTAL (PERCENTS)	100.0	100.0	100.0	100.0	100.0	100.0	100.0
SOLID FUELS	3.8	2.4	2.0	1.8	1.7	14.5	12.6
OIL	50.9	39.8	40.9	42.6	27.6	22.2	21.5
GAS	12.2	15.0	16.3	11.1	23.3	21.4	22.2
ELECTRICITY	33.1	42.8	40.7	43.1	46.2	41.9	43.7
HEAT [2]	na	na	na	1.4	1.3	-	-

Source: IEA/OECD Energy Balances and IEA Country Submissions (1982).

1. Includes use in agricultural, commercial, public services and residential sectors.
2. Includes only heat from combined heat and power plants.

NEW ZEALAND

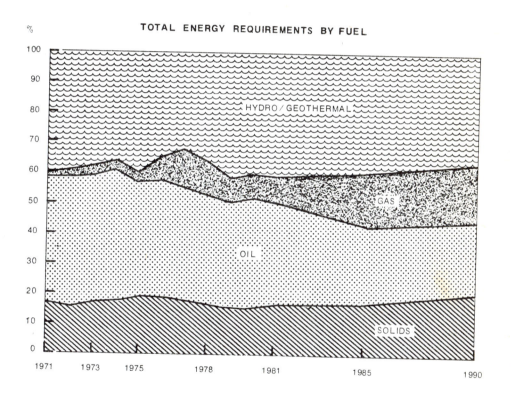

NEW ZEALAND

4/A1 TOTAL PRIMARY ENERGY REQUIREMENTS (TPER) BY FUEL

	1973	1978	1979	1980	1981	1985	1990
TPER (MTOE) [1]	9.6	11.4	11.1	11.7	11.8	15.2	18.7
SOLID FUELS	1.6	1.9	1.7	1.8	2.0	2.6	3.9
OIL	4.0	4.1	3.9	4.2	4.0	4.0	4.5
GAS	.3	1.3	.9	.9	1.0	2.6	3.5
NUCLEAR	-	-	-	-	-	-	-
HYDRO/GEOTHERMAL	3.7	4.1	4.6	4.7	4.8	6.0	6.8
OTHER [2]	-	-	-	-	-	-	-
TPER (PERCENTS) [1]	100.0	100.0	100.0	100.0	100.0	100.0	100.0
SOLID FUELS	17.0	16.3	15.3	15.3	16.6	16.9	21.0
OIL	41.9	36.2	35.3	36.3	33.8	26.3	23.9
GAS	3.0	11.5	7.7	8.0	8.4	17.2	18.9
NUCLEAR	-	-	-	-	-	-	-
HYDRO/GEOTHERMAL	38.1	36.0	41.7	40.4	41.1	39.7	36.2
OTHER [2]	-	-	-	-	-	-	-

Source: IEA/OECD Energy Balances and IEA Country Submissions (1982).

1. Net imports of electricity are only included in totals.
2. Includes other electricity generation (solar, wind, etc.) and traded synthetic fuels.

4/A2 TOTAL FINAL ENERGY CONSUMPTION (TFC) BY FUEL

	1973	1978	1979	1980	1981	1985	1990
TFC (MTOE)	6.5	7.2	7.1	7.7	7.7	9.3	11.1
SOLID FUELS	1.1	1.4	1.4	1.5	1.7	2.0	2.3
OIL	3.9	3.8	3.8	4.2	3.9	3.9	4.7
GAS	.1	.3	.3	.3	.4	1.2	1.5
ELECTRICITY	1.4	1.6	1.6	1.7	1.7	2.2	2.6
HEAT [1]	-	-	-	-		-	-
TFC (PERCENTS)	100.0	100.0	100.0	100.0	100.0	100.0	100.0
SOLID FUELS	16.7	19.4	19.9	19.6	21.8	21.2	20.8
OIL	60.1	53.4	53.0	54.2	51.0	41.8	42.0
GAS	2.2	4.5	4.2	4.3	4.7	13.2	13.3
ELECTRICITY	21.0	22.7	22.9	21.9	22.5	23.7	23.9
HEAT [1]	-	-	-	-	-	-	-

Source: IEA/OECD Energy Balances and IEA Country Submissions (1982).

1. Includes only heat production from combined heat and power plants.

NEW ZEALAND

4/A3 TOTAL SOLID FUELS BALANCE
(MTOE)

	1973	1978	1979	1980	1981	1985	1990
PRODUCTION	1.7	1.9	1.7	2.0	2.1	2.8	4.3
IMPORTS	-	.0	.0	-	-	-	-
EXPORTS	.0	.0	.0	.0	-.2	-.3	-.4
MARINE BUNKERS	-	-	-	-	-	-	-
STOCK CHANGE	.0	.0	.0	-.1	.0	-	-
PRIMARY REQUIREMENTS	1.6	1.9	1.7	1.8	2.0	2.6	3.9
ELECTRICITY GENERATION	-.5	-.4	-.2	-.2	-.2	-.4	-1.5
GAS MANUFACTURE	.0	.0	.0	.0	.0	.0	-
LIQUEFACTION	-	-	-	-	-	-	-
ENERGY SECTOR OWN USE + LOSSES [1]	.0	.0	.0	.0	.0	-.1	-.1
FINAL CONSUMPTION	1.1	1.4	1.4	1.5	1.7	2.0	2.3
INDUSTRY SECTOR	.9	1.0	1.1	1.0	1.0	1.4	1.7
IRON AND STEEL	.1	.1	.1	.1	.1	.4	.6
CHEMICAL/ PETROCHEMICAL	-	-	-	.1	.2	-	-
OTHER INDUSTRY	.8	.9	1.0	.7	.8	1.0	1.1
TRANSPORTATION SECTOR	-	-	-	.0	.0	-	-
OTHER SECTORS	.2	.3	.3	.5	.6	.6	.6
AGRICULTURE	.0	-	-	.0	.0	.0	.0
COMMERCIAL AND PUBLIC SERVICE	-	.2	.2	.2	.2	.1	.1
RESIDENTIAL	.2	.1	.1	.4	.5	.4	.4

Source: IEA/OECD Energy Balances and IEA Country Submissions (1982).

1. Includes statistical difference.

NEW ZEALAND

4/B1 SOLID FUELS INDIGENOUS PRODUCTION BY FUEL TYPE
(MTOE)

	1973	1978	1979	1980	1981	1985	1990
TOTAL SOLID FUELS	1.66	1.88	1.73	1.95	2.08	2.84	4.28
HARD COAL	1.63	1.58	1.22	1.46	1.51	2.25e	3.65e
COKING COAL	na	.15	.17	.17	.23	.45e	1.00e
STEAM COAL	na	1.42	1.05	1.29	1.28	1.80e	2.65e
BROWN COAL/LIGNITE	.03	.03	.04	.04	.04	.06e	.10e
OTHER SOLID FUELS	-	.27	.47	.45	.53	.53e	.53e

Source: IEA/OECD Coal Statistics, IEA/OECD Energy Balances, IEA Country Submissions (1982).

4/B2 COAL PRODUCTION BY TYPE
(THOUSAND METRIC TONS)

	1973	1978	1979	1980	1981
TOTAL COAL	2468	2183	1948	2291	2362
HARD COAL	2325	2032	1739	2083	2150
COKING COAL	na	218	236	243	323
STEAM COAL	na	1814	1503	1840	1827
BROWN COAL/LIGNITE	143	151	209	208	212
DERIVED PRODUCTS:					
PATENT FUEL	15	10	10	9	7
COKE OVEN COKE	20	28	3	3	3
GAS COKE	59	34	36	36	27
BROWN COAL BRIQUETTES	-	-	-	-	-
COKE OVEN GAS(TCAL)	-	-	-	-	-
BLAST FURNACE GAS(TCAL)	-	-	-	-	-

Source: IEA/OECD Coal Statistics.

NEW ZEALAND

4/C1 SOLID FUELS TRADE BY TYPE
(MTOE)

	1973	1978	1979	1980	1981	1985	1990
IMPORTS							
TOTAL SOLID FUELS	-	-	-	-	-	-	-
HARD COAL	-	-	-	-	-	-	-
COKING COAL	-	-	-	-	-	-	-
STEAM COAL	-	-	-	-	-	-	-
BROWN COAL/LIGNITE	-	-	-	-	-	-	-
OTHER SOLID FUELS	-	-	-	-	-	-	-
EXPORTS							
TOTAL SOLID FUELS	.02	-	.01	.05	.16	.28	.36
HARD COAL	-	-	.01	.05	.16	.28e	.36e
COKING COAL	-	-	.01	.05	.13	.28e	.36e
STEAM COAL	-	-	-	-	.04	-	-
BROWN COAL/LIGNITE	-	-	-	-	-	-	-
OTHER SOLID FUELS [1]	.02	-	-	-	-	-	-

Source: IEA/OECD Coal Statistics, IEA/OECD Energy Balances, IEA Country Submissions (1982).

1. Includes derived coal products, e.g. coke.

NEW ZEALAND

4/D1 FUEL INPUT IN ELECTRICITY GENERATION

	1973	1978	1979	1980	1981	1985	1990
TOTAL (MTOE)	4.6	5.6	5.4	5.4	5.5	7.5	9.1
SOLID FUELS	.5	.4	.2	.2	.2	.4	1.5
OIL	.4	.1	.0	.0	.0	.0	.1
GAS	.1	.9	.6	.4	.4	1.0	.7
NUCLEAR	-	-	-	-	-	-	-
HYDRO/GEOTHERMAL	3.7	4.1	4.6	4.7	4.8	6.0	6.8
OTHER [1]	-	-	-	-	-	-	-
TOTAL (PERCENTS)	100.0	100.0	100.0	100.0	100.0	100.0	100.0
SOLID FUELS	10.6	7.4	3.9	3.9	4.2	5.8	16.2
OIL	8.0	2.5	.4	.2	.3	.4	1.5
GAS	1.8	16.7	10.1	8.1	7.9	13.8	7.8
NUCLEAR	-	-	-	-	-	-	-
HYDRO/GEOTHERMAL	79.6	73.3	85.5	87.8	87.7	80.0	74.4
OTHER [1]	-	-	-	-	-	-	-

Source: IEA/OECD Energy Balances and IEA Country Submissions (1982).

1. Solar, wind, etc.

4/D2 ELECTRICTY PRODUCTION BY FUEL TYPE

	1973	1978	1979	1980	1981	1985	1990
TOTAL (TWH)	18.5	21.9	21.7	22.3	23.1	28.9	34.2
SOLID FUELS	1.6	.6	.3	.4	.3	.9	4.8
OIL	1.4	.1	.0	.0	.0	.1	.5
GAS	-	3.7	1.6	1.7	2.2	4.5	2.9
NUCLEAR	-	-	-	-	-	-	-
HYDRO/GEOTHERMAL	15.6	17.4	19.8	20.1	20.6	23.3	26.1
OTHER [1]	-	-	-	-	-	-	-
TOTAL (PERCENTS)	100.0	100.0	100.0	100.0	100.0	100.0	100.0
SOLID FUELS	8.5	2.9	1.6	1.9	1.2	3.3	13.9
OIL	7.5	.6	.1	.1	.1	.4	1.5
GAS	-	16.8	7.3	7.7	9.4	15.5	8.5
NUCLEAR	-	-	-	-	-	-	-
HYDRO/GEOTHERMAL	84.0	79.6	91.0	90.4	89.3	80.8	76.1
OTHER [1]	-	-	-	-	-	-	-

Source: IEA/OECD Energy Balances and IEA Country Submissions (1982).

1. Solar, wind, etc.

NEW ZEALAND

4/D3 PROJECTIONS OF ELECTRICITY GENERATION CAPACITY BY FUEL
(GW)

	Nuclear	Hydro/ Geoth.	Solid Fuels	Oil	Gas	Multi Firing [1]	Total
I 1981							
Operating [2]	-	4.1	0.2	0.5	0.2	0.9	5.9
II 1981-1985							
Capacity:							
Under Construction	-	0.7	-	-	-	0.9	1.5
Authorised	-	-	-	-	-	-	-
Other Planned	-	-	-	-	-	-	-
Conversion	-	-	-	-	-	-	-
Decommissioning	-	-	-	-	-	-	-
Total Operating 1985	-	4.8	0.2	0.5	0.2	1.8	7.5
III 1986-1990							
Capacity:							
Under Construction	-	0.4	-	-	-	-	0.4
Authorised	-	0.3	-	-	-	-	0.3
Other Planned	-	-	-	-	-	-	-
Conversion	-	-	0.2	-0.2e	-	-	-
Decommissioning	-	-	-	-	-	-	-
Total Operating 1990	-	5.5	0.4	0.3	0.2	1.8	8.2

Source: IEA/OECD Electricity Statistics and IEA Country Submissions (1982).

1.   Includes "Other" capacity.
2.   1981 data are for the fiscal year ending 31 March 1982.

NEW ZEALAND

4/D6 ENERGY USE IN INDUSTRY [1] BY FUEL

	1973	1978	1979	1980	1981	1985	1990
TOTAL (MTOE)	2.0	2.8	2.8	2.6	2.7	4.1	5.4
SOLID FUELS	.9	1.0	1.1	1.0	1.0	1.4	1.7
OIL	.6	.9	.9	.8	.7	.6	1.1
GAS	.0	.2	.2	.2	.2	1.0	1.2
ELECTRICITY	.5	.6	.6	.7	.7	1.1	1.4
HEAT [2]	-	-	-	-	-	-	-
TOTAL (PERCENTS)	100.0	100.0	100.0	100.0	100.0	100.0	100.0
SOLID FUELS	45.2	37.2	38.0	37.6	39.4	33.5	31.9
OIL	28.5	32.2	32.4	28.5	27.3	15.5	20.6
GAS	1.7	8.3	7.0	9.0	8.2	25.0	21.7
ELECTRICITY	24.6	22.3	22.7	25.0	25.1	26.0	25.8
HEAT [2]	-	-	-	-	-	-	-

Source: IEA/OECD Energy Balances and IEA Country Submissions (1982).

(1) Includes non-energy use of petroleum products.
(2) Includes only heat from combined heat and power plants.

4/D7 ENERGY USE IN OTHER SECTORS [1] BY FUEL

	1973	1978	1979	1980	1981	1985	1990
TOTAL (MTOE)	2.3	1.8	1.8	2.4	2.4	1.9	2.0
SOLID FUELS	.2	.3	.3	.5	.6	.6	.6
OIL	1.1	.4	.4	.8	.6	.1	.1
GAS	.1	.1	.1	.1	.1	.1	.1
ELECTRICITY	.9	1.0	1.0	1.0	1.1	1.1	1.3
HEAT [2]	-	-	-	-	-	-	-
TOTAL (PERCENTS)	100.0	100.0	100.0	100.0	100.0	100.0	100.0
SOLID FUELS	8.7	18.7	19.3	21.1	25.5	30.9	28.8
OIL	48.3	22.2	20.3	33.0	25.2	5.2	3.4
GAS	4.7	4.9	5.6	3.9	5.9	5.2	6.3
ELECTRICITY	38.2	54.3	54.8	42.1	43.4	58.8	61.5
HEAT [2]	-	-	-	-	-	-	-

Source: IEA/OECD Energy Balances and IEA Country Submissions (1982).

1. Includes use in agricultural, commercial, public services and residential sectors.
2. Includes only heat from combined heat and power plants.

JAPAN

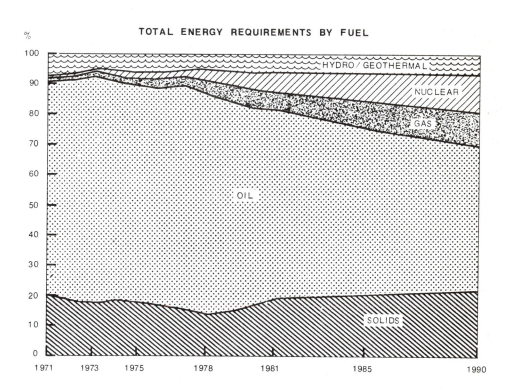

TOTAL ENERGY REQUIREMENTS BY FUEL

JAPAN

4/A1 TOTAL PRIMARY ENERGY REQUIREMENTS (TPER) BY FUEL

	1973	1978	1979	1980	1981	1985	1990
TPER (MTOE) [1]	340.4	362.3	377.0	365.1	362.1	430.9	519.2
SOLID FUELS	59.4	50.6	55.8	63.0	69.1	88.4	111.8
OIL	255.6	262.6	263.8	237.2	226.7	232.6	242.5
GAS	5.4	16.1	19.2	21.9	22.4	38.3	57.9
NUCLEAR	2.4	14.5	17.2	20.2	21.5	39.4	62.5
HYDRO/GEOTHERMAL	17.6	18.4	21.1	22.8	22.4	27.5	33.9
OTHER [2]	-	-	-	-	-	4.7	10.6
TPER (PERCENTS) [1]	100.0	100.0	100.0	100.0	100.0	100.0	100.0
SOLID FUELS	17.5	14.0	14.8	17.3	19.1	20.5	21.5
OIL	75.1	72.5	70.0	65.0	62.6	54.0	46.7
GAS	1.6	4.4	5.1	6.0	6.2	8.9	11.2
NUCLEAR	.7	4.0	4.6	5.5	5.9	9.1	12.0
HYDRO/GEOTHERMAL	5.2	5.1	5.6	6.2	6.2	6.4	6.5
OTHER [2]	-	-	-	-	-	1.1	2.0

Source: IEA/OECD Energy Balances and IEA Country Submissions (1982).

1. Net imports of electricity are only included in totals.
2. Includes other electricity generation (solar, wind, etc.), traded synthetic fuels and direct use of solar, wind and other non-solid renewable energy sources.

4/A2 TOTAL FINAL ENERGY CONSUMPTION (TFC) BY FUEL

	1973	1978	1979	1980	1981	1985	1990
TFC (MTOE)	249.7	269.6	293.0	261.2	254.5	297.3	353.3
SOLID FUELS	37.8	30.7	32.9	37.8	39.5	48.7	58.5
OIL	169.0	187.3	205.7	170.1	161.0	174.0	194.7
GAS	7.2	8.9	9.6	9.2	9.6	16.1	24.0
ELECTRICITY	35.7	42.8	44.9	44.1	44.4	55.1	68.4
HEAT [1]	-	-	-	-	-	3.4	7.7
TFC (PERCENTS)	100.0	100.0	100.0	100.0	100.0	100.0	100.0
SOLID FUELS	15.2	11.4	11.2	14.5	15.5	16.4	16.6
OIL	67.7	69.5	70.2	65.1	63.3	58.5	55.1
GAS	2.9	3.3	3.3	3.5	3.8	5.4	6.8
ELECTRICITY	14.3	15.9	15.3	16.9	17.4	18.5	19.4
HEAT [1]	-	-	-	-	-	1.1	2.2

Source: IEA/OECD Energy Balances and IEA Country Submissions (1982).

1. Includes heat production from combined heat and power plants and direct use of solar, wind and other non-solid renewable energy sources.

JAPAN

4/A3 TOTAL SOLID FUELS BALANCE
(MTOE)

	1973	1978	1979	1980	1981	1985	1990
PRODUCTION	18.4	13.6	13.1	13.2	12.9	13.4	13.9
IMPORTS	42.4	38.6	43.4	50.1	57.0	75.0	97.9
EXPORTS	-.4	-.7	-1.5	-1.4	-1.3	-	-
MARINE BUNKERS	-	-	-	-	-	-	-
STOCK CHANGE	-1.0	-1.0	.8	1.1	.5	-	-
PRIMARY REQUIREMENTS	59.4	50.6	55.8	63.0	69.1	88.4	111.8
ELECTRICITY GENERATION	-12.2	-12.8	-13.6	-11.7	-15.2	-26.3	-40.3
GAS MANUFACTURE	-2.4	-1.7	-1.7	-1.4	-1.7	-1.2	-1.2
LIQUEFACTION	-	-	-	-	-	-	-
ENERGY SECTOR OWN USE + LOSSES [1]	-7.1	-5.5	-7.6	-12.1	-12.6	-12.2	-11.8
FINAL CONSUMPTION	37.8	30.7	32.9	37.8	39.5	48.7	58.5
INDUSTRY SECTOR	35.8	29.9	32.0	36.4	38.2	47.8	57.9
IRON AND STEEL	33.9	27.4	29.0	30.5	29.1	na	na
CHEMICAL/ PETROCHEMICAL	-	.3	.4	.3	.4	na	na
OTHER INDUSTRY	1.9	2.2	2.5	5.6	8.7	na	na
TRANSPORTATION SECTOR	.2	-	-	-	-	-	-
OTHER SECTORS	1.8	.7	.9	1.4	1.3	.9	.6
AGRICULTURE	-	-	-	-	-	-	-
COMMERCIAL AND PUBLIC SERVICE	-	-	-	-	-	-	-
RESIDENTIAL	1.8	.7	.9	1.4	1.3	-	-

Source: IEA/OECD Energy Balances and IEA Country Submissions (1982).

1. Includes statistical difference.

JAPAN

4/B1 SOLID FUELS INDIGENOUS PRODUCTION BY FUEL TYPE
(MTOE)

	1973	1978	1979	1980	1981	1985	1990
TOTAL SOLID FUELS	18.38	13.64	13.05	13.19	12.93	13.40	13.90
HARD COAL	18.34	13.56	12.98	13.18	12.93	13.40e	13.90e
COKING COAL	na	6.33	5.60	5.08	4.56	4.40e	4.60e
STEAM COAL	na	7.23	7.38	8.10	8.37	9.00e	9.30e
BROWN COAL/LIGNITE	.04	.01	.01	.01	-	-	-
OTHER SOLID FUELS	-	.07	.06	-	-	-	-

Source: IEA/OECD Coal Statistics, IEA/OECD Energy Balances, IEA Country Submissions (1982).

4/B2 COAL PRODUCTION BY TYPE
(THOUSAND METRIC TONS)

	1973	1978	1979	1980	1981
TOTAL COAL	25190	18586	17789	18054	17695
HARD COAL	25090	18549	17760	18027	17687
COKING COAL	na	8659	7663	6943	6236
STEAM COAL	na	9890	10097	11084	11451
BROWN COAL/LIGNITE	100	37	29	27	8
DERIVED PRODUCTS:					
PATENT FUEL	1200	420	520	432	331
COKE OVEN COKE	52300	43970	47630	47463	47241
GAS COKE	4772	3253	3282	3497	3450
BROWN COAL BRIQUETTES	-	-	-	-	-
COKE OVEN GAS(TCAL)	74330	64420	69730	75067	73642
BLAST FURNACE GAS(TCAL)	114490	88550	91870	93858	89621

Source: IEA/OECD Coal Statistics.

JAPAN

4/C1 SOLID FUELS TRADE BY TYPE
(MTOE)

	1973	1978	1979	1980	1981	1985	1990
IMPORTS							
TOTAL SOLID FUELS	42.40	38.64	43.41	50.12	56.99	75.00	97.90
HARD COAL	42.40	38.64	43.41	50.12	56.99	75.00e	97.90e
COKING COAL	na	37.19	41.42	45.49	47.76	53.00e	59.00e
STEAM COAL	na	1.45	1.99	4.64	9.23	22.00e	38.90e
BROWN COAL/LIGNITE	-	-	-	-	-	-	-
OTHER SOLID FUELS	-	-	-	-	-	-	-
EXPORTS							
TOTAL SOLID FUELS	.39	.71	1.48	1.38	1.30	-	-
HARD COAL	-	.04	.03	.04	.03	-	-
COKING COAL	-	.03	.01	.02	.03	-	-
STEAM COAL	-	.01	.02	.02	.01	-	-
BROWN COAL/LIGNITE	-	-	-	-	-	-	-
OTHER SOLID FUELS [1]	.39	.67	1.45	1.34	1.27	-	-

Source: IEA/OECD Coal Statistics, IEA/OECD Energy Balances, IEA Country Submissions (1982).

1. Includes derived coal products, e.g. coke.

JAPAN

4/C2 COAL IMPORTS BY ORIGIN
(THOUSAND METRIC TONS)

	1978	1979	1980	1981
HARD COAL IMPORTS	52858	59386	68570	77958
COKING COAL	50876	56660	62227	65333
IMPORTS FROM:				
AUSTRALIA	24149	27013	25917	29148
CANADA	10895	9777	10759	9252
GERMANY	398	223	-	-
UNITED KINGDOM	-	-	-	-
UNITED STATES·	9956	13987	19414	21761
OTHER OECD	-	-	70	124
CZECHOSLOVAKIA	-	-	-	-
POLAND	429	489	387	65
USSR	2244	1932	1898	1112
CHINA	-	-	979	1148
SOUTH AFRICA	2360	2389	2803	2723
OTHER NON-OECD	445	850	-	-
STEAM COAL	1982	2726	6343	12625
IMPORTS FROM:				
AUSTRALIA	669	1287	3462	5427
CANADA	105	62	352	1199
GERMANY	-	-	-	-
UNITED KINGDOM	-	-	-	-
UNITED STATES	-	8	385	2079
OTHER OECD	-	-	-	19
CZECHOSLOVAKIA	-	-	-	-
POLAND	-	-	-	-
USSR	149	147	233	286
CHINA	-	-	1086	1614
SOUTH AFRICA	157	142	385	1625
OTHER NON-OECD	902	1080	440	376

Source: IEA/OECD Coal Statistics.

JAPAN

4/D1 FUEL INPUT IN ELECTRICITY GENERATION

	1973	1978	1979	1980	1981	1985	1990
TOTAL (MTOE)	106.2	116.7	124.5	121.6	123.6	162.7	211.0
SOLID FUELS [1]	12.2	12.8	13.6	11.7	15.2	26.3	40.3
OIL	72.0	60.8	59.5	50.4	47.6	43.9	37.3
GAS	2.0	10.1	13.0	16.4	16.9	25.9	37.7
NUCLEAR	2.4	14.5	17.2	20.2	21.5	39.4	62.5
HYDRO/GEOTHERMAL	17.6	18.4	21.1	22.8	22.4	27.2	33.2
OTHER [2]	-	-	-	-	-	-	-
TOTAL (PERCENTS)	100.0	100.0	100.0	100.0	100.0	100.0	100.0
SOLID FUELS	11.4	11.0	10.9	9.7	12.3	16.2	19.1
OIL	67.8	52.1	47.8	41.5	38.5	27.0	17.7
GAS	1.9	8.7	10.5	13.5	13.6	15.9	17.9
NUCLEAR	2.2	12.5	13.9	16.6	17.4	24.2	29.6
HYDRO/GEOTHERMAL	16.6	15.8	16.9	18.7	18.1	16.7	15.7
OTHER [2]	-	-	-	-	-	-	-

Source: IEA/OECD Energy Balances and IEA Country Submissions (1982).

1. Beginning with 1980, a break in series because of change to calendar year data and revision of use of blast furnace gas in autogeneration. Previous years' data include use of blast furnace gas in industrial process heating boilers and should be reduced by about 1.5 Mtoe. Later information from Japanese authoroties indicates further that the figure for 1980 should be 12.5 Mtoe instead of 11.7 Mtoe.
2. Solar, wind, etc.

4/D2 ELECTRICTY PRODUCTION BY FUEL TYPE [1]

	1973	1978	1979	1980	1981	1985	1990
TOTAL (TWH)	470.3	501.5	529.1	520.4	529.3	665.0	840.0
SOLID FUELS	37.3	30.9	33.2	39.0	47.1	80.3	123.0
OIL	340.8	281.8	266.6	224.7	221.2	199.1	160.0
GAS	10.5	54.2	73.0	81.1	81.7	123.0	179.0
NUCLEAR	9.7	59.3	70.4	82.6	87.8	159.7	253.0
HYDRO/GEOTHERMAL	71.9	75.2	85.9	93.0	91.5	102.9	125.0
OTHER [2]	-	-	-	-	-	-	-
TOTAL (PERCENTS)	100.0	100.0	100.0	100.0	100.0	100.0	100.0
SOLID FUELS	7.9	6.2	6.3	7.5	8.9	12.1	14.6
OIL	72.5	56.2	50.4	43.2	41.8	29.9	19.0
GAS	2.2	10.8	13.8	15.6	15.4	18.5	21.3
NUCLEAR	2.1	11.8	13.3	15.9	16.6	24.0	30.1
HYDRO/GEOTHERMAL	15.3	15.0	16.2	17.9	17.3	15.5	14.9
OTHER [2]	-	-	-	-	-	-	-

Source: IEA/OECD Energy Balances and IEA Country Submissions (1982).

1. Includes only electricity generated by public utilities.
2. Solar, wind, etc.

JAPAN

4/D3 PROJECTIONS OF ELECTRICITY GENERATION CAPACITY BY FUEL
(GW)

	Nuclear	Hydro/ Geoth.	Coal	Oil	Gas	Multi Firing [1]	Total
I 1981							
Operating [2]	16.1	30.6	6.0	62.0	20.8	-	135.5
II 1981-1985							
Capacity:							
Under Construction	9.5	4.5	2.6	3.3	5.2	-	25.1
Authorised	-	0.5	0.5	1.9	2.5	-	5.4
Other Planned	-	0.1	0.3	1.2	-	-	1.6
Conversion	-	-	3.3	-8.4	5.1	-	-
Decommissioning	-	-0.1	-0.4	-0.9	-	-	-1.4
Total Operating 1985	25.6	35.6	12.3	59.1	33.6	-	166.2
III 1986-1990							
Capacity:							
Under Construction	-	0.3	-	0.4	-	-	0.7
Authorised	7.3	3.0	5.2	0.5	4.4	-	20.4
Other Planned [3]	13.1	7.8	5.5	-13.0	8.3	-	21.7
Conversion	-	-	-	-	-	-	-
Decommissioning	-	-	-	-	-	-	-
Total Operating 1990	46.0	46.7	23.0	47.0	46.3	-	209.0

Source: IEA/OECD Electricity Statistics and IEA Country Submissions (1982).

1. Includes "Other" capacity.
2. 1981 data are for the fiscal year ending 31 March 1982.
3. Includes "Conversion" and "Decommissioning" Capacity.

Note: Generation capacity of autoproducers is not included

JAPAN

4/D4 ENERGY USE IN IRON AND STEEL INDUSTRY BY FUEL

	1973	1978	1979	1980	1981	1985	1990
TOTAL (MTOE)	51.8	40.4	42.9	42.3	38.3	na	na
SOLID FUELS	33.9	27.4	29.0	30.5	29.1	na	na
OIL	11.7	6.9	7.3	5.5	3.2	na	na
GAS	-	-	-	-	-	na	na
ELECTRICITY	6.2	6.1	6.6	6.3	6.1	na	na
HEAT [1]	-	-	-	-	-	-	-
TOTAL (PERCENTS)	100.0	100.0	100.0	100.0	100.0	na	na
SOLID FUELS	65.4	67.8	67.7	72.1	75.9	na	na
OIL	22.7	17.0	17.0	13.0	8.2	na	na
GAS	-	-	-	-	-	na	na
ELECTRICITY	11.9	15.2	15.3	15.0	15.8	na	na
HEAT [1]	-	-	-	-	-	-	-

Source: IEA/OECD Energy Balances and IEA Country Submissions (1982).

1. Includes only heat from combined heat and power plants.

4/D5 ENERGY USE IN OTHER INDUSTRIES BY FUEL

	1973	1978	1979	1980	1981	1985	1990
TOTAL (MTOE)	101.5	117.7	135.2	106.1	103.2	na	na
SOLID FUELS	1.9	2.5	3.0	5.9	9.2	na	na
OIL	78.6	91.6	107.6	76.7	70.8	na	na
GAS	2.1	2.6	2.7	2.1	2.1	na	na
ELECTRICITY	18.9	21.0	21.9	21.4	21.0	na	na
HEAT [1]	-	-	-	-	-	-	-
TOTAL (PERCENTS)	100.0	100.0	100.0	100.0	100.0	na	na
SOLID FUELS	1.9	2.2	2.2	5.6	8.9	na	na
OIL	77.4	77.8	79.6	72.3	68.6	na	na
GAS	2.1	2.2	2.0	1.9	2.1	na	na
ELECTRICITY	18.6	17.8	16.2	20.2	20.4	na	na
HEAT [1]	-	-	-	-	-	-	-

Source: IEA/OECD Energy Balances and IEA Country Submissions (1982).

1. Includes only heat from combined heat and power plants.

JAPAN

4/D6 ENERGY USE IN INDUSTRY [1] BY FUEL

	1973	1978	1979	1980	1981	1985	1990
TOTAL (MTOE)	153.3	158.1	178.1	148.4	141.4	162.3	189.3
SOLID FUELS	35.8	29.9	32.0	36.4	38.2	47.8	57.9
OIL	90.3	98.4	114.9	82.2	74.0	78.9	86.8
GAS	2.1	2.6	2.7	2.1	2.1	6.3	11.4
ELECTRICITY	25.1	27.1	28.5	27.7	27.1	29.3	33.2
HEAT [2]	-	-	-	-	-	-	-
TOTAL (PERCENTS)	100.0	100.0	100.0	100.0	100.0	100.0	100.0
SOLID FUELS	23.3	18.9	18.0	24.5	27.0	29.5	30.6
OIL	58.9	62.3	64.5	55.4	52.3	48.6	45.9
GAS	1.4	1.7	1.5	1.4	1.5	3.9	6.0
ELECTRICITY	16.3	17.1	16.0	18.7	19.2	18.1	17.5
HEAT [2]	-	-	-	-	-	-	-

Source: IEA/OECD Energy Balances and IEA Country Submissions (1982).

1. Includes non-energy use of petroleum products.
2. Includes only heat from combined heat and power plants.

4/D7 ENERGY USE IN OTHER SECTORS [1] BY FUEL

	1973	1978	1979	1980	1981	1985	1990
TOTAL (MTOE)	51.2	58.1	59.8	64.2	65.0	72.6	93.0
SOLID FUELS	1.8	.7	.9	1.4	1.3	.9	.6
OIL	34.8	36.8	37.0	40.6	40.3	34.1	38.4
GAS	5.0	6.3	6.9	7.1	7.5	9.8	12.6
ELECTRICITY	9.5	14.4	15.0	15.1	15.9	24.4	33.7
OTHER [2]	-	-	-	-	-	3.4	7.7
TOTAL (PERCENTS)	100.0	100.0	100.0	100.0	100.0	100.0	100.0
SOLID FUELS	3.6	1.3	1.5	2.2	2.0	1.2	.6
OIL	68.0	63.3	61.9	63.2	62.0	47.0	41.3
GAS	9.9	10.8	11.5	11.1	11.5	13.5	13.5
ELECTRICITY	18.6	24.7	25.1	23.5	24.5	33.6	36.2
OTHER [2]	-	-	-	-	-	4.7	8.3

Source: IEA/OECD Energy Balances and IEA Country Submissions (1982).

1. Includes use in agricultural, commercial, public services and residential sectors.
2. Includes heat from combined heat and power plants and direct use of solar, wind and other non-solid renewable energy sources.

5. OECD EUROPE

TOTAL ENERGY REQUIREMENTS BY FUEL

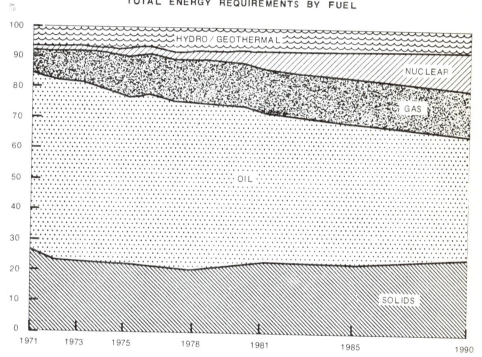

OECD EUROPE

5/A1 TOTAL PRIMARY ENERGY REQUIREMENTS (TPER) BY FUEL

	1973	1978	1979	1980	1981	1985	1990
TPER (MTOE) [1]	1200.5	1237.8	1290.2	1251.0	1218.0	1349.5	1499.2
SOLID FUELS	278.7	256.4	279.2	285.3	285.7	310.7	374.2
OIL	704.6	677.7	688.4	643.8	590.6	612.6	592.0
GAS	122.6	171.4	182.1	178.6	175.2	200.4	221.6
NUCLEAR	18.7	40.5	44.3	51.4	71.7	119.2	182.9
HYDRO/GEOTHERMAL	75.9	91.7	96.0	91.5	94.5	97.2	108.0
OTHER [2]	-	-	-	-	-	8.1	20.6
TPER (PERCENTS) [1]	100.0	100.0	100.0	100.0	100.0	100.0	100.0
SOLID FUELS	23.2	20.7	21.6	22.8	23.5	23.0	25.0
OIL	58.8	54.8	53.4	51.6	48.6	45.4	39.5
GAS	10.2	13.8	14.1	14.3	14.4	14.9	14.8
NUCLEAR	1.6	3.3	3.4	4.0	5.7	8.8	12.2
HYDRO/GEOTHERMAL	6.2	7.4	7.4	7.2	7.6	7.2	7.2
OTHER [2]	-	-	-	-	-	.6	1.4

Source: IEA/OECD Energy Balances and IEA Country Submissions (1982).

1. Net imports of electricity are only included in totals.
2. Includes other electricity generation (solar, wind, etc.), traded synthetic fuels and direct use of solar, wind and other non-solid renewable energy sources.

5/A2 TOTAL FINAL ENERGY CONSUMPTION (TFC) BY FUEL

	1973	1978	1979	1980	1981	1985	1990
TFC (MTOE)	906.4	908.9	948.7	903.9	869.1	963.2	1048.6
SOLID FUELS	135.0	98.9	104.2	100.1	101.2	124.7	144.7
OIL	571.1	554.3	571.2	523.2	487.4	501.4	500.4
GAS	99.1	136.1	147.4	147.0	146.6	174.9	200.2
ELECTRICITY	101.2	119.7	125.8	126.7	127.2	144.6	172.7
HEAT/OTHER [1]	na	na	na	7.0	6.7	17.5	30.7
TFC (PERCENTS)	100.0	100.0	100.0	100.0	100.0	100.0	100.0
SOLID FUELS	14.9	10.9	11.0	11.1	11.6	13.0	13.8
OIL	63.0	61.0	60.2	57.9	56.1	52.1	47.7
GAS	10.9	15.0	15.5	16.3	16.9	18.2	19.1
ELECTRICITY	11.2	13.2	13.3	14.0	14.6	15.0	16.5
HEAT/OTHER [1]	na	na	na	.8	.8	1.8	2.9

Source: IEA/OECD Energy Balances and IEA Country Submissions (1982).

1. Includes heat production from combined heat and power plants and direct use of solar, wind and other non-solid renewable energy sources.

OECD EUROPE

5/A3 TOTAL SOLID FUELS BALANCE

	1973	1978	1979	1980	1981	1985	1990
PRODUCTION	244.6	216.9	221.3	231.1	235.3	234.5	252.7
IMPORTS	53.8	61.5	75.6	86.2	87.0	90.5	133.6
EXPORTS	-23.8	-24.6	-25.1	-21.6	-23.9	-14.1	-12.1
MARINE BUNKERS	.0	-	-	-	-	-	-
STOCK CHANGE	4.2	2.7	7.5	-10.4	-12.7	-.1	.1
PRIMARY REQUIREMENTS	278.7	256.4	279.2	285.3	285.7	310.7	374.2
ELECTRICITY GENERATION	-126.4	-146.3	-159.4	-168.9	-169.4	-162.5	-200.3
GAS MANUFACTURE	-2.3	-.3	-.3	-.3	-.1	-1.0	-4.1
LIQUEFACTION	-	-	-	-	-	-5.0	-7.7
ENERGY SECTOR OWN USE + LOSSES [1]	-15.1	-10.9	-15.1	-15.7	-14.7	-17.5	-17.4
FINAL CONSUMPTION	135.0	98.9	104.2	100.1	101.2	124.7	144.7
INDUSTRY SECTOR	84.4	63.4	67.0	63.3	65.5	85.8	102.9
IRON AND STEEL	53.1	41.5	43.2	39.1	39.3	na	na
CHEMICAL/ PETROCHEMICAL	6.9	4.2	4.0	4.4	5.0	na	na
OTHER INDUSTRY	24.4	17.8	19.8	19.8	21.2	na	na
TRANSPORTATION SECTOR	1.8	.8	.6	.5	.6	.3	.6
OTHER SECTORS	48.9	34.6	36.6	36.3	35.2	37.2	39.7
AGRICULTURE	.1	.0	.0	.0	.1	.2	.2
COMMERCIAL AND PUBLIC SERVICE	2.2	1.5	1.5	1.1	1.0	.0	.2
RESIDENTIAL	43.9	32.2	33.9	34.2	33.1	20.7	20.9

Source: IEA/OECD Energy Balances and IEA Country Submissions (1982).

1. Includes statistical difference.

Note: See United Kingdom, Table 5/A3 for explanation of changes to IEA Country Submissions (1982) as published in "Energy Policies and Programmes of IEA Countries, 1982 Review".

5/B1 SOLID FUELS INDIGENOUS PRODUCTION BY FUEL TYPE
(MTOE)

	1973	1978	1979	1980	1981	1985	1990
TOTAL SOLID FUELS	244.59	216.90	221.28	231.11	235.33	234.48	252.69
HARD COAL	194.80	161.02	162.87	168.34	168.64	159.69	164.91
COKING COAL	na	53.87	54.29	53.40	53.21	52.63	52.74
STEAM COAL	na	107.15	108.59	114.94	115.44	107.06	112.17
BROWN COAL/LIGNITE	28.70	34.75	38.09	40.51	43.92	54.23	65.77
OTHER SOLID FUELS	21.09	21.13	20.32	22.26	22.77	20.56	22.01

Source: IEA/OECD Coal Statistics, IEA/OECD Energy Balances, IEA Country Submissions (1982).

Note: See Table 5/A3

5/B2 COAL PRODUCTION BY TYPE
(THOUSAND METRIC TONS)

	1973	1978	1979	1980	1981
TOTAL COAL	433679	424392	439596	455659	468728
HARD COAL	285580	254531	257003	266148	266138
COKING COAL	na	81158	81306	79678	76897
STEAM COAL	na	173373	175697	186470	189241
BROWN COAL/LIGNITE	148099	169861	182593	189511	202590
DERIVED PRODUCTS:					
PATENT FUEL	7620	4884	4998	4293	4032
COKE OVEN COKE	90744	72292	75896	75346	73342
GAS COKE	2055	892	1216	912	325
BROWN COAL BRIQUETTES	6862	5298	6727	6916	7054
COKE OVEN GAS(TCAL)	173234	134714	142961	141236	142878
BLAST FURNACE GAS(TCAL)	203604	155251	171374	148172	149314

Source: IEA/OECD Coal Statistics.

OECD EUROPE

5/C1 SOLID FUELS TRADE BY TYPE
(MTOE)

	1973	1978	1979	1980	1981	1985	1990
IMPORTS							
TOTAL SOLID FUELS	53.79	61.51	75.57	86.23	87.01	90.48	133.58
HARD COAL	41.97	53.41	64.95	74.99	76.52	85.79	129.42
COKING COAL	na	27.08	31.89	32.99	32.20	30.16	33.41
STEAM COAL	na	26.33	33.06	42.01	44.32	55.63	96.01
BROWN COAL/LIGNITE	.37	.42	.49	.62	.79	1.24	1.12
OTHER SOLID FUELS [1]	11.45	7.68	10.13	10.62	9.70	3.45	3.04
EXPORTS							
TOTAL SOLID FUELS	23.84	24.65	25.14	21.63	23.87	14.12	12.12
HARD COAL	13.46	15.06	13.33	12.47	15.18	13.10	11.10
COKING COAL	na	9.33	7.29	6.13	7.20	9.80	7.70
STEAM COAL	na	5.72	6.04	6.34	7.98	3.30	3.40
BROWN COAL/LIGNITE	.01	.01	.01	.01	.01	.01	.01
OTHER SOLID FUELS [1]	10.37	9.58	11.80	9.15	8.68	1.01	1.01

Source: IEA/OECD Coal Statistics, IEA/OECD Energy Balances, IEA Country Submissions (1982).

1. Includes derived coal products, e.g. coke.

Note: See Table 5/A3

OECD EUROPE

5/C2 COAL IMPORTS BY ORIGIN
(THOUSAND METRIC TONS)

	1978	1979	1980	1981
HARD COAL IMPORTS	77801	94849	109944	111819
COKING COAL	39098	46202	47718	46141
IMPORTS FROM:				
AUSTRALIA	4622	4950	4326	4892
CANADA	434	522	227	424
GERMANY	9987	9129	8281	8398
UNITED KINGDOM	110	258	75	1678
UNITED STATES	8715	15314	20903	23687
OTHER OECD	405	326	740	574
CZECHOSLOVAKIA	832	970	1065	860
POLAND	10517	10671	8808	3877
USSR	2722	2744	2712	1030
CHINA	-	-	-	-
SOUTH AFRICA	211	619	545	377
OTHER NON-OECD	543	699	36	344
STEAM COAL	38703	48647	62226	65678
IMPORTS FROM:				
AUSTRALIA	2953	4532	4874	3604
CANADA	816	665	691	1236
GERMANY	6642	6098	4134	3087
UNITED KINGDOM	2201	2367	4175	7886
UNITED STATES	86	1526	13583	24503
OTHER OECD	1327	1532	1346	1450
CZECHOSLOVAKIA	243	203	255	217
POLAND	11020	12652	10645	2765
USSR	2204	2263	1783	829
CHINA	-	70	417	195
SOUTH AFRICA	10830	16399	20017	19784
OTHER NON-OECD	381	330	306	122

Source: IEA/OECD Coal Statistics.

OECD EUROPE

5/C3 HARD COAL EXPORTS BY DESTINATION
(THOUSAND METRIC TONS)

	1973	1978	1979	1980
TOTAL EXPORTS	22371	19816	18798	23603
EXPORTS TO:				
AUSTRALIA	1	-	-	-
AUSTRIA	228	238	100	308
BELGIUM	4726	4015	2928	2804
CANADA	-	-	-	-
DENMARK	1104	809	932	1939
FINLAND	-	1	177	20
FRANCE	7586	7619	6632	7258
GERMANY	1236	1609	2602	2513
GREECE	-	-	-	-
ICELAND	-	-	-	14
IRELAND	182	161	210	358
ITALY	2574	1996	2550	2682
JAPAN	375	247	1	306
LUXEMBOURG	38	58	60	45
NETHERLANDS	1612	1323	1453	1683
NEW ZEALAND	-	-	-	-
NORWAY	165	180	101	202
PORTUGAL	15	3	7	43
SPAIN	478	82	33	119
SWEDEN	306	284	72	209
SWITZERLAND	92	159	253	257
TURKEY	163	50	-	15
UNITED KINGDOM	305	263	183	122
UNITED STATES	103	-	-	-
TOTAL NON-OECD	1082	719	504	2706

Source: IEA/OECD Statistics.

OECD EUROPE

5/D1 FUEL INPUT IN ELECTRICITY GENERATION

	1973	1978	1979	1980	1981	1985	1990
TOTAL (MTOE)	330.4	386.7	407.7	413.6	422.3	468.5	564.5
SOLID FUELS	126.4	146.3	159.4	168.9	169.4	162.5	200.3
OIL	85.3	77.7	79.9	77.2	65.7	62.7	45.2
GAS	24.1	30.5	28.0	24.5	21.0	25.1	22.3
NUCLEAR	18.7	40.5	44.3	51.4	71.7	119.2	182.9
HYDRO/GEOTHERMAL	75.9	91.7	96.0	91.5	94.5	97.2	108.0
OTHER [1]	-	-	-	-	-	1.9	5.7
TOTAL (PERCENTS)	100.0	100.0	100.0	100.0	100.0	100.0	100.0
SOLID FUELS	38.4	37.8	39.1	41.1	40.5	34.7	35.5
OIL	25.9	20.1	19.6	18.8	15.7	13.4	8.0
GAS	7.3	7.9	6.9	6.0	5.0	5.3	4.0
NUCLEAR	5.7	10.5	10.9	12.2	16.7	25.5	32.4
HYDRO/GEOTHERMAL	22.7	23.7	23.6	21.9	22.2	20.7	19.1
OTHER [1]	-	-	-	-	-	.4	1.0

Source: IEA/OECD Energy Balances and IEA Country Submissions (1982).
1. Solar, wind, etc.

5/D2 ELECTRICTY PRODUCTION BY FUEL TYPE

	1973	1978	1979	1980	1981	1985	1990
TOTAL (TWH)	1392.4	1637.2	1725.4	1742.7	1759.2	1938.8	2313.5
SOLID FUELS	467.0	559.4	602.6	638.8	638.5	640.1	801.0
OIL	394.3	351.0	355.0	347.5	292.4	253.4	181.5
GAS	109.7	140.6	141.5	122.4	103.7	114.4	98.4
NUCLEAR	73.8	167.9	184.4	213.9	294.0	484.1	740.9
HYDRO/GEOTHERMAL	347.6	418.4	442.0	420.1	430.7	441.9	484.8
OTHER [1]	-	-	-	-	-	5.0	6.7
TOTAL (PERCENTS)	100.0	100.0	100.0	100.0	100.0	100.0	100.0
SOLID FUELS	33.5	34.2	34.9	36.7	36.3	33.0	34.6
OIL	28.3	21.4	20.6	19.9	16.6	13.1	7.8
GAS	7.9	8.6	8.2	7.0	5.9	5.9	4.3
NUCLEAR	5.3	10.3	10.7	12.3	16.7	25.0	32.0
HYDRO/GEOTHERMAL	25.0	25.6	25.6	24.1	24.5	22.8	21.0
OTHER [1]	-	-	-	-	-	.3	.3

Source: IEA/OECD Energy Balances and IEA Country Submissions (1982).

2. Solar, wind, etc.

5/D4 ENERGY USE IN IRON AND STEEL INDUSTRY BY FUEL

	1973	1978	1979	1980	1981	1985	1990
TOTAL (MTOE)	92.0	71.7	74.5	66.4	63.9	na	na
SOLID FUELS	53.1	41.5	43.2	39.1	39.3	na	na
OIL	18.8	11.7	11.5	8.4	7.0	na	na
GAS	11.4	9.2	9.7	9.4	8.4	na	na
ELECTRICITY	8.7	9.4	10.0	9.5	9.3	na	na
HEAT [1]	-	-	-	-	-	na	na
TOTAL (PERCENTS)	100.0	100.0	100.0	100.0	100.0	na	na
SOLID FUELS	57.7	57.8	58.0	58.8	61.5	na	na
OIL	20.4	16.3	15.5	12.6	10.9	na	na
GAS	12.4	12.9	13.1	14.2	13.1	na	na
ELECTRICITY	9.4	13.1	13.5	14.3	14.5	na	na
HEAT [1]	-	-	-	-	-	na	na

Source: IEA/OECD Energy Balances and IEA Country Submissions (1982).

1. Includes only heat from combined heat and power plants.

5/D5 ENERGY USE IN OTHER INDUSTRIES BY FUEL

	1973	1978	1979	1980	1981	1985	1990
TOTAL (MTOE)	319.7	301.4	321.1	299.4	280.1	na	na
SOLID FUELS	31.3	22.0	23.8	24.2	26.2	na	na
OIL	200.6	174.1	186.2	162.8	142.5	na	na
GAS	42.1	55.3	58.3	57.7	57.3	na	na
ELECTRICITY	45.8	50.0	52.7	53.2	52.6	na	na
HEAT [1]	na	na	na	1.5	1.5	na	na
TOTAL (PERCENTS)	100.0	100.0	100.0	100.0	100.0	na	na
SOLID FUELS	9.8	7.3	7.4	8.1	9.3	na	na
OIL	62.7	57.8	58.0	54.4	50.9	na	na
GAS	13.2	18.3	18.2	19.3	20.5	na	na
ELECTRICITY	14.3	16.6	16.4	17.8	18.8	na	na
HEAT [1]	na	na	na	.5	.5	na	na

Source: IEA/OECD Energy Balances and IEA Country Submissions (1982).

1. Includes only heat from combined heat and power plants.

OECD EUROPE

5/D6 ENERGY USE IN INDUSTRY [1] BY FUEL

	1973	1978	1979	1980	1981	1985	1990
TOTAL (MTOE)	411.7	373.1	395.6	365.8	344.0	407.3	453.4
SOLID FUELS	84.4	63.4	67.0	63.3	65.5	87.3	104.3
OIL	219.4	185.8	197.8	171.2	149.5	162.4	162.5
GAS	53.5	64.5	68.0	67.1	65.6	86.0	99.4
ELECTRICITY	54.5	59.4	62.8	62.7	61.9	69.0	81.9
HEAT/OTHER [2]	na	na	na	1.5	1.5	2.6	5.3
TOTAL (PERCENTS)	100.0	100.0	100.0	100.0	100.0	100.0	100.0
SOLID FUELS	20.5	17.0	16.9	17.3	19.0	21.4	23.0
OIL	53.3	49.8	50.0	46.8	43.4	39.9	35.8
GAS	13.0	17.3	17.2	18.3	19.1	21.1	21.9
ELECTRICITY	13.2	15.9	15.9	17.1	18.0	16.9	18.1
HEAT/OTHER [2]	na	na	na	.4	.4	.6	1.2

Source: IEA/OECD Energy Balances and IEA Country Submissions (1982).

1. Includes non-energy use of petroleum products.
2. Includes heat from combined heat and power plants and direct use of solar, wind and other non-solid renewable energy sources.

5/D7 ENERGY USE IN OTHER SECTORS [1] BY FUEL

	1973	1978	1979	1980	1981	1985	1990
TOTAL (MTOE)	327.4	341.9	352.9	337.1	327.1	345.6	369.1
SOLID FUELS	48.9	34.6	36.6	36.3	35.2	37.2	39.7
OIL	189.1	178.8	177.5	155.2	144.2	132.9	118.2
GAS	45.5	71.3	79.1	79.5	80.7	88.4	100.2
ELECTRICITY	44.0	57.2	59.8	60.6	61.9	72.6	87.1
HEAT/OTHER [2]	na	na	na	5.5	5.2	14.5	23.9
TOTAL (PERCENTS)	100.0	100.0	100.0	100.0	100.0	100.0	100.0
SOLID FUELS	14.9	10.1	10.4	10.8	10.8	10.8	10.8
OIL	57.8	52.3	50.3	46.0	44.1	38.5	32.0
GAS	13.9	20.9	22.4	23.6	24.7	25.6	27.2
ELECTRICITY	13.4	16.7	16.9	18.0	18.9	21.0	23.6
HEAT/OTHER [2]	na	na	na	1.6	1.6	4.2	6.5

Source: IEA/OECD Energy Balances and IEA Country Submissions (1982).

1. Includes use in agricultural, commercial, public services and residential sectors.
2. Includes heat from combined heat and power plants and direct use of solar, wind and other non-solid renewable energy sources.

AUSTRIA

TOTAL ENERGY REQUIREMENTS BY FUEL

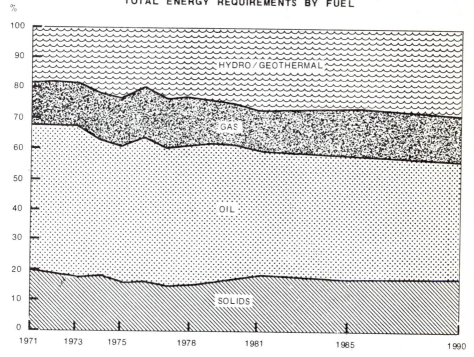

AUSTRIA

5/A1 TOTAL PRIMARY ENERGY REQUIREMENTS (TPER) BY FUEL

	1973	1978	1979	1980	1981	1985	1990
TPER (MTOE)[1]	24.3	25.1	27.1	27.2	26.3	29.1	31.6
SOLID FUELS	4.3	3.9	4.4	4.8	4.9	5.0	5.5
OIL	12.3	11.6	12.5	12.2	11.0	12.0	12.4
GAS	3.4	4.0	3.9	3.7	3.5	4.4	4.7
NUCLEAR	-	-	-	-	-	-	-
HYDRO/GEOTHERMAL	4.5	5.8	6.6	6.8	7.2	7.8	9.1
OTHER [2]	-	-	-	-	-	-	-
TPER (PERCENTS) [1]	100.0	100.0	100.0	100.0	100.0	100.0	100.0
SOLID FUELS	17.5	15.4	16.4	17.6	18.6	17.1	17.3
OIL	50.5	46.3	46.2	44.8	41.9	41.3	39.3
GAS	14.0	15.9	14.3	13.7	13.4	15.2	14.9
NUCLEAR	-	-	-	-	-	-	-
HYDRO/GEOTHERMAL	18.5	23.3	24.3	25.2	27.5	26.7	28.7
OTHER [2]	-	-	-	-	-	-	-

Source: IEA/OECD Energy Balances and IEA Country Submissions (1982).

1. Net imports of electricity are only included in totals.
2. Includes other electricity generation (solar, wind, etc.) and traded synthetic fuels.

5/A2 TOTAL FINAL ENERGY CONSUMPTION (TFC) BY FUEL

	1973	1978	1979	1980	1981	1985	1990
TFC (MTOE)	18.6	18.6	19.5	19.5	18.7	20.7	22.1
SOLID FUELS	3.2	3.0	3.4	3.7	3.7	3.7	3.6
OIL	10.9	10.3	10.6	10.0	9.2	10.0	10.6
GAS	2.4	2.7	2.7	2.8	2.6	3.1	3.4
ELECTRICITY	2.2	2.6	2.7	2.8	2.9	3.3	3.9
HEAT [1]	na	na	na	.2	.2	.5	.6
TFC (PERCENTS)	100.0	100.0	100.0	100.0	100.0	100.0	100.0
SOLID FUELS	17.2	16.1	17.5	18.8	20.0	18.1	16.1
OIL	58.3	55.3	54.3	51.3	49.6	48.5	47.9
GAS	12.7	14.5	14.1	14.4	14.1	14.9	15.5
ELECTRICITY	11.7	14.1	14.1	14.5	15.3	16.1	17.7
HEAT [1]	na	na	na	1.0	1.0	2.4	2.8

Source: IEA/OECD Energy Balances and IEA Country Submissions (1982).

1. Includes only heat production from combined heat and power plants.

AUSTRIA

5/A3 TOTAL SOLID FUELS BALANCE
(MTOE)

	1973	1978	1979	1980	1981	1985	1990
PRODUCTION	1.3	1.5	1.5	1.9	2.0	2.0	1.9
IMPORTS	3.0	2.4	3.0	2.9	2.9	3.0	3.6
EXPORTS	-.1	-.1	.0	.0	.0	.0	.0
MARINE BUNKERS	-	-	-	-	-	-	-
STOCK CHANGE	.0	-.1	.0	.0	.0	-.1	-
PRIMARY REQUIREMENTS	4.3	3.9	4.4	4.8	4.9	5.0	5.5
ELECTRICITY GENERATION	-.7	-.7	-.8	-.9	-1.0	-1.0	-1.7
GAS MANUFACTURE	.0	-	-	-	-	-	-
LIQUEFACTION	-	-	-	-	-	-	-
ENERGY SECTOR							
OWN USE + LOSSES [1]	-.3	-.2	-.3	-.2	-.2	-.2	-.2
FINAL CONSUMPTION	3.2	3.0	3.4	3.7	3.7	3.7	3.6
INDUSTRY SECTOR	1.4	1.4	1.7	1.8	1.9	1.8	1.8
IRON AND STEEL	1.2	1.2	1.5	1.4	1.5	na	na
CHEMICAL/							
PETROCHEMICAL	.0	.0	.0	.0	.0	na	na
OTHER INDUSTRY	.2	.2	.2	.4	.4	na	na
TRANSPORTATION SECTOR	.3	.2	.2	.2	.2	.0	.0
OTHER SECTORS	1.5	1.4	1.6	1.7	1.6	1.9	1.7
AGRICULTURE	-	-	-	-	-	-	-
COMMERCIAL AND							
PUBLIC SERVICE	-	-	-	-	-	-	-
RESIDENTIAL	1.5	1.4	1.6	1.7	1.6	-	-

Source: IEA/OECD Energy Balances and IEA Country Submissions (1982).

1. Includes statistical difference.

AUSTRIA

5/B1 SOLID FUELS INDIGENOUS PRODUCTION BY FUEL TYPE
(MTOE)

	1973	1978	1979	1980	1981	1985	1990
TOTAL SOLID FUELS	1.31	1.53	1.46	1.89	1.96	2.01	1.90
HARD COAL	-	-	-	-	-	-	-
COKING COAL	-	-	-	-	-	-	-
STEAM COAL	-	-	-	-	-	-	-
BROWN COAL/LIGNITE	.73	.90	.80	.84	.90	.95	.76
OTHER SOLID FUELS	.58	.63	.66	1.05	1.06	1.06	1.14

Source: IEA/OECD Coal Statistics, IEA/OECD Energy Balances, IEA Country Submissions (1982).

5/B2 COAL PRODUCTION BY TYPE
(THOUSAND METRIC TONS)

	1973	1978	1979	1980	1981
TOTAL COAL	3634	3076	2741	2865	3061
HARD COAL	-	-	-	-	-
COKING COAL	-	-	-	-	-
STEAM COAL	-	-	-	-	-
BROWN COAL/LIGNITE	3634	3076	2741	2865	3061
DERIVED PRODUCTS:					
PATENT FUEL	-	-	-	-	-
COKE OVEN COKE	1719	1484	1689	1729	1652
GAS COKE	-	-	-	-	-
BROWN COAL BRIQUETTES	-	-	-	-	-
COKE OVEN GAS(TCAL)	3820	2838	3255	3335	3224
BLAST FURNACE GAS(TCAL)	4936	3959	4713	4387	4101

Source: IEA/OECD Coal Statistics.

AUSTRIA

5/C1 SOLID FUELS TRADE BY TYPE
(MTOE)

	1973	1978	1979	1980	1981	1985	1990
IMPORTS							
TOTAL SOLID FUELS	2.99	2.44	2.98	2.91	2.94	3.05	3.59
HARD COAL	1.96	1.50	1.86	1.91	1.80	2.71	2.53
COKING COAL	na	1.50	1.86	1.91	1.80	1.52	1.34
STEAM COAL	na	-	-	-	-	1.19	1.19
BROWN COAL/LIGNITE	.09	.07	.06	.07	.15	.24	.12
OTHER SOLID FUELS [1]	.94	.87	1.06	.93	.99	.10	.94

Source: IEA/OECD Coal Statistics, IEA/OECD Energy Balances, IEA Country Submissions (1982).

1. Includes derived coal products, e.g. coke.

5/C2 COAL IMPORTS BY ORIGIN
(THOUSAND METRIC TONS)

	1978	1979	1980	1981
HARD COAL IMPORTS	2277	2803	2898	2727
COKING COAL IMPORTS FROM:	2277	2803	2898	2727
AUSTRALIA	-	-	-	-
CANADA	73	89	-	-
GERMANY	187	230	83	205
UNITED KINGDOM	-	-	-	85
UNITED STATES	-	-	189	942
OTHER OECD	-	-	6	82
CZECHOSLOVAKIA	585	721	802	745
POLAND	831	1023	1005	546
USSR	592	728	782	79
CHINA	-	-	-	-
SOUTH AFRICA	1	2	1	-
OTHER NON-OECD	8	10	30	43
STEAM COAL IMPORTS	-	-	-	-

Source: IEA/OECD Coal Statistics.

AUSTRIA

5/D1 FUEL INPUT IN ELECTRICITY GENERATION

	1973	1978	1979	1980	1981	1985	1990
TOTAL (MTOE)	6.8	8.6	9.3	9.7	9.9	11.1	12.8
SOLID FUELS	.7	.7	.8	.9	1.0	1.0	1.7
OIL	.8	1.1	1.4	1.5	1.1	1.4	1.2
GAS	.8	.9	.6	.5	.5	.9	.8
NUCLEAR	-	-	-	-	-	-	-
HYDRO/GEOTHERMAL	4.5	5.8	6.6	6.8	7.2	7.8	9.1
OTHER [1]	-	-	-	-	-	-	-
TOTAL (PERCENTS)	100.0	100.0	100.0	100.0	100.0	100.0	100.0
SOLID FUELS	10.2	8.3	8.1	8.8	10.1	9.3	13.4
OIL	12.1	13.0	14.5	15.3	11.1	12.5	9.4
GAS	11.5	10.3	6.9	5.5	5.4	8.3	6.2
NUCLEAR	-	-	-	-	-	-	-
HYDRO/GEOTHERMAL	66.1	68.4	70.5	70.4	73.5	69.9	71.1
OTHER [1]	-	-	-	-	-	-	-

Source: IEA/OECD Energy Balances and IEA Country Submissions (1982).

1. Solar, wind, etc.

5/D2 ELECTRICTY PRODUCTION BY FUEL TYPE

	1973	1978	1979	1980	1981	1985	1990
TOTAL (TWH)	31.3	38.1	40.6	42.0	42.9	45.9	53.2
SOLID FUELS	3.4	3.1	3.1	3.3	3.6	3.9	6.9
OIL	4.3	4.7	4.6	5.7	4.6	5.1	4.4
GAS	4.4	5.4	4.8	3.8	3.8	3.8	3.3
NUCLEAR	-	-	-	-	-	-	-
HYDRO/GEOTHERMAL	19.2	24.9	28.0	29.1	30.8	33.0	38.6
OTHER [1]	-	-	-	-	-	-	-
TOTAL (PERCENTS)	100.0	100.0	100.0	100.0	100.0	100.0	100.0
SOLID FUELS	10.8	8.1	7.7	8.0	8.5	8.6	13.0
OIL	13.9	12.3	11.4	13.6	10.7	11.2	8.3
GAS	14.1	14.2	11.8	9.1	9.0	8.3	6.2
NUCLEAR	-	-	-	-	-	-	-
HYDRO/GEOTHERMAL	61.2	65.4	69.0	69.3	71.9	72.0	72.5
OTHER [1]	-	-	-	-	-	-	-

Source: IEA/OECD Energy Balances and IEA Country Submissions (1982).

1. Solar, wind, etc.

AUSTRIA

5/D3 PROJECTIONS OF ELECTRICITY GENERATION CAPACITY BY FUEL
(GW)

	Nuclear	Hydro/ Geoth.	Solid Fuels	Oil	Gas	Multi Firing [1]	Total
I 1980							
Operating		9.2	0.7	0.1	0.5	2.3 [3]	13.9[2]
II 1980-1985							
Capacity:							
Under Construction [4]	-	1.8	0.3	-	-	-	2.1
Authorised	-	-	-	-	-	-	-
Other Planned	-	-	-	-	-	-	-
Conversion	-	-	-	-	-	-	-
Decommissioning	-	-	-	-	-	-	-
Total Operating 1985	-	11.0	1.0	0.1	0.5	2.3	16.0 [2]
III 1986-1990							
Capacity:							
Under Construction [4]	-	1.4	1.4	-	-	0.4	3.2
Authorised	-	-	-	-	-	-	-
Other Planned	-	-	-	-	-	-	-
Conversion	-	-	-	-	-	-	-
Decommissioning	-	-	-	-	-	-	-
Total Operating 1990	-	12.4	2.4	0.1	0.5	2.7	19.2 [2]

Source: IEA/OECD Electricity Statistics and IEA Country Submissions (1981).

1. Includes "Other" capacity.
2. Total includes industry and federal railway-owned capacity which are not reflected in the Table for Solid Fuels, oil, gas and multifiring capacity.
3. These plants are mainly fueled with coal and gas: the use of oil is technically possible.
4. Includes authorised and other planned.

AUSTRIA

5/D4 ENERGY USE IN IRON AND STEEL INDUSTRY BY FUEL

	1973	1978	1979	1980	1981	1985	1990
TOTAL (MTOE)	2.3	2.0	2.4	2.3	2.1	na	na
SOLID FUELS	1.2	1.2	1.5	1.4	1.5	na	na
OIL	.6	.3	.4	.4	.1	na	na
GAS	.4	.3	.4	.3	.3	na	na
ELECTRICITY	.2	.2	.2	.2	.2	na	na
HEAT [1]	-	-	-	-	-	na	na
TOTAL (PERCENTS)	100.0	100.0	100.0	100.0	100.0	na	na
SOLID FUELS	52.0	62.5	60.6	59.7	69.8	na	na
OIL	24.9	13.6	17.4	18.3	6.8	na	na
GAS	16.1	15.3	14.5	14.5	15.2	na	na
ELECTRICITY	7.0	8.6	7.5	7.5	8.2	na	na
HEAT [1]	-	-	-	-	-	na	na

Source: IEA/OECD Energy Balances and IEA Country Submissions (1982).

1. Includes only heat from combined heat and power plants.

5/D5 ENERGY USE IN OTHER INDUSTRIES BY FUEL

	1973	1978	1979	1980	1981	1985	1990
TOTAL (MTOE)	6.5	5.5	5.3	5.6	5.4	na	na
SOLID FUELS	.3	.2	.2	.4	.4	na	na
OIL	3.9	2.7	2.6	2.6	2.5	na	na
GAS	1.4	1.6	1.5	1.5	1.5	na	na
ELECTRICITY	.9	1.0	1.0	1.0	1.0	na	na
HEAT [1]	-	-	-	-	-	na	na
TOTAL (PERCENTS)	100.0	100.0	100.0	100.0	100.0	na	na
SOLID FUELS	4.1	3.5	3.4	7.2	8.1	na	na
OIL	60.8	50.0	48.3	46.5	46.1	na	na
GAS	21.6	28.8	29.2	27.4	26.6	na	na
ELECTRICITY	13.6	17.7	19.0	18.8	19.1	na	na
HEAT [1]	-	-	-	-	-	na	na

Source: IEA/OECD Energy Balances and IEA Country Submissions (1982).

1. Includes only heat from combined heat and power plants.

AUSTRIA

5/D6 ENERGY USE IN INDUSTRY [1] BY FUEL

	1973	1978	1979	1980	1981	1985	1990
TOTAL (MTOE)	8.8	7.5	7.7	7.9	7.5	7.7	8.1
SOLID FUELS	1.4	1.4	1.7	1.8	1.9	1.8	1.8
OIL	4.5	3.0	3.0	3.0	2.7	2.7	2.9
GAS	1.8	1.9	1.9	1.9	1.8	1.8	2.0
ELECTRICITY	1.0	1.1	1.2	1.2	1.2	1.3	1.4
HEAT [2]	-	-	-	-	-	.0	.0
TOTAL (PERCENTS)	100.0	100.0	100.0	100.0	100.0	100.0	100.0
SOLID FUELS	16.5	19.1	21.4	22.8	25.2	23.7	22.0
OIL	51.5	40.4	38.6	38.2	35.2	35.3	35.5
GAS	20.2	25.3	24.6	23.6	23.5	23.8	24.3
ELECTRICITY	11.9	15.3	15.4	15.4	16.1	16.8	17.6
HEAT [2]	-	-	-	-	-	.5	.6

Source: IEA/OECD Energy Balances and IEA Country Submissions (1982).

1. Includes non-energy use of petroleum products.
2. Includes only heat from combined heat and power plants.

5/D7 ENERGY USE IN OTHER SECTORS [1] BY FUEL

	1973	1978	1979	1980	1981	1985	1990
TOTAL (MTOE)	6.2	6.9	7.2	7.1	6.6	8.3	9.1
SOLID FUELS	1.5	1.4	1.6	1.7	1.6	1.9	1.7
OIL	3.1	3.4	3.4	2.8	2.5	3.0	3.1
GAS	.6	.8	.8	1.0	.9	1.2	1.4
ELECTRICITY	1.0	1.3	1.4	1.4	1.5	1.8	2.2
HEAT [2]	na	na	na	.2	.2	.4	.6
TOTAL (PERCENTS)	100.0	100.0	100.0	100.0	100.0	100.0	100.0
SOLID FUELS	24.6	20.4	22.0	23.8	24.3	22.6	19.3
OIL	49.7	48.8	47.4	39.9	37.8	35.9	34.3
GAS	9.8	11.8	11.7	13.5	13.1	14.5	15.7
ELECTRICITY	15.9	19.0	18.9	20.2	22.0	21.8	24.6
HEAT [2]	na	na	na	2.6	2.8	5.3	6.1

Source: IEA/OECD Energy Balances and IEA Country Submissions (1982).

1. Includes use in agricultural, commercial, public services and residential sectors.
2. Includes only heat from combined heat and power plants.

BELGIUM

TOTAL ENERGY REQUIREMENTS BY FUEL

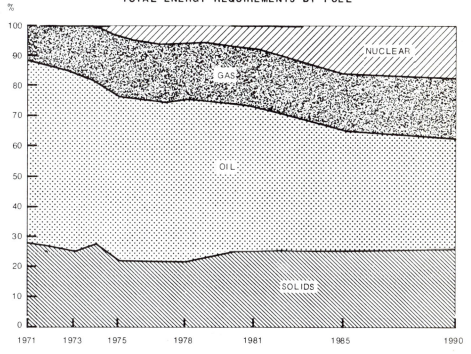

BELGIUM

5/A1 TOTAL PRIMARY ENERGY REQUIREMENTS (TPER) BY FUEL

	1973	1978	1979	1980	1981	1985	1990
TPER (MTOE) [1]	46.7	46.5	48.7	46.8	43.6	45.1	47.6
SOLID FUELS	11.7	10.0	11.3	11.6	11.1	11.2	12.3
OIL	27.6	25.1	25.2	23.1	20.7	18.1	17.5
GAS	7.3	8.7	9.5	9.1	8.4	8.5	9.5
NUCLEAR	.0	2.8	2.6	3.0	3.1	7.4	8.4
HYDRO/GEOTHERMAL	.2	.1	.1	.2	.3	.1	.1
OTHER [2]	-	-	-	-	-	-	-
TPER (PERCENTS) [1]	100.0	100.0	100.0	100.0	100.0	100.0	100.0
SOLID FUELS	25.1	21.6	23.3	24.8	25.5	24.8	25.8
OIL	59.0	54.0	51.7	49.4	47.5	40.1	36.8
GAS	15.7	18.7	19.5	19.4	19.2	18.8	20.0
NUCLEAR	.0	5.9	5.4	6.5	7.2	16.4	17.6
HYDRO/GEOTHERMAL	.3	.2	.3	.4	.6	.2	.2
OTHER [2]	-	-	-	-	-	-	-

Source: IEA/OECD Energy Balances and IEA Country Submissions (1982).

1. Net imports of electricity are only included in totals.
2. Includes other electricity generation (solar, wind, etc.), traded synthetic fuels.

5/A2 TOTAL FINAL ENERGY CONSUMPTION (TFC) BY FUEL

	1973	1978	1979	1980	1981	1985	1990
TFC (MTOE)	37.6	36.7	38.2	36.1	33.4	33.7	35.0
SOLID FUELS	8.1	6.0	6.3	5.9	5.4	5.9	6.0
OIL	21.2	19.9	20.2	17.7	16.1	15.4	14.8
GAS	5.2	7.3	7.9	7.8	7.3	8.1	9.1
ELECTRICITY	3.0	3.6	3.8	3.7	3.7	4.3	5.1
HEAT [1]	na	na	na	1.0	.9	-	-
TFC (PERCENTS)	100.0	100.0	100.0	100.0	100.0	100.0	100.0
SOLID FUELS	21.6	16.4	16.5	16.4	16.1	17.5	17.1
OIL	56.5	54.1	52.9	48.9	48.2	45.7	42.3
GAS	13.8	19.8	20.7	21.5	21.8	24.0	26.0
ELECTRICITY	8.1	9.7	9.9	10.4	11.2	12.8	14.6
HEAT [1]	na	na	na	2.8	2.7	-	-

Source: IEA/OECD Energy Balances and IEA Country Submissions (1982).

1. Includes only heat production from combined heat and power plants.

BELGIUM

5/A3 TOTAL SOLID FUELS BALANCE
(MTOE)

	1973	1978	1979	1980	1981	1985	1990
PRODUCTION	6.2	4.6	4.3	4.4	4.3	4.2	4.2
IMPORTS	5.9	5.4	7.8	8.0	7.9	7.3	8.4
EXPORTS	-.6	-.3	-.8	-.9	-1.1	-.5	-.5
MARINE BUNKERS	-	-	-	-	-	-	-
STOCK CHANGE	.2	.3	.1	.0	.0	.2	.2
PRIMARY REQUIREMENTS	11.7	10.0	11.3	11.6	11.1	11.2	12.3
ELECTRICITY GENERATION	-2.8	-3.3	-4.0	-4.6	-4.9	-4.1	-5.1
GAS MANUFACTURE	-	-	-	-	-	-	-
LIQUEFACTION	-	-	-	-	-	-	-
ENERGY SECTOR OWN USE + LOSSES [1]	-.8	-.7	-1.1	-1.1	-.9	-1.2	-1.2
FINAL CONSUMPTION	8.1	6.0	6.3	5.9	5.4	5.9	6.0
INDUSTRY SECTOR	5.7	4.7	5.0	4.9	4.4	5.1	5.2
IRON AND STEEL	5.0	3.6	4.0	3.8	3.5	4.2	4.2
CHEMICAL/ PETROCHEMICAL	.1	.1	.1	.1	.0	-	-
OTHER INDUSTRY	.7	1.0	.9	1.0	.9	.9	1.0
TRANSPORTATION SECTOR	.0	.0	.0	.0	.0	-	-
OTHER SECTORS	2.4	1.3	1.3	1.1	1.0	.9	.8
AGRICULTURE	-	-	-	-	-	-	-
COMMERCIAL AND PUBLIC SERVICE	-	-	-	-	-	-	-
RESIDENTIAL	2.4	1.3	1.3	1.1	1.0	-	-

Source: IEA/OECD Energy Balances and IEA Country Submissions (1982).

1. Includes statistical difference.

BELGIUM

5/B1 SOLID FUELS INDIGENOUS PRODUCTION BY FUEL TYPE
(MTOE)

	1973	1978	1979	1980	1981	1985	1990
TOTAL SOLID FUELS	6.19	4.64	4.31	4.44	4.34	4.20	4.20
HARD COAL	6.19	•4.61	4.29	4.43	4.33	4.20	4.20
COKING COAL	na	2.66	2.64	2.83	2.67	2.50	2.20
STEAM COAL	na	1.95	1.64	1.60	1.66	1.70	2.00
BROWN COAL/LIGNITE	-	-	-	-	-	-	-
OTHER SOLID FUELS	-	.03	.02	.01	.01	-	-

Source: IEA/OECD Coal Statistics, IEA/OECD Energy Balances, IEA Country Submissions (1982).

5/B2 COAL PRODUCTION BY TYPE
(THOUSAND METRIC TONS)

	1973	1978	1979	1980	1981
TOTAL COAL	8842	6590	6125	6324	6186
HARD COAL	8842	6590	6125	6324	6186
COKING COAL	na	3805	3777	4036	3810
STEAM COAL	na	2785	2348	2288	2376
BROWN COAL/LIGNITE	-	-	-	-	-
DERIVED PRODUCTS:					
PATENT FUEL	453	124	153	82	54
COKE OVEN COKE	7774	5747	6450	6048	6004
GAS COKE	-	-	-	-	-
BROWN COAL BRIQUETTES	-	-	-	-	-
COKE OVEN GAS(TCAL)	14471	10302	11474	10742	10796
BLAST FURNACE GAS(TCAL)	18709	14372	15190	13333	12965

Source: IEA/OECD Coal Statistics.

BELGIUM

5/C1 SOLID FUELS TRADE BY TYPE
(MTOE)

	1973	1978	1979	1980	1981	1985	1990
IMPORTS							
TOTAL SOLID FUELS	5.90	5.40	7.78	8.05	7.91	7.30	8.40
HARD COAL	5.03	4.90	6.74	7.10	7.04	6.50e	7.60e
COKING COAL	na	2.44	3.39	2.92	2.95	2.20e	2.10e
STEAM COAL	na	2.46	3.35	4.18	4.09	4.30e	5.50e
BROWN COAL/LIGNITE	-	-	-	.02	.02	-	-
OTHER SOLID FUELS [1]	.87	.50	1.04	.93	.85	.80	.80
EXPORTS							
TOTAL SOLID FUELS	.59	.33	.84	.87	1.11	.50	.50
HARD COAL	.24	.16	.22	.34	.55	.50	.50
COKING COAL	na	.07	.10	.12	.22	.10	-
STEAM COAL	na	.09	.12	.21	.33	.40	.50
BROWN COAL/LIGNITE	-	-	-	-	-	-	-
OTHER SOLID FUELS [1]	.35	.17	.62	.53	.56	-	-

Source: IEA/OECD Coal Statistics, IEA/OECD Energy Balances, IEA Country Submissions (1982).

1. Includes derived coal products, e.g. coke.

BELGIUM

5/C2 COAL IMPORTS BY ORIGIN
(THOUSAND METRIC TONS)

	1978	1979	1980	1981
HARD COAL IMPORTS	7007	9622	10139	10051
COKING COAL	3490	4841	4168	4215
IMPORTS FROM:				
AUSTRALIA	209	265	206	311
CANADA	148	56	-	-
GERMANY	1751	1479	783	1000
UNITED KINGDOM	22	-	-	171
UNITED STATES	833	2431	2719	2504
OTHER OECD	15	13	13	10
CZECHOSLOVAKIA	59	61	25	30
POLAND	392	448	422	181
USSR	46	31	-	6
CHINA	-	-	-	-
SOUTH AFRICA	15	57	-	1
OTHER NON-OECD	-	-	-	1
STEAM COAL	3517	4781	5971	5836
IMPORTS FROM:				
AUSTRALIA	-	10	51	231
CANADA	26	-	-	56
GERMANY	2316	1937	1563	1283
UNITED KINGDOM	129	242	217	168
UNITED STATES	-	110	1436	1018
OTHER OECD	72	63	239	185
CZECHOSLOVAKIA	-	3	14	6
POLAND	105	242	162	-
USSR	233	333	216	88
CHINA	-	-	-	32
SOUTH AFRICA	606	1830	2048	2769
OTHER NON-OECD	30	11	25	-

Source: IEA/OECD Coal Statistics.

BELGIUM

5/D1 FUEL INPUT IN ELECTRICITY GENERATION

	1973	1978	1979	1980	1981	1985	1990
TOTAL (MTOE)	9.8	11.1	12.0	12.9	12.2	13.1	15.1
SOLID FUELS	2.8	3.3	4.0	4.6	4.9	4.1	5.1
OIL	4.9	3.6	3.7	3.9	3.0	1.2	1.2
GAS	1.9	1.3	1.5	1.3	1.0	.3	.3
NUCLEAR	.0	2.8	2.6	3.0	3.1	7.4	8.4
HYDRO/GEOTHERMAL	.2	.1	.1	.2	.3	.1	.1
OTHER [1]	-	-	-	-	-	-	-
TOTAL (PERCENTS)	100.0	100.0	100.0	100.0	100.0	100.0	100.0
SOLID FUELS	28.7	29.8	33.3	35.4	39.7	31.3	33.8
OIL	49.7	32.6	31.0	29.9	24.5	9.2	7.9
GAS	19.8	11.7	12.5	9.7	8.3	2.3	2.0
NUCLEAR	.2	24.8	22.1	23.4	25.4	56.5	55.6
HYDRO/GEOTHERMAL	1.6	1.0	1.1	1.5	2.1	.8	.7
OTHER [1]	-	-	-	-	-	-	-

Source: IEA/OECD Energy Balances and IEA Country Submissions (1982).

1. Solar, wind, etc.

5/D2 ELECTRICTY PRODUCTION BY FUEL TYPE

	1973	1978	1979	1980	1981	1985	1990
TOTAL (TWH)	40.6	50.4	51.7	53.6	50.8	55.3	65.0
SOLID FUELS	8.9	13.8	15.1	16.2	18.1	15.4	22.9
OIL	21.3	17.2	17.8	18.2	13.9	4.5	4.5
GAS	9.6	6.3	6.8	5.9	4.8	1.1	1.1
NUCLEAR	.1	12.5	11.4	12.5	12.9	33.3	35.5
HYDRO/GEOTHERMAL	.6	.5	.6	.8	1.1	1.0	1.0
OTHER [1]	-	-	-	-	-	-	-
TOTAL (PERCENTS)	100.0	100.0	100.0	100.0	100.0	100.0	100.0
SOLID FUELS	22.0	27.4	29.2	30.3	35.6	27.8	35.2
OIL	52.5	34.2	34.5	33.9	27.4	8.1	6.9
GAS	23.7	12.5	13.2	10.9	9.4	2.0	1.7
NUCLEAR	.2	24.8	22.1	23.4	25.4	60.2	54.6
HYDRO/GEOTHERMAL	1.6	1.0	1.1	1.5	2.1	1.8	1.5
OTHER [1]	-	-	-	-	-	-	-

Source: IEA/OECD Energy Balances and IEA Country Submissions (1982).

1. Solar, wind, etc.

BELGIUM

5/D3 PROJECTIONS OF ELECTRICITY GENERATION CAPACITY BY FUEL
(GW)

	Nuclear	Hydro/ Geoth.	Solid Fuels	Oil	Gas	Multi Firing[1]	Total
I 1980							
Operating	1.7	1.3	0.9	2.3	-	4.9	11.1
II 1980-1985							
Capacity:							
Under Construction	3.8	-	-	-	-	-	3.8
Authorised	-	-	-	-	-	-	-
Other Planned	-	-	-	-	-	-	-
Conversion	-	-	-	-	-	-	-
Decommissioning	-	-	-	-0.3	-	-	-0.3
Total Operating 1985	5.5	1.3	0.9	2.0	-	4.9	14.6
III 1986-1990							
Capacity:							
Under Construction	-	-	-	-	-	-	-
Authorised	-	-	-	-	-	-	-
Other Planned	-	-	-	-	-	-	-
Conversion	-	-	-	-	-	-	-
Decommissioning	-	-	-	-0.1	-	-	-0.1
Total Operating 1990	5.5	1.3	0.9	1.9	-	4.9	14.5

Source: IEA/OECD Electricity Statistics and IEA Country Submissions (1981).

1. Includes "Other" capacity.

BELGIUM

5/D4 ENERGY USE IN IRON AND STEEL INDUSTRY BY FUEL

	1973	1978	1979	1980	1981	1985	1990
TOTAL (MTOE)	7.1	5.4	5.8	5.3	4.9	na	na
SOLID FUELS	5.0	3.6	4.0	3.8	3.5	4.2	4.2
OIL	.8	.4	.3	.2	.2	na	na
GAS	.9	.9	1.0	.9	.8	na	na
ELECTRICITY	.5	.4	.5	.4	.4	na	na
HEAT [1]	-	-	-	-	-	-	-
TOTAL (PERCENTS)	100.0	100.0	100.0	100.0	100.0	na	na
SOLID FUELS	69.7	66.8	69.4	71.7	71.5	na	na
OIL	10.9	7.9	5.8	3.6	4.2	na	na
GAS	13.0	17.4	16.9	16.6	15.5	na	na
ELECTRICITY	6.3	7.9	7.9	8.1	8.9	na	na
HEAT [1]	-	-	-	-	-	-	-

Source: IEA/OECD Energy Balances and IEA Country Submissions (1982).

1. Includes only heat from combined heat and power plants.

5/D5 ENERGY USE IN OTHER INDUSTRIES BY FUEL

	1973	1978	1979	1980	1981	1985	1990
TOTAL (MTOE)	12.2	11.4	11.3	11.3	10.3	na	na
SOLID FUELS	.8	1.1	.9	1.1	.9	na	na
OIL	7.1	5.6	5.3	4.5	4.0	na	na
GAS	2.8	3.2	3.4	3.2	2.8	na	na
ELECTRICITY	1.6	1.6	1.7	1.6	1.6	na	na
HEAT [1]	na	na	na	1.0	.9	-	-
TOTAL (PERCENTS)	100.0	100.0	100.0	100.0	100.0	na	na
SOLID FUELS	6.2	9.6	8.2	9.3	9.2	na	na
OIL	58.1	48.8	47.1	39.5	38.7	na	na
GAS	22.8	27.7	29.6	28.1	27.6	na	na
ELECTRICITY	12.9	13.9	15.0	14.5	15.6	na	na
HEAT [1]	na	na	na	8.6	8.9	-	-

Source: IEA/OECD Energy Balances and IEA Country Submissions (1982).

1. Includes only heat from combined heat and power plants.

BELGIUM

5/D6 ENERGY USE IN INDUSTRY [1] BY FUEL

	1973	1978	1979	1980	1981	1985	1990
TOTAL (MTOE)	19.4	16.9	17.2	16.6	15.1	15.8	16.7
SOLID FUELS	5.7	4.7	5.0	4.9	4.4	5.1	5.2
OIL	7.9	6.0	5.7	4.7	4.2	4.2	4.0
GAS	3.7	4.1	4.3	4.1	3.6	4.1	4.6
ELECTRICITY	2.0	2.0	2.2	2.1	2.0	2.4	2.9
HEAT [2]	na	na	na	1.0	.9	-	-
TOTAL (PERCENTS)	100.0	100.0	100.0	100.0	100.0	100.0	100.0
SOLID FUELS	29.6	28.0	29.0	29.2	29.2	32.3	31.1
OIL	40.7	35.6	33.1	28.0	27.6	26.6	24.0
GAS	19.2	24.4	25.3	24.4	23.7	25.9	27.5
ELECTRICITY	10.5	12.0	12.6	12.5	13.5	15.2	17.4
HEAT [2]	na	na	na	5.9	6.0	-	-

Source: IEA/OECD Energy Balances and IEA Country Submissions (1982).

1. Includes non-energy use of petroleum products.
2. Includes only heat from combined heat and power plants.

5/D7 ENERGY USE IN OTHER SECTORS [1] BY FUEL

	1973	1978	1979	1980	1981	1985	1990
TOTAL (MTOE)	13.2	14.1	15.0	13.6	12.5	12.6	12.8
SOLID FUELS	2.4	1.3	1.3	1.1	1.0	.9	.8
OIL	8.4	8.2	8.5	7.2	6.2	5.9	5.4
GAS	1.5	3.2	3.6	3.7	3.7	4.0	4.5
ELECTRICITY	.9	1.5	1.5	1.6	1.6	1.8	2.1
HEAT [2]	na	na	na	.0	-	-	-
TOTAL (PERCENTS)	100.0	100.0	100.0	100.0	100.0	100.0	100.0
SOLID FUELS	18.0	9.2	8.8	7.9	7.6	7.1	6.3
OIL	63.6	58.1	57.1	52.8	49.6	46.8	42.2
GAS	11.2	22.4	23.9	27.3	29.8	31.7	35.2
ELECTRICITY	7.2	10.3	10.2	11.7	13.0	14.3	16.4
HEAT [2]	na	na	na	.3	-	-	-

Source: IEA/OECD Energy Balances and IEA Country Submissions (1982).

1. Includes use in agricultural, commercial, public services and residential sectors.
2. Includes only heat from combined heat and power plants.

DENMARK

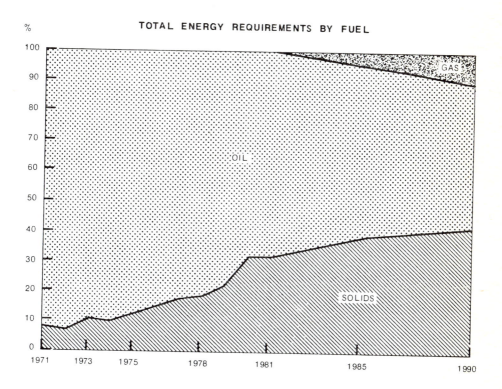

TOTAL ENERGY REQUIREMENTS BY FUEL

%

100
90
80
70
60
50
40
30
20
10
0

OIL

GAS

SOLIDS

1971   1973   1975   1978   1981   1985   1990

DENMARK

5/A1 TOTAL PRIMARY ENERGY REQUIREMENTS (TPER) BY FUEL

	1973	1978	1979	1980	1981	1985	1990
TPER (MTOE) [1]	19.5	20.3	20.8	19.2	17.3	19.2	20.2
SOLID FUELS	2.1	3.7	4.6	6.1	5.4	7.3	8.2
OIL	17.5	16.3	15.9	13.2	11.5	10.7	9.2
GAS	-	-	-	-	-	.9	2.1
NUCLEAR	-	-	-	-	-	-	-
HYDRO/GEOTHERMAL	.0	.0	.0	.0	.0	-	-
OTHER [2]	-	-	-	-	-	.3	.7
TPER (PERCENTS) [1]	100.0	100.0	100.0	100.0	100.0	100.0	100.0
SOLID FUELS	10.5	18.1	22.1	31.9	31.1	38.0	40.6
OIL	89.5	80.3	76.6	68.6	66.1	55.7	45.5
GAS	-	-	-	-	-	4.7	10.4
NUCLEAR	-	-	-	-	-	-	-
HYDRO/GEOTHERMAL	.0	.0	.0	.0	.0	-	-
OTHER [2]	-	-	-	-	-	1.6	3.5

Source: IEA/OECD Energy Balances and IEA Country Submissions (1982).

1. Net imports of electricity are only included in totals.
2. Includes other electricity generation (solar, wind, etc.), traded synethetic fuels and direct use of solar, wind and other non-solid renewable energy sources.

5/A2 TOTAL FINAL ENERGY CONSUMPTION (TFC) BY FUEL

	1973	1978	1979	1980	1981	1985	1990
TFC (MTOE)	16.2	15.9	16.3	15.3	14.2	14.6	15.1
SOLID FUELS	.4	.8	.8	.8	.8	1.2	1.5
OIL	14.3	13.2	13.4	11.8	10.6	9.3	7.8
GAS	.1	.1	.1	.1	.1	.7	1.8
ELECTRICITY	1.4	1.8	1.9	1.9	1.9	2.1	2.3
HEAT [1]	na	na	na	.7	.8	1.3	1.7
TFC (PERCENTS)	100.0	100.0	100.0	100.0	100.0	100.0	100.0
SOLID FUELS	2.4	4.9	5.1	5.3	5.8	8.2	9.9
OIL	88.4	83.0	82.5	76.7	74.8	63.7	51.7
GAS	.7	.7	.7	.7	.7	4.8	11.9
ELECTRICITY	8.4	11.4	11.8	12.5	13.5	14.4	15.2
HEAT [1]	na	na	na	4.8	5.3	8.9	11.3

Source: IEA/OECD Energy Balances and IEA Country Submissions (1982).

1. Includes heat production from combined heat and power plants and direct use of solar, wind and other non-solid renewable energy sources.

DENMARK

5/A3 TOTAL SOLID FUELS BALANCE
(MTOE)

	1973	1978	1979	1980	1981	1985	1990
PRODUCTION	-	.2	.3	.3	.4	.3	.3
IMPORTS	2.0	3.7	4.6	6.1	6.6	7.0	7.9
EXPORTS	.0	.0	.0	.0	.0	-	-
MARINE BUNKERS	-	-	-	-	-	-	-
STOCK CHANGE	.1	-.3	-.3	-.2	-1.6	-	-
PRIMARY REQUIREMENTS	2.1	3.7	4.6	6.1	5.4	7.3	8.2
ELECTRICITY GENERATION	-1.7	-3.0	-3.7	-5.5	-4.3	-6.1	-6.7
GAS MANUFACTURE	.0	.0	.0	.0	.0	-	-
LIQUEFACTION	-	-	-	-	-	-	-
ENERGY SECTOR OWN USE + LOSSES [1]	.0	.1	-.1	.2	-.2	-	-
FINAL CONSUMPTION	.4	.8	.8	.8	.8	1.2	1.5
INDUSTRY SECTOR	.3	.5	.6	.5	.4	.7	.9
IRON AND STEEL	.0	.0	.0	.0	.0	-	-
CHEMICAL/ PETROCHEMICAL	-	-	.0	-	-	-	-
OTHER INDUSTRY	.2	.5	.5	.5	.4	.7	.9
TRANSPORTATION SECTOR	.0	-	-	-	-	-	-
OTHER SECTORS	.1	.3	.3	.3	.4	.5	.6
AGRICULTURE	-	-	-	.0	.1	-	-
COMMERCIAL AND PUBLIC SERVICE	-	-	-	-	-	-	-
RESIDENTIAL	.1	.3	.3	.3	.3	-	-

Source: IEA/OECD Energy Balances and IEA Country Submissions (1982).

1.   Includes statistical difference.

DENMARK

5/B1 SOLID FUELS INDIGENOUS PRODUCTION BY FUEL TYPE
(MTOE)

	1973	1978	1979	1980	1981	1985	1990
TOTAL SOLID FUELS	-	.24	.26	.26	.36	.30	.30
HARD COAL	-	-	-	-	-	-	-
COKING COAL	-	-	-	-	-	-	-
STEAM COAL	-	-	-	-	-	-	-
BROWN COAL/LIGNITE	-	-	-	-	-	-	-
OTHER SOLID FUELS	-	.24	.26	.26	.36	.30	.30

Source: IEA/OECD Coal Statistics, IEA/OECD Energy Balances, IEA Country Submissions (1982).

5/C1 SOLID FUELS TRADE BY TYPE
(MTOE)

	1973	1978	1979	1980	1981	1985	1990
IMPORTS							
TOTAL SOLID FUELS	2.03	3.75	4.63	6.10	6.61	7.00	7.90
HARD COAL	1.93	3.67	4.53	5.98	6.53	7.00e	7.90e
COKING COAL	na	-	-	-	.01	-	-
STEAM COAL	na	3.67	4.53	5.98	6.52	7.00e	7.90e
BROWN COAL/LIGNITE	-	-	-	-	-	-	-
OTHER SOLID FUELS [1]	.10	.08	.10	.12	.08	-	-
EXPORTS							
TOTAL SOLID FUELS	.04	.04	.04	.04	.03	-	-
HARD COAL	-	-	-	-	-	-	-
COKING COAL	-	-	-	-	-	-	-
STEAM COAL	-	-	-	-	-	-	-
BROWN COAL/LIGNITE	-	-	-	-	-	-	-
OTHER SOLID FUELS [1]	.04	.04	.04	.04	.03	-	-

Source: IEA/OECD Coal Statistics, IEA/OECD Energy Balances, IEA Country Submissions (1982).

1. Includes derived coal products, e.g. coke.

DENMARK

5/C2 COAL IMPORTS BY ORIGIN
(THOUSAND METRIC TONS)

	1978	1979	1980	1981
HARD COAL IMPORTS	6110	7552	9969	10886
COKING COAL	-	-	7	24
STEAM COAL	6110	7552	9962	10862
IMPORTS FROM:				
AUSTRALIA	177	609	436	279
CANADA	307	133	204	495
GERMANY	928	601	275	6
UNITED KINGDOM	145	234	592	1939
UNITED STATES	2	178	1376	4265
OTHER OECD	21	30	44	58
CZECHOSLOVAKIA	6	4	4	4
POLAND	3078	2982	3422	636
USSR	528	472	395	86
CHINA	-	-	-	-
SOUTH AFRICA	868	2290	3118	3094
OTHER NON-OECD	50	19	96	-

Source: IEA/OECD Coal Statistics.

DENMARK

5/D1 FUEL INPUT IN ELECTRICITY GENERATION

	1973	1978	1979	1980	1981	1985	1990
TOTAL (MTOE)	4.5	5.1	5.7	6.7	5.0	7.4	8.2
SOLID FUELS	1.7	3.0	3.7	5.5	4.3	6.1	6.7
OIL	2.8	2.1	2.0	1.2	.7	1.0	1.0
GAS	-	-	-	-	-	.2	.2
NUCLEAR	-	-	-	-	-	-	-
HYDRO/GEOTHERMAL	.0	.0	.0	.0	.0	-	-
OTHER [1]	-	-	-	-	-	.1	.3
TOTAL (PERCENTS)	100.0	100.0	100.0	100.0	100.0	100.0	100.0
SOLID FUELS	36.8	58.1	65.2	82.1	86.7	82.4	81.7
OIL	63.0	41.8	34.7	17.8	13.1	13.5	12.2
GAS	-	-	-	-	-	2.7	2.4
NUCLEAR	-	-	-	-	-	-	-
HYDRO/GEOTHERMAL	.1	.1	.1	.1	.2	-	-
OTHER [1]	-	-	-	-	-	1.4	3.7

Source: IEA/OECD Energy Balances and IEA Country Submissions (1982).

1. Solar, wind, etc.

5/D2 ELECTRICTY PRODUCTION BY FUEL TYPE

	1973	1978	1979	1980	1981	1985	1990
TOTAL (TWH)	19.1	20.8	22.5	27.1	19.8	26.4	30.0
SOLID FUELS	6.8	11.2	13.8	22.1	17.2	22.0	24.6
OIL	12.3	9.6	8.6	5.0	2.5	3.1	3.4
GAS	-	-	-	-	-	.8	1.0
NUCLEAR	-	-	-	-	-	-	-
HYDRO/GEOTHERMAL	.0	.0	.0	.0	.0	-	-
OTHER [1]	-	-	-	-	-	.5	1.0
TOTAL (PERCENTS)	100.0	100.0	100.0	100.0	100.0	100.0	100.0
SOLID FUELS	35.8	53.7	61.5	81.6	87.1	83.3	82.0
OIL	64.1	46.2	38.4	18.3	12.8	11.7	11.3
GAS	-	-	-	-	-	3.0	3.3
NUCLEAR	-	-	-	-	-	-	-
HYDRO/GEOTHERMAL	.1	.1	.1	.1	.2	-	-
OTHER [1]	-	-	-	-	-	1.9	3.3

Source: IEA/OECD Energy Balances and IEA Country Submissions (1982).

2. Solar, wind, etc.

DENMARK

5/D3 PROJECTIONS OF ELECTRICITY GENERATION CAPACITY BY FUEL
(GW)

	Nuclear	Hydro/ Geoth.	Solid Fuels	Oil	Gas	Multi Firing [1]	Total
I 1981							
Operating	-	-	0.6	2.1	-	4.8	7.5
II 1981-1985							
Capacity:							
Under Construction	-	-	-	-	-	0.9	0.9
Authorised	-	-	-	-	-	-	-
Other Planned	-	-	-	-	-	-	-
Conversion	-	-	-	-0.3	-	0.3	-
Decommissioning	-	-	-	-	-	-0.1	-0.1
Total Operating 1985	-	-	0.6	1.8 [2]	-	5.9	8.3
III 1986-1990							
Capacity:							
Under Construction	-	-	-	-	-	-	-
Authorised	-	-	-	-	-	-	-
Other Planned	-	-	-	-	-	1.3	1.3
Conversion	-	-	-	-	-	-	-
Decommissioning	-	-	-0.1	-0.1	-	-0.3	-0.5
Total Operating 1990	-	-	0.5	1.7	-	6.9	9.1

Source: IEA/OECD Electricity Statistics and IEA Country Submissions (1982).

1. Includes "Other" capacity.
2. Some of this capacity (0.3 Mtoe) may be fired with gas.

DENMARK

5/D6 ENERGY USE IN INDUSTRY [1] BY FUEL

	1973	1978	1979	1980	1981	1985	1990
TOTAL (MTOE)	4.0	3.7	3.7	3.5	3.0	3.5	3.9
SOLID FUELS	.3	.5	.6	.5	.4	.7	.9
OIL	3.4	2.7	2.6	2.4	2.0	2.0	1.7
GAS	.0	.0	.0	.0	.0	.2	.6
ELECTRICITY	.4	.5	.5	.5	.5	.6	.7
HEAT [2]	-	-	-	-	-	-	-
TOTAL (PERCENTS)	100.0	100.0	100.0	100.0	100.0	100.0	100.0
SOLID FUELS	6.5	14.2	15.4	14.7	12.8	20.0	23.1
OIL	84.1	72.0	70.6	69.8	68.6	57.1	43.6
GAS	.4	.6	.6	.6	.3	5.7	15.4
ELECTRICITY	9.0	13.2	13.4	14.9	18.3	17.1	17.9
HEAT [2]	-	-	-	-	-	-	-

Source: IEA/OECD Energy Balances and IEA Country Submissions (1982).

1. Includes non-energy use of petroleum products.
2. Includes only heat from combined heat and power plants.

5/D7 ENERGY USE IN OTHER SECTORS [1] BY FUEL

	1973	1978	1979	1980	1981	1985	1990
TOTAL (MTOE)	8.6	8.5	8.9	8.5	7.8	8.1	8.1
SOLID FUELS	.1	.3	.3	.3	.4	.5	.6
OIL	7.4	6.9	7.2	6.0	5.2	4.3	3.0
GAS	.1	.1	.1	.1	.1	.5	1.2
ELECTRICITY	1.0	1.3	1.4	1.4	1.4	1.5	1.6
OTHER [2]	na	na	na	.7	.8	1.3	1.7
TOTAL (PERCENTS)	100.0	100.0	100.0	100.0	100.0	100.0	100.0
SOLID FUELS	1.5	3.0	2.9	3.7	5.6	6.2	7.4
OIL	85.8	80.7	80.5	70.3	66.4	53.1	37.0
GAS	1.2	1.0	1.0	.9	1.1	6.2	14.8
ELECTRICITY	11.5	15.4	15.7	16.4	17.3	18.5	19.8
OTHER [2]	na	na	na	8.7	9.6	16.0	21.0

Source: IEA/OECD Energy Balances and IEA Country Submissions (1982).

1. Includes use in agricultural, commercial, public services and residential sectors.
2. Includes heat from combined heat and power plants and direct use of solar, wind and other non-solid renewable energy sources.

FINLAND

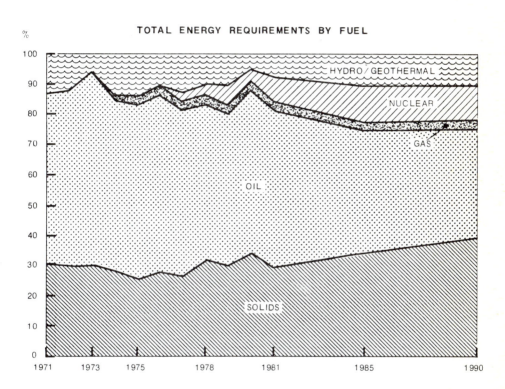

TOTAL ENERGY REQUIREMENTS BY FUEL

FINLAND

5/A1 TOTAL PRIMARY ENERGY REQUIREMENTS (TPER) BY FUEL

	1973	1978	1979	1980	1981	1985	1990
TPER (MTOE) [1]	22.6	24.2	25.8	26.4	25.9	28.0	29.8
SOLID FUELS	6.4	7.7	7.7	8.4	6.6	9.5	11.6
OIL	13.6	12.4	12.8	13.0	11.7	11.1	10.5
GAS	-	.8	.8	.8	.6	.8	1.0
NUCLEAR	-	.8	1.7	1.7	3.6	3.4	3.4
HYDRO/GEOTHERMAL	2.6	2.4	2.7	2.5	3.4	3.0	3.0
OTHER [2]	-	-	-	-	-	-	-
TPER (PERCENTS) [1]	100.0	100.0	100.0	100.0	100.0	100.0	100.0
SOLID FUELS	28.3	31.6	29.9	31.8	25.5	33.9	38.9
OIL	60.2	51.0	49.8	49.2	45.2	39.7	35.1
GAS	-	3.4	3.2	3.0	2.3	2.8	3.3
NUCLEAR	-	3.5	6.5	6.4	13.9	12.1	11.3
HYDRO/GEOTHERMAL	11.5	10.0	10.4	9.5	13.1	10.6	10.2
OTHER [2]	-	-	-	-	-	-	-

Source: IEA/OECD Energy Balances and IEA Country Submissions (1982).

1.   Net imports of electricity are only included in totals.
2.   Includes other electricity generation (solar, wind, etc.) and traded synthetic fuels.

5/A2 TOTAL FINAL ENERGY CONSUMPTION (TFC) BY FUEL

	1973	1978	1979	1980	1981	1985	1990
TFC (MTOE)	18.9	18.8	18.7	18.5	18.4	20.6	21.6
SOLID FUELS	4.9	4.2	4.3	4.6	4.9	5.9	6.4
OIL	11.6	11.2	10.9	10.2	9.7	9.7	9.2
GAS	.0	.5	.4	.4	.5	.6	.8
ELECTRICITY	2.3	2.8	3.0	3.2	3.3	4.0	4.7
HEAT [1]	na	na	na	na	na	.3	.5
TFC (PERCENTS)	100.0	100.0	100.0	100.0	100.0	100.0	100.0
SOLID FUELS	26.2	22.4	23.2	25.0	26.8	28.7	29.6
OIL	61.5	59.8	58.2	55.4	52.7	47.3	42.6
GAS	.1	2.8	2.3	2.3	2.5	3.1	3.7
ELECTRICITY	12.3	15.0	16.3	17.3	18.0	19.3	21.9
HEAT [1]	na	na	na	na	na	1.7	2.3

Source: IEA/OECD Energy Balances and IEA Country Submissions (1982).

1.   Includes only heat production from combined heat and power plants.

FINLAND

5/A3 TOTAL SOLID FUELS BALANCE
(MTOE)

	1973	1978	1979	1980	1981	1985	1990
PRODUCTION	4.0	3.6	3.7	4.2	3.9	5.1	5.8
IMPORTS	2.4	3.7	3.9	3.8	4.3	4.3	5.8
EXPORTS	.0	-	-	-	-	-	-
MARINE BUNKERS	-	-	-	-	-	-	-
STOCK CHANGE	.1	.4	.1	.4	-1.7	-	-
PRIMARY REQUIREMENTS	6.4	7.7	7.7	8.4	6.6	9.5	11.6
ELECTRICITY GENERATION [1]	-1.5	-3.0	-2.9	-3.3	-1.3	-2.9	-4.4
GAS MANUFACTURE	.0	-	-	-	-	-.6	-.6
LIQUEFACTION	-	-	-	-	-	-	-
ENERGY SECTOR OWN USE + LOSSES [2]	-	-.4	-.5	-.5	-.4	-.1	-.1
FINAL CONSUMPTION	4.9	4.2	4.3	4.6	4.9	5.9	6.4
INDUSTRY SECTOR	3.0	2.0	2.6	2.9	3.2	4.0	4.4
IRON AND STEEL	.5	.3	.4	.3	.3	na	na
CHEMICAL/ PETROCHEMICAL	.2	.2	.1	.1	.1	na	na
OTHER INDUSTRY	2.4	1.5	2.2	2.5	2.7	na	na
TRANSPORTATION SECTOR	.0	-	-	-	-	-	-
OTHER SECTORS	1.9	2.2	1.7	1.7	1.7	1.8	1.9
AGRICULTURE	-	-	-	-	-	.2	.2
COMMERCIAL AND PUBLIC SERVICE	-	-	-	-	-	-	-
RESIDENTIAL	1.9	2.2	1.7	1.7	1.7	1.6	1.8

Source: IEA/OECD Energy Balances and IEA Country Submissions (1982).

1.   Includes also solid fuels inputs to district heat production.
2.   Includes statistical difference.

FINLAND

5/B1 SOLID FUELS INDIGENOUS PRODUCTION BY FUEL TYPE
(MTOE)

	1973	1978	1979	1980	1981	1985	1990
TOTAL SOLID FUELS	3.98	3.63	3.72	4.18	3.94	5.14	5.79
HARD COAL	-	-	-	-	-	-	-
COKING COAL	-	-	-	- .	-	-	-
STEAM COAL	-	-	-	-	-	-	-
BROWN COAL/LIGNITE	-	-	-	-	-	-	-
OTHER SOLID FUELS	3.98	3.63	3.72	4.18	3.94	5.14	5.79

Source: IEA/OECD Coal Statistics, IEA/OECD Energy Balances, IEA Country Submissions (1982).

5/C1 SOLID FUELS TRADE BY TYPE
(MTOE)

	1973	1978	1979	1980	1981	1985	1990
IMPORTS							
TOTAL SOLID FUELS	2.37	3.66	3.88	3.79	4.33	4.35	5.80
HARD COAL	1.81	3.02	3.01	2.94	3.56	3.60	5.10
COKING COAL	na	2.96	2.93	2.86	3.49	3.60	5.10
STEAM COAL	na	.05	.08	.08	.07	.00	.00
BROWN COAL/LIGNITE	-	-	-	-	-	-	-
OTHER SOLID FUELS [1]	.56	.64	.87	.85	.77	.77	.70
EXPORTS							
TOTAL SOLID FUELS	.02	-	-	-	-	-	-
HARD COAL	-	-	-	-	-	-	-
COKING COAL	-	-	-	-	-	-	-
STEAM COAL	-	-	-	-	-	-	-
BROWN COAL/LIGNITE	-	-	-	-	-	-	-
OTHER SOLID FUELS [1]	.02	-	-	-	-	-	-

Source: IEA/OECD Coal Statistics, IEA/OECD Energy Balances, IEA Country Submissions (1982).

1. Includes derived coal products, e.g. coke.

FINLAND

5/C2 COAL IMPORTS BY ORIGIN
(THOUSAND METRIC TONS)

	1978	1979	1980	1981
HARD COAL IMPORTS	4789	4771	4669	5650
COKING COAL IMPORTS FROM:	4703	4647	4542	5538
AUSTRALIA	-	3	-	-
CANADA	-	-	4	-
GERMANY	-	-	-	-
UNITED KINGDOM	3	12	4	1138
UNITED STATES	-	-	511	2264
OTHER OECD	-	1	4	5
CZECHOSLOVAKIA	-	-	-	-
POLAND	4089	4013	3535	1422
USSR	611	618	484	709
CHINA	-	-	-	-
SOUTH AFRICA	-	-	-	-
OTHER NON-OECD	-	-	-	-
STEAM COAL IMPORTS FROM:	86	124	127	112
AUSTRALIA	-	-	-	-
CANADA	-	-	-	-
GERMANY	-	-	-	-
UNITED KINGDOM	-	-	-	-
UNITED STATES	-	-	22	3
OTHER OECD	-	-	-	1
CZECHOSLOVAKIA	-	-	-	-
POLAND	-	-	-	-
USSR	86	124	105	97
CHINA	-	-	-	-
SOUTH AFRICA	-	-	-	11
OTHER NON-OECD	-	-	-	-

Source: IEA/OECD Coal Statistics.

FINLAND

5/D1 FUEL INPUT IN ELECTRICITY GENERATION

	1973	1978	1979	1980	1981	1985	1990
TOTAL (MTOE)	5.1	7.5	9.0	9.3	9.7	10.3	11.9
SOLID FUELS [2]	1.5	3.0	2.9	3.3	1.3	2.9	4.4
OIL	1.0	.8	1.3	1.4	1.2	.6	.6
GAS	-	.3	.4	.4	.2	.4	.4
NUCLEAR	-	.8	1.7	1.7	3.6	3.4	3.4
HYDRO/GEOTHERMAL	2.6	2.4	2.7	2.5	3.4	3.0	3.0
OTHER [1]	-	-	-	-	-	-	-
TOTAL (PERCENTS)	100.0	100.0	100.0	100.0	100.0	100.0	100.0
SOLID FUELS	29.4	40.7	32.3	35.5	13.4	27.8	37.1
OIL	19.6	11.3	14.7	15.1	12.4	6.2	5.4
GAS	-	4.3	4.3	4.3	2.1	4.0	3.8
NUCLEAR	-	11.3	18.6	18.3	37.1	32.9	28.3
HYDRO/GEOTHERMAL	51.0	32.4	30.0	26.9	35.1	29.0	25.5
OTHER [1]	-	-	-	-	-	-	-

Source: IEA/OECD Energy Balances and IEA Country Submissions (1982).

1. Solar, wind, etc.
2. Includes solid fuels inputs to district heating.

5/D2 ELECTRICTY PRODUCTION BY FUEL TYPE

	1973	1978	1979	1980	1981	1985	1990
TOTAL (TWH)	26.1	35.8	39.2	40.6	41.0	46.6	52.5
SOLID FUELS	7.3	15.7	15.3	17.3	8.3	14.8	21.1
OIL	8.3	4.9	4.3	4.5	3.3	3.9	2.7
GAS	-	1.9	1.9	1.7	1.0	2.5	2.9
NUCLEAR	-	3.4	6.8	7.0	14.8	13.5	13.5
HYDRO/GEOTHERMAL	10.5	9.8	10.9	10.2	13.7	11.9	12.2
OTHER [1]	-	-	-	-	-	-	-
TOTAL (PERCENTS)	100.0	100.0	100.0	100.0	100.0	100.0	100.0
SOLID FUELS	28.1	43.8	39.0	42.6	20.3	31.8	40.3
OIL	31.6	13.8	11.0	11.0	8.0	8.3	5.1
GAS	-	5.3	4.8	4.2	2.5	5.4	5.6
NUCLEAR	-	9.6	17.3	17.1	36.0	29.0	25.7
HYDRO/GEOTHERMAL	40.3	27.5	27.9	25.1	33.3	25.5	23.2
OTHER [1]	-	-	-	-	-	-	-

Source: IEA/OECD Energy Balances and IEA Country Submissions (1982).

1. Solar, wind, etc.

FINLAND

5/D3 PROJECTIONS OF ELECTRICITY GENERATION CAPACITY BY FUEL
(GW)

	Nuclear	Hydro/ Geoth.	Solid Fuels	Oil	Gas	Multi Firing	Total
I 1981							
Operating	2.2	2.5	2.5	2.2	0.5	1.2	11.1
II 1981-1985							
Capacity:							
Under Construction	-	0.1	0.1	-	-	0.1	0.3
Authorised	-	-	0.2	0.1	-	-	0.3
Other Planned	-	-	-	-	-	0.1	0.1
Conversion	-	-	0.6	-0.6	-	-	-
Decommissioning	-	-	-0.2	-	-	-	-0.2
Total Operating 1985	2.2	2.6	3.2	1.7	0.5	1.4	11.6
III 1986-1990							
Capacity:							
Under Construction	-	-	0.1	-	-	0.2	0.3
Authorised	-	0.1	-	-	-	0.2	0.3
Other Planned	-	0.1	-	-	-	-	0.1
Conversion	-	-	-	-	-	-	-
Decommissioning	-	-	-	-0.1	-	-	-0.1
Total Operating 1990	2.2	2.8	3.3	1.6	0.5	1.8	12.2

Source: IEA/OECD Electricity Statistics and IEA Country Submissions (1982).

FINLAND

5/D6 ENERGY USE IN INDUSTRY [1] BY FUEL

	1973	1978	1979	1980	1981	1985	1990
TOTAL (MTOE)	9.4	7.9	8.8	9.0	9.3	10.4	11.3
SOLID FUELS	3.0	2.0	2.6	2.9	3.2	4.0	4.4
OIL	4.8	3.7	3.9	3.6	3.6	3.3	3.3
GAS	.0	.5	.4	.4	.5	.6	.7
ELECTRICITY	1.6	1.7	1.9	2.0	2.0	2.4	2.8
HEAT [2]	na	na	na	na	na	.0	.0
TOTAL (PERCENTS)	100.0	100.0	100.0	100.0	100.0	100.0	100.0
SOLID FUELS	32.1	25.8	30.0	32.9	34.5	38.8	39.3
OIL	51.2	46.7	43.9	40.6	39.0	31.9	28.9
GAS	.1	6.1	4.9	4.7	4.9	5.7	6.5
ELECTRICITY	16.6	21.4	21.2	21.9	21.7	23.2	24.8
HEAT [2]	na	na	na	na	na	.4	.4

Source: IEA/OECD Energy Balances and IEA Country Submissions (1982).

1. Includes non-energy use of petroleum products.
2. Includes only heat from combined heat and power plants.

5/D7 ENERGY USE IN OTHER SECTORS [1] BY FUEL

	1973	1978	1979	1980	1981	1985	1990
TOTAL (MTOE)	7.0	8.1	7.0	6.6	6.2	7.0	7.0
SOLID FUELS	1.9	2.2	1.7	1.7	1.7	1.8	1.9
OIL	4.3	4.8	4.1	3.7	3.2	3.3	2.6
GAS	.0	.0	.0	.0	.0	.0	.0
ELECTRICITY	.8	1.1	1.2	1.2	1.3	1.5	1.9
HEAT [2]	na	na	na	na	na	.3	.4
TOTAL (PERCENTS)	100.0	100.0	100.0	100.0	100.0	100.0	100.0
SOLID FUELS	27.5	26.5	24.3	25.3	28.0	26.5	28.0
OIL	61.6	59.6	59.1	56.2	51.2	46.8	37.9
GAS	.1	.5	.1	.1	.1	.6	.7
ELECTRICITY	10.8	13.5	16.6	18.4	20.7	21.8	27.1
HEAT [2]	na	na	na	na	na	4.3	6.3

Source: IEA/OECD Energy Balances and IEA Country Submissions (1982).

1. Includes use in agricultural, commercial, public services and residential sectors.
2. Includes only heat from combined heat and power plants.

FRANCE

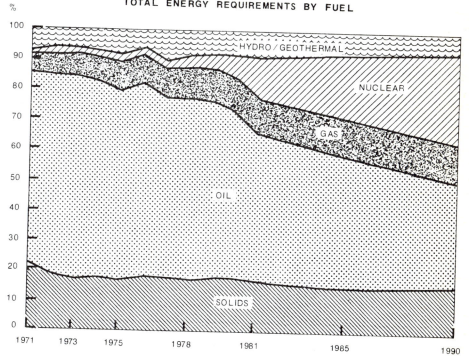

TOTAL ENERGY REQUIREMENTS BY FUEL

%

HYDRO / GEOTHERMAL

NUCLEAR

GAS

OIL

SOLIDS

1971 1973 1975 1978 1981 1985 1990

FRANCE

5/A1 TOTAL PRIMARY ENERGY REQUIREMENTS (TPER) BY FUEL

	1973	1978	1979	1980	1981	1985	1990
TPER (MTOE) [1]	182.2	191.8	198.0	198.1	196.7	213.7	250.1
SOLID FUELS	30.9	32.8	35.2	35.7	32.7	31.0	37.8
OIL	122.5	114.9	115.6	109.3	97.4	90.4	82.6
GAS	13.9	19.6	21.3	21.9	22.3	26.5	30.8
NUCLEAR	3.6	7.4	9.4	14.3	26.3	44.2	72.4
HYDRO/GEOTHERMAL	11.6	16.8	16.0	16.6	18.3	16.6	16.5
OTHER [2]	-	-	-	-	-	5.0	11.0
TPER (PERCENTS) [1]	100.0	100.0	100.0	100.0	100.0	100.0	100.0
SOLID FUELS	16.9	17.1	17.8	18.0	16.6	14.5	15.1
OIL	67.3	59.9	58.4	55.2	49.5	42.3	33.0
GAS	7.6	10.2	10.7	11.1	11.3	12.4	12.3
NUCLEAR	2.0	3.8	4.8	7.2	13.4	20.7	28.9
HYDRO/GEOTHERMAL	6.4	8.8	8.1	8.4	9.3	7.8	6.6
OTHER [2]	-	-	-	-	-	2.3	4.4

Source: IEA/OECD energy balances and country submissions (1981).

1. Net imports of electricity are only included in totals.
2. Includes other electricity generation (solar, wind, etc.), traded synthetic fuels and direct use of solar, wind and other non-solid renewable energy sources.

5/A2 TOTAL FINAL ENERGY CONSUMPTION (TFC) BY FUEL

	1973	1978	1979	1980	1981	1985	1990
TFC (MTOE)	142.7	143.9	148.1	143.6	136.4	146.8	164.3
SOLID FUELS	18.4	13.5	12.5	12.1	11.7	17.2	23.6
OIL	99.8	95.9	98.6	93.0	85.2	80.5	73.0
GAS	11.2	17.5	19.2	20.4	21.0	26.2	32.6
ELECTRICITY	13.3	17.0	17.7	18.2	18.5	18.4	24.6
OTHER [1]	-	-	-	-	-	4.5	10.5
TFC (PERCENTS)	100.0	100.0	100.0	100.0	100.0	100.0	100.0
SOLID FUELS	12.9	9.4	8.4	8.4	8.6	11.7	14.4
OIL	69.9	66.6	66.6	64.7	62.5	54.8	44.4
GAS	7.9	12.2	13.0	14.2	15.4	17.8	19.8
ELECTRICITY	9.3	11.8	12.0	12.7	13.5	12.6	15.0
OTHER [1]	-	-	-	-	-	3.1	6.4

Source: IEA/OECD Energy Balances and Country Submissions (1981).

1. Includes heat production from combined heat and power plants and direct use of solar, wind and other non-solid renewable energy sources.

FRANCE

5/A3 TOTAL SOLID FUELS BALANCE
(MTOE)

	1973	1978	1979	1980	1981	1985	1990
PRODUCTION	19.0	15.4	14.9	14.6	14.8	14.7	20.0
IMPORTS	11.4	17.8	20.8	22.8	21.0	16.3	17.8
EXPORTS	-1.3	-.9	-1.4	-.9	-1.2	-	-
MARINE BUNKERS	-	-	-	-	-	-	-
STOCK CHANGE	1.8	.5	.9	-.8	-1.9	-	-
PRIMARY REQUIREMENTS	30.9	32.8	35.2	35.7	32.7	31.0	37.8
ELECTRICITY GENERATION	-10.7	-17.4	-19.0	-19.3	-17.4	-11.3	-8.5
GAS MANUFACTURE	-.1	.0	-.1	.0	.0	-	-3.0
LIQUEFACTION	-	-	-	-	-	-	-2.7
ENERGY SECTOR							
OWN USE + LOSSES [1]	-1.6	-1.8	-3.7	-4.3	-3.6	-2.5	-
FINAL CONSUMPTION	18.4	13.5	12.5	12.1	11.7	17.2	23.6
INDUSTRY SECTOR	12.3	9.3	8.5	8.7	8.6	13.6	18.8
IRON AND STEEL	9.4	7.5	6.5	6.3	5.5	na	na
CHEMICAL/							
PETROCHEMICAL	.9	.5	.6	.7	.8	na	na
OTHER INDUSTRY	2.1	1.3	1.3	1.7	2.3	na	na
TRANSPORTATION SECTOR	.1	.0	.0	.0	.0	.2	.5
OTHER SECTORS	6.0	4.1	4.0	3.4	3.0	3.4	4.3
AGRICULTURE	-	-	-	-	-	-	-
COMMERCIAL AND							
PUBLIC SERVICE	-	-	-	-	-	-	-
RESIDENTIAL	6.0	4.1	4.0	3.4	3.0	-	-

Source: IEA/OECD Energy Balances and Country Submissions (1981).

1. Includes statistical difference.

FRANCE

5/B1 SOLID FUELS INDIGENOUS PRODUCTION BY FUEL TYPE
(MTOE)

	1973	1978	1979	1980	1981	1985	1990
TOTAL SOLID FUELS	19.00	15.36	14.93	14.65	14.80	14.70	20.00
HARD COAL	18.45	14.81	14.44	14.13	14.21	14.11e	19.41e
COKING COAL	na	4.16	4.52	4.31	4.33	4.23e	4.23e
STEAM COAL	na	10.65	9.92	9.82	9.88	9.88e	15.18e
BROWN COAL/LIGNITE	.55	.55	.49	.51	.59	.59e	.59e
OTHER SOLID FUELS	-	-	-	.01	-	-	-

Source: IEA/OECD Coal Statistics, IEA/OECD Energy Balances, Country Submissions (1981).

5/B2 COAL PRODUCTION BY TYPE
(THOUSAND METRIC TONS)

	1973	1978	1979	1980	1981
TOTAL COAL	28446	22420	23080	22750	23240
HARD COAL	25682	19690	20630	20190	20300
COKING COAL	na	5946	6457	6158	6191
STEAM COAL	na	13744	14173	14032	14109
BROWN COAL/LIGNITE	2764	2730	2450	2560	2940
DERIVED PRODUCTS:					
PATENT FUEL	3233	2180	2130	1760	1600
COKE OVEN COKE	11881	10682	11610	11120	10720
GAS COKE	-	-	-	-	-
BROWN COAL BRIQUETTES	-	-	-	-	-
COKE OVEN GAS(TCAL)	25153	22474	24022	23278	26646
BLAST FURNACE GAS(TCAL)	39020	30022	31010	29191	30769

Source: IEA/OECD Coal Statistics.

FRANCE

5/C1 SOLID FUELS TRADE BY TYPE
(MTOE)

	1973	1978	1979	1980	1981	1985	1990
IMPORTS							
TOTAL SOLID FUELS	11.38	17.81	20.84	22.77	21.00	16.30	17.80
HARD COAL	8.75	16.41	19.03	20.59	19.25	16.30e	17.80e
COKING COAL	na	5.82	7.02	7.32	7.15	6.30e	6.20e
STEAM COAL	na	10.59	12.01	13.27	12.10	10.00e	11.60e
BROWN COAL/LIGNITE	-	-	.01	-	-	-	-
OTHER SOLID FUELS [1]	2.63	1.40	1.80	2.18	1.75	-	-
EXPORTS							
TOTAL SOLID FUELS	1.34	.85	1.44	.89	1.16	-	-
HARD COAL	.61	.31	.36	.30	.49	-	-
COKING COAL	na	-	-	-	-	-	-
STEAM COAL	na	.31	.36	.30	.49	-	-
BROWN COAL/LIGNITE	.01	-	-	-	-	-	-
OTHER SOLID FUELS [1]	.72	.54	1.08	.59	.67	-	-

Source: IEA/OECD Coal Statistics, IEA/OECD Energy Balances, Country Submissions (1981).

1. Includes derived coal products, e.g. coke.

FRANCE

5/C2 COAL IMPORTS BY ORIGIN
(THOUSAND METRIC TONS)

	1978	1979	1980	1981
HARD COAL IMPORTS	23441	27190	29410	27500
COKING COAL IMPORTS FROM:	8316	10029	10456	10220
AUSTRALIA	987	1230	1093	1045
CANADA	-	50	18	20
GERMANY	4136	4340	4175	4000
UNITED KINGDOM	1	-	18	6
UNITED STATES	1462	2950	3742	4340
OTHER OECD	228	64	50	53
CZECHOSLOVAKIA	-	-	-	-
POLAND	1311	905	816	390
USSR	-	-	-	-
CHINA	-	-	-	-
SOUTH AFRICA	191	490	544	366
OTHER NON-OECD	-	-	-	-
STEAM COAL IMPORTS FROM:	15125	17161	18954	17280
AUSTRALIA	785	1180	557	715
CANADA	-	20	-	-
GERMANY	2217	2410	1185	550
UNITED KINGDOM	891	790	1442	2754
UNITED STATES	36	420	3600	5070
OTHER OECD	98	36	10	7
CZECHOSLOVAKIA	-	-	-	-
POLAND	3441	3555	2634	280
USSR	853	740	780	310
CHINA	-	70	83	-
SOUTH AFRICA	6643	7920	8596	7504
OTHER NON-OECD	161	20	67	90

Source: IEA/OECD Coal Statistics.

FRANCE

5/C3 HARD COAL EXPORTS BY DESTINATION
(THOUSAND METRIC TONS)

	1978	1979	1980	1981
TOTAL EXPORTS	443	520	430	700
EXPORTS TO:				
AUSTRALIA	-	-	-	-
AUSTRIA	-	-	10	90
BELGIUM	51	60	40	20
CANADA	-	-	-	-
DENMARK	-	-	-	-
FINLAND	-	-	-	-
FRANCE	-	-	-	-
GERMANY	264	310	300	360
GREECE	-	-	-	-
ICELAND	-	-	-	-
IRELAND	-	-	-	10
ITALY	8	10	-	20
JAPAN	-	-	-	-
LUXEMBOURG	17	20	20	-
NETHERLANDS	-	-	-	-
NEW ZEALAND	-	-	-	
NORWAY	53	60	-	80
PORTUGAL	-	-	-	-
SPAIN	-	-	-	-
SWEDEN	-	-	-	10
SWITZERLAND	8	10	10	60
TURKEY	42	50	-	-
UNITED KINGDOM	-	-	-	-
UNITED STATES	-	-	-	-
TOTAL NON-OECD	-	-	50	50

Source: IEA/OECD Coal Statistics.

FRANCE

5/D1 FUEL INPUT IN ELECTRICITY GENERATION

	1973	1978	1979	1980	1981	1985	1990
TOTAL (MTOE)	43.8	54.9	57.0	60.4	69.0	75.5	100.6
SOLID FUELS	10.7	17.4	19.0	19.3	17.4	11.3	8.5
OIL	15.8	11.9	11.2	9.2	6.4	2.2	2.4
GAS	2.1	1.4	1.4	1.0	.6	.7	.3
NUCLEAR	3.6	7.4	9.4	14.3	26.3	44.2	72.4
HYDRO/GEOTHERMAL	11.6	16.8	16.0	16.6	18.3	16.6	16.5
OTHER [1]	-	-	-	-	-	.5	.5
TOTAL (PERCENTS)	100.0	100.0	100.0	100.0	100.0	100.0	100.0
SOLID FUELS	24.5	31.7	33.3	32.0	25.2	15.0	8.4
OIL	36.0	21.7	19.6	15.3	9.2	2.9	2.4
GAS	4.9	2.6	2.5	1.6	.9	.9	.3
NUCLEAR	8.1	13.5	16.6	23.7	38.1	58.5	72.0
HYDRO/GEOTHERMAL	26.6	30.6	28.1	27.4	26.5	22.0	16.4
OTHER [1]	-	-	-	-	-	.7	.5

Source: IEA/OECD Energy Balances and Country Submissions (1981).

1. Solar, wind, etc.

5/D2 ELECTRICTY PRODUCTION BY FUEL TYPE

	1973	1978	1979	1980	1981	1985	1990
TOTAL (TWH)	181.7	226.6	241.4	258.0	276.1	284.2	380.7
SOLID FUELS	36.1	64.6	71.8	70.5	60.1	40.9	35.1
OIL	72.5	55.5	53.1	48.4	32.1	8.7	9.4
GAS	10.1	6.7	8.7	7.1	5.4	3.2	1.5
NUCLEAR	14.7	30.5	40.0	61.3	105.2	164.0	271.0
HYDRO/GEOTHERMAL	48.3	69.3	67.8	70.7	73.3	66.0	62.5
OTHER [1]	-	-	-	-	-	1.4	1.2
TOTAL (PERCENTS)	100.0	100.0	100.0	100.0	100.0	100.0	100.0
SOLID FUELS	19.9	28.5	29.8	27.3	21.8	14.4	9.2
OIL	39.9	24.5	22.0	18.8	11.6	3.1	2.5
GAS	5.6	3.0	3.6	2.8	2.0	1.1	.4
NUCLEAR	8.1	13.5	16.6	23.7	38.1	57.7	71.2
HYDRO/GEOTHERMAL	26.6	30.6	28.1	27.4	26.5	23.2	16.4
OTHER [1]	-	-	-	-	-	.5	.3

Source: IEA/OECD Energy Balances and Country Submissions (1981).

1. Solar, wind, etc.

FRANCE

5/D3 PROJECTIONS OF ELECTRICITY GENERATION CAPACITY BY FUEL
(GW)

	Nuclear	Hydro/ Geoth.	Solid Fuels	Oil	Gas	Multi Firing	Total
I 1981							
Operating	21.6	19.5	7.5	11.9	0.6	9.6	70.7
II 1981-1985							
Capacity:							
Under Construction							
Authorised							
Other Planned							
Conversion			na				
Decommissioning							
Total Operating 1985							
III 1986-1990							
Capacity:							
Under Construction							
Authorised							
Other Planned							
Conversion			na				
Decommissioning							
Total Operating 1990							

Source: IEA/OECD Electricity Statistics.

FRANCE

5/D4 ENERGY USE IN IRON AND STEEL INDUSTRY BY FUEL

	1973	1978	1979	1980	1981	1985	1990
TOTAL (MTOE)	13.9	12.0	10.9	10.4	8.8	na	na
SOLID FUELS	9.4	7.5	6.5	6.3	5.5	na	na
OIL	2.4	2.0	1.7	1.3	.5	na	na
GAS	1.0	1.3	1.3	1.6	1.5	na	na
ELECTRICITY	1.2	1.3	1.3	1.3	1.3	na	na
OTHER [1]	-	-	-	-	-	na	na
TOTAL (PERCENTS)	100.0	100.0	100.0	100.0	100.0	na	na
SOLID FUELS	67.0	62.1	59.7	60.0	62.3	na	na
OIL	17.4	16.8	16.0	12.4	5.8	na	na
GAS	7.1	10.6	12.2	15.0	16.6	na	na
ELECTRICITY	8.4	10.5	12.1	12.6	15.2	na	na
OTHER [1]	-	-	-	-	-	na	na

Source: IEA/OECD Energy Balances and Country Submissions (1981).

1. Includes heat from combined heat and power plants and direct use of solar, wind and other non-solid renewable energy sources.

5/D5 ENERGY USE IN OTHER INDUSTRIES BY FUEL

	1973	1978	1979	1980	1981	1985	1990
TOTAL (MTOE)	47.1	46.2	51.5	49.9	45.0	na	na
SOLID FUELS	3.0	1.9	2.0	2.4	3.1	na	na
OIL	32.8	29.9	33.7	31.0	24.5	na	na
GAS	4.7	7.7	8.4	8.6	9.4	na	na
ELECTRICITY	6.6	6.8	7.4	8.0	8.0	na	na
OTHER [1]	-	-	-	-	-	na	na
TOTAL (PERCENTS)	100.0	100.0	100.0	100.0	100.0	na	na
SOLID FUELS	6.3	4.0	3.8	4.8	6.9	na	na
OIL	69.7	64.6	65.5	62.0	54.5	na	na
GAS	10.0	16.6	16.3	17.2	20.8	na	na
ELECTRICITY	14.0	14.8	14.3	16.0	17.8	na	na
OTHER [1]	-	-	-	-	-	na	na

Source: IEA/OECD Energy Balances and Country Submissions (1981).

1. Includes heat from combined heat and power plants and direct use of solar, wind and other non-solid renewable energy sources.

FRANCE

5/D6 ENERGY USE IN INDUSTRY [1] BY FUEL

	1973	1978	1979	1980	1981	1985	1990
TOTAL (MTOE)	61.0	58.2	62.4	60.4	53.9	61.4	71.7
SOLID FUELS	12.3	9.3	8.5	8.7	8.6	13.6	18.8
OIL	35.3	31.9	35.4	32.3	25.1	24.7	20.4
GAS	5.7	9.0	9.7	10.2	10.8	14.3	18.9
ELECTRICITY	7.7	8.1	8.7	9.3	9.3	7.8	10.7
OTHER [2]	-	-	-	-	-	1.0	2.9
TOTAL (PERCENTS)	100.0	100.0	100.0	100.0	100.0	100.0	100.0
SOLID FUELS	20.2	16.0	13.6	14.3	16.0	22.1	26.2
OIL	57.8	54.7	56.8	53.4	46.5	40.2	28.5
GAS	9.4	15.4	15.6	16.8	20.1	23.3	26.4
ELECTRICITY	12.7	13.9	14.0	15.4	17.3	12.7	14.9
OTHER [2]	-	-	-	-	-	1.6	4.0

Source: IEA/OECD Energy Balances and Country Submissions (1981).

1. Includes non-energy use of petroleum products.
2. Includes heat from combined heat and power plants and direct use of solar, wind and other non-solid renewable energy sources.

5/D7 ENERGY USE IN OTHER SECTORS [1] BY FUEL

	1973	1978	1979	1980	1981	1985	1990
TOTAL (MTOE)	54.5	54.2	53.1	50.3	49.1	48.6	50.5
SOLID FUELS	6.0	4.1	4.0	3.4	3.0	3.4	4.3
OIL	38.0	33.2	31.2	28.4	27.4	20.1	13.1
GAS	5.5	8.6	9.5	10.2	10.2	11.9	13.7
ELECTRICITY	5.0	8.3	8.4	8.3	8.5	10.1	13.3
OTHER [2]	-	-	-	-	-	3.1	6.1
TOTAL (PERCENTS)	100.0	100.0	100.0	100.0	100.0	100.0	100.0
SOLID FUELS	11.0	7.6	7.5	6.8	6.2	7.0	8.5
OIL	69.8	61.3	58.8	56.5	55.8	41.4	26.0
GAS	10.1	15.8	17.9	20.3	20.8	24.5	27.2
ELECTRICITY	9.2	15.2	15.8	16.5	17.3	20.7	26.3
OTHER [2]	-	-	-	-	-	6.4	12.1

Source: IEA/OECD Energy Balances and Country Submissions (1981).

1. Includes use in agricultural, commercial, public services and residential sectors.
2. Includes heat from combined heat and power plants and direct use of solar, wind and other non-solid renewable energy sources.

GERMANY

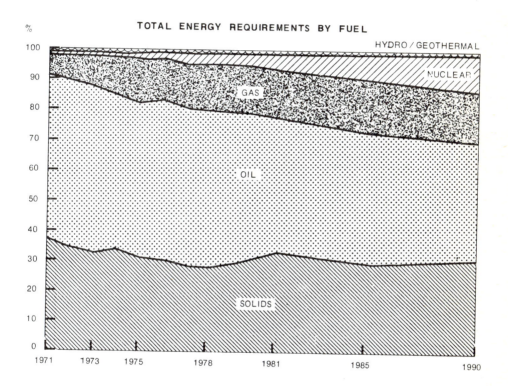

TOTAL ENERGY REQUIREMENTS BY FUEL

%

HYDRO / GEOTHERMAL

NUCLEAR

GAS

OIL

SOLIDS

GERMANY

5/A1 TOTAL PRIMARY ENERGY REQUIREMENTS (TPER) BY FUEL

	1973	1978	1979	1980	1981	1985	1990
TPER (MTOE) [1]	269.6	275.2	287.4	274.7	263.2	289.0	306.0
SOLID FUELS	87.1	76.8	83.3	84.9	86.4	85.0	95.0
OIL	147.8	143.3	144.7	130.9	116.6	125.0	119.0
GAS	27.6	41.5	44.9	43.3	41.7	50.0	51.0
NUCLEAR	2.7	8.8	10.1	10.6	12.9	24.0	37.0
HYDRO/GEOTHERMAL	3.5	4.5	4.4	4.5	4.8	4.0	4.0
OTHER [2]	-	-	-	-	-	-	-
TPER (PERCENTS) [1]	100.0	100.0	100.0	100.0	100.0	100.0	100.0
SOLID FUELS	32.3	27.9	29.0	30.9	32.9	29.4	31.0
OIL	54.8	52.1	50.4	47.7	44.4	43.3	38.9
GAS	10.2	15.1	15.6	15.8	15.9	17.3	16.7
NUCLEAR	1.0	3.2	3.5	3.8	4.8	8.3	12.1
HYDRO/GEOTHERMAL	1.3	1.6	1.5	1.6	1.8	1.4	1.3
OTHER [2]	-	-	-	-	-	-	-

Source: IEA/OECD Energy Balances and IEA Country Submissions (1982).

1. Net imports of electricity are only included in totals.
2. Includes other electricity generation (solar, wind, etc.) and traded synthetic fuels.

5/A2 TOTAL FINAL ENERGY CONSUMPTION (TFC) BY FUEL

	1973	1978	1979	1980	1981	1985	1990
TFC (MTOE)	207.3	199.9	211.2	200.5	191.3	208.0	215.0
SOLID FUELS	34.9	21.7	24.2	24.4	24.3	20.0	21.0
OIL	128.6	125.8	129.6	116.2	106.0	111.0	108.0
GAS	21.7	26.1	29.9	29.0	30.2	40.0	43.0
ELECTRICITY	22.1	26.3	27.4	27.4	27.3	32.0	37.0
HEAT [1]	na	na	na	3.5	3.5	5.0	6.0
TFC (PERCENTS)	100.0	100.0	100.0	100.0	100.0	100.0	100.0
SOLID FUELS	16.8	10.8	11.5	12.2	12.7	9.6	9.8
OIL	62.0	62.9	61.4	57.9	55.4	53.4	50.2
GAS	10.5	13.1	14.2	14.5	15.8	19.2	20.0
ELECTRICITY	10.7	13.1	13.0	13.6	14.3	15.4	17.2
HEAT [1]	na	na	na	1.8	1.8	2.4	2.8

Source: IEA/OECD Energy Balances and IEA Country Submissions (1982).

1. Includes only heat production from combined heat and power plants.

GERMANY

5/A3 TOTAL SOLID FUELS BALANCE
(MTOE)

	1973	1978	1979	1980	1981	1985	1990
PRODUCTION	96.3	87.1	91.2	91.9	93.0	87.0	89.0
IMPORTS	7.0	6.1	7.2	8.3	9.2	7.0	13.0
EXPORTS	-17.4	-19.8	-19.1	-14.5	-13.4	-9.0	-7.0
MARINE BUNKERS	-	-	-	-	-	-	-
STOCK CHANGE	1.2	3.3	4.0	-.8	-2.5	-	-
PRIMARY REQUIREMENTS	87.1	76.8	83.3	84.9	86.4	85.0	95.0
ELECTRICITY GENERATION	-48.0	-52.7	-55.6	-57.0	-58.5	-52.0	-60.0
GAS MANUFACTURE	-1.6	-.2	-.2	-.1	.0	-	-
LIQUEFACTION	-	-	-	-	-	-5.0	-5.0
ENERGY SECTOR OWN USE + LOSSES [1]	-2.7	-2.1	-3.1	-3.3	-3.5	-8.0	-9.0
FINAL CONSUMPTION	34.9	21.7	24.2	24.4	24.3	20.0	21.0
INDUSTRY SECTOR	24.1	16.3	18.0	18.6	19.0	15.0	16.0
IRON AND STEEL	15.1	11.1	12.2	12.4	12.4	na	na
CHEMICAL/ PETROCHEMICAL	3.9	1.9	2.0	2.0	2.3	na	na
OTHER INDUSTRY	5.1	3.4	3.8	4.2	4.4	na	na
TRANSPORTATION SECTOR	.7	.1	.1	.1	.1	-	-
OTHER SECTORS	10.1	5.3	6.1	5.7	5.2	4.0	4.0
AGRICULTURE	-	-	-	-	-	-	-
COMMERCIAL AND PUBLIC SERVICE	.0	-	-	-	-	-	-
RESIDENTIAL	9.1	4.4	5.1	5.1	4.6	-	-

Source: IEA/OECD Energy Balances and IEA Country Submissions (1982).

1. Includes statistical difference.
Note: Solid fuels import projections for 1985 should be 10 Mtoe, according to new information from German authorities.

GERMANY

5/B1 SOLID FUELS INDIGENOUS PRODUCTION BY FUEL TYPE
(MTOE)

	1973	1978	1979	1980	1981	1985	1990
TOTAL SOLID FUELS	96.26	87.14	91.16	91.93	93.03	87.00	89.00
HARD COAL	72.56	61.54	63.73	64.54	65.26	61.00	62.00
COKING COAL	na	35.68	37.48	38.27	39.04	37.00e	37.00e
STEAM COAL	na	25.86	26.25	26.27	26.22	24.00e	25.00e
BROWN COAL/LIGNITE	22.55	24.35	26.12	25.97	26.13	26.00	27.00
OTHER SOLID FUELS	1.15	1.25	1.31	1.42	1.64	-	-

Source: IEA/OECD Coal Statistics, IEA/OECD Energy Balances, IEA Country Submissions (1982).

5/B2 COAL PRODUCTION BY TYPE
(THOUSAND METRIC TONS)

	1973	1978	1979	1980	1981
TOTAL COAL	222312	213691	223919	224354	226194
HARD COAL	103654	90104	93311	94492	95545
COKING COAL	na	52239	54875	56036	57158
STEAM COAL	na	37865	38436	38456	38387
BROWN COAL/LIGNITE	118658	123587	130608	129862	130649
DERIVED PRODUCTS:					
PATENT FUEL	2271	1453	1673	1455	1332
COKE OVEN COKE	33997	25593	26697	28669	28216
GAS COKE	1547	782	937	678	84
BROWN COAL BRIQUETTES	6747	4868	6261	6480	6515
COKE OVEN GAS(TCAL)	66213	49592	51531	55453	54670
BLAST FURNACE GAS(TCAL)	54960	42783	50574	47439	44181

Source: IEA/OECD Coal Statistics.

GERMANY

5/C1 SOLID FUELS TRADE BY TYPE
(MTOE)

	1973	1978	1979	1980	1981	1985	1990
IMPORTS							
TOTAL SOLID FUELS	7.04	6.13	7.19	8.32	9.25	7.00	13.00
HARD COAL	5.33	4.73	5.68	6.58	7.37	6.00	12.00
COKING COAL	na	.48	.80	.90	.45	1.00e	1.00e
STEAM COAL	na	4.25	4.88	5.69	6.92	5.00e	11.00e
BROWN COAL/LIGNITE	.24	.29	.32	.43	.54	1.00e	1.00e
OTHER SOLID FUELS [1]	1.47	1.11	1.19	1.31	1.34	-	-
EXPORTS							
TOTAL SOLID FUELS	17.38	19.79	19.08	14.49	13.35	9.00	7.00
HARD COAL	9.81	12.83	10.65	8.45	8.18	9.00e	7.00e
COKING COAL	na	9.12	7.07	5.89	6.30	9.00e	7.00e
STEAM COAL	na	3.71	3.57	2.56	1.89	-	-
BROWN COAL/LIGNITE	-	-	-	-	-	-	-
OTHER SOLID FUELS [1]	7.57	6.96	8.43	6.04	5.17	-	-

Source: IEA/OECD Coal Statistics, IEA/OECD Energy Balances, IEA Country Submissions (1982).

1. Includes derived coal products, e.g. coke.
Note: Solid fuels import projections for 1985 should be 10 Mtoe, according to new information from German authorities. The difference of 3 Mtoe is projected to be steam coal.

GERMANY

5/C2 COAL IMPORTS BY ORIGIN
(THOUSAND METRIC TONS)

	1978	1979	1980	1981
HARD COAL IMPORTS	6931	8321	9641	10786
COKING COAL IMPORTS FROM:	710	1173	1312	653
AUSTRALIA	-	-	-	26
CANADA	-	-	-	-
GERMANY	-	-	-	-
UNITED KINGDOM	48	10	48	30
UNITED STATES	553	997	1050	422
OTHER OECD	89	152	158	170
CZECHOSLOVAKIA	-	-	-	-
POLAND	-	-	-	-
USSR	20	14	56	5
CHINA	-	-	-	-
SOUTH AFRICA	-	-	-	-
OTHER NON-OECD	-	-	-	-
STEAM COAL IMPORTS FROM:	6221	7148	8329	10133
AUSTRALIA	763	623	578	580
CANADA	428	512	449	629
GERMANY	-	-	-	-
UNITED KINGDOM	554	693	1354	1823
UNITED STATES	32	213	1218	2988
OTHER OECD	1004	1196	822	934
CZECHOSLOVAKIA	152	129	146	171
POLAND	2041	2402	1918	1012
USSR	96	196	143	16
CHINA	-	-	209	155
SOUTH AFRICA	1108	1052	1492	1825
OTHER NON-OECD	43	132	-	-

Source: IEA/OECD Coal Statistics.

GERMANY

5/C3 HARD COAL EXPORTS BY DESTINATION
(THOUSAND METRIC TONS)

	1978	1979	1980	1981
TOTAL EXPORTS	18788	15586	12369	11982
EXPORTS TO:				
AUSTRALIA	-	-	-	-
AUSTRIA	228	238	90	218
BELGIUM	4320	3375	2241	2180
CANADA	-	-	-	-
DENMARK	944	559	257	1
FINLAND	-	-	-	-
FRANCE	6594	6743	5254	4417
GERMANY	-	-	-	-
GREECE	-	-	-	-
ICELAND	-	-	-	-
IRELAND	7	3	-	5
ITALY	2512	1967	2544	2591
JAPAN	375	247	1	-
LUXEMBOURG	-	-	-	-
NETHERLANDS	1442	1261	1320	1137
NEW ZEALAND	-	-	-	-
NORWAY	18	23	30	30
PORTUGAL	6	-	-	2
SPAIN	447	63	20	22
SWEDEN	272	163	1	12
SWITZERLAND	78	111	118	119
TURKEY	121	-	-	-
UNITED KINGDOM	249	188	74	115
UNITED STATES	103	-	-	-
TOTAL NON-OECD	1072	645	419	1133

Source: IEA/OECD Coal Statistics.

GERMANY

5/C4 COKING COAL EXPORTS BY DESTINATION
(THOUSAND METRIC TONS)

	1978	1979	1980	1981
TOTAL EXPORTS	13354	10352	8617	9218
EXPORTS TO:				
AUSTRALIA	-	-	-	-
AUSTRIA	213	226	73	159
BELGIUM	2183	1518	787	862
CANADA	-	-	-	-
DENMARK	-	-	-	-
FINLAND	-	-	-	-
FRANCE	5020	4999	4198	3961
GERMANY	-	-	-	-
GREECE	-	-	-	-
ICELAND	-	-	-	-
IRELAND	-	-	-	-
ITALY	2480	1937	2527	2580
JAPAN	375	247	1	-
LUXEMBOURG	-	-	-	-
NETHERLANDS	975	736	682	819
NEW ZEALAND	-	-	-	-
NORWAY	-	-	-	-
PORTUGAL	-	-	-	-
SPAIN	447	63	20	22
SWEDEN	234	161	-	-
SWITZERLAND	18	15	-	-
TURKEY	70	-	-	-
UNITED KINGDOM	198	141	-	-
UNITED STATES	103	-	-	-
TOTAL NON-OECD	1038	309	329	815

Source: IEA/OECD Coal Statistics.

GERMANY

5/C5 STEAM COAL EXPORTS BY DESTINATION
(THOUSAND METRIC TONS)

	1978	1979	1980	1981
TOTAL EXPORTS	5434	5234	3752	2764
EXPORTS TO:				
AUSTRALIA	-	-	-	-
AUSTRIA	15	12	17	59
BELGIUM	2137	1857	1454	1318
CANADA	-	-	-	-
DENMARK	944	559	257	1
FINLAND	-	-	-	-
FRANCE	1574	1744	1056	456
GERMANY	-	-	-	-
GREECE	-	-	-	-
ICELAND	-	-	-	-
IRELAND	7	3	-	5
ITALY	32	30	17	11
JAPAN	-	-	-	-
LUXEMBOURG	-	-	-	-
NETHERLANDS	467	525	638	318
NEW ZEALAND	-	-	-	-
NORWAY	18	23	30	30
PORTUGAL	6	-	-	2
SPAIN	-	-	-	-
SWEDEN	38	2	1	12
SWITZERLAND	60	96	118	119
TURKEY	51	-	-	-
UNITED KINGDOM	51	47	74	115
UNITED STATES	-	-	-	-
TOTAL NON-OECD	34	336	90	318

Source: IEA/OECD Coal Statistics.

GERMANY

5/D1 FUEL INPUT IN ELECTRICITY GENERATION

	1973	1978	1979	1980	1981	1985	1990
TOTAL (MTOE)	68.1	86.3	88.6	89.7	89.0	97.0	115.0
SOLID FUELS	48.0	52.7	55.6	57.0	58.5	52.0	60.0
OIL	6.4	6.1	5.5	5.4	3.3	4.0	3.0
GAS	7.5	14.2	13.0	12.2	9.5	13.0	11.0
NUCLEAR	2.7	8.8	10.1	10.6	12.9	24.0	37.0
HYDRO/GEOTHERMAL	3.5	4.5	4.4	4.5	4.8	4.0	4.0
OTHER [1]	-	-	-	-	-	-	-
TOTAL (PERCENTS)	100.0	100.0	100.0	100.0	100.0	100.0	100.0
SOLID FUELS	70.4	61.0	62.8	63.7	66.0	53.6	52.2
OIL	9.4	7.1	6.2	6.0	3.7	4.1	2.6
GAS	11.0	16.5	14.7	13.7	10.7	13.4	9.6
NUCLEAR	3.9	10.2	11.4	11.6	14.4	24.7	32.2
HYDRO/GEOTHERMAL	5.2	5.2	5.0	5.0	5.3	4.1	3.5
OTHER [1]	-	-	-	-	-	-	-

Source: IEA/OECD Energy Balances and IEA Country Submissions (1982).

1. Solar, wind, etc.

5/D2 ELECTRICTY PRODUCTION BY FUEL TYPE

	1973	1978	1979	1980	1981	1985	1990
TOTAL (TWH)	297.8	353.4	372.2	368.8	368.8	427.0	498.0
SOLID FUELS	192.1	203.3	214.1	219.7	228.2	218.0	253.0
OIL	42.8	31.1	27.7	25.7	20.0	22.0	16.0
GAS	35.6	64.5	69.5	61.0	47.0	62.0	49.0
NUCLEAR	11.8	35.9	42.3	43.7	53.6	106.0	161.0
HYDRO/GEOTHERMAL	15.5	18.5	18.5	18.6	20.0	19.0	19.0
OTHER [1]	-	-	-	-	-	-	-
TOTAL (PERCENTS)	100.0	100.0	100.0	100.0	100.0	100.0	100.0
SOLID FUELS	64.5	57.5	57.5	59.6	61.9	51.1	50.8
OIL	14.4	8.8	7.5	7.0	5.4	5.2	3.2
GAS	12.0	18.3	18.7	16.6	12.7	14.5	9.8
NUCLEAR	3.9	10.2	11.4	11.9	14.5	24.8	32.3
HYDRO/GEOTHERMAL	5.2	5.2	5.0	5.1	5.4	4.4	3.8
OTHER [1]	-	-	-	-	-	-	-

Source: IEA/OECD Energy Balances and IEA Country Submissions (1982).

1. Solar, wind, etc.

GERMANY

5/D3 PROJECTIONS OF ELECTRICITY GENERATION CAPACITY BY FUEL
(GW)

	Nuclear	Hydro/ Geoth.	Coal	Oil	Gas	Multi Firing [1]	Total
I 1981							
Operating	9.8	6.5	32.8	11.7	8.5	15.5	85.4
II 1981-1985							
Capacity:							
Under Construction	7.2	0.3	5.8	0.3	-	0.3	13.9
Authorised	-	-	-	-	-	-	-
Other Planned	-	-	-	-	-	-	-
Conversion	-	-	-	-	-	-	-
Decommissioning	-	-0.1	-1.9	-	-	-	-2.0
Total Operating 1985	17.0	6.7	27.7	12.0	8.5	15.8	97.3
III 1986-1990							
Capacity:							
Under Construction	2.7	0.1	0.8	-	-	-	3.6
Authorised	6.2	-	9.5	-	-	-	15.7
Other Planned	-	-	-	-	-	-	-
Conversion	-	-	-	-	-	-	-
Decommissioning	-	-0.1	-2.6	-4.6	-2.7	-	-10.0
Total Operating 1990	25.9	6.7	35.4	7.3	5.8	15.8	106.6

Source: IEA/OECD Electricity Statistics and IEA Country Submissions (1981).

1.   Includes "Other" capacity.

GERMANY

5/D4 ENERGY USE IN IRON AND STEEL INDUSTRY BY FUEL

	1973	1978	1979	1980	1981	1985	1990
TOTAL (MTOE)	27.3	18.2	19.7	18.9	17.5	na	na
SOLID FUELS	15.1	11.1	12.2	12.4	12.4	na	na
OIL	4.5	2.3	2.4	1.5	.7	na	na
GAS	5.6	2.8	3.0	2.8	2.4	na	na
ELECTRICITY	2.1	2.0	2.2	2.1	2.0	na	na
HEAT [1]	-	-	-	-	-	na	na
TOTAL (PERCENTS)	100.0	100.0	100.0	100.0	100.0	na	na
SOLID FUELS	55.3	60.9	62.0	65.9	70.6	na	na
OIL	16.4	12.9	12.0	7.8	4.2	na	na
GAS	20.7	15.1	15.1	15.1	13.7	na	na
ELECTRICITY	7.7	11.0	10.9	11.3	11.5	na	na
HEAT [1]	-	-	-	-	-	na	na

Source: IEA/OECD Energy Balances and IEA Country Submissions (1982).

1. Includes only heat from combined heat and power plants.

5/D5 ENERGY USE IN OTHER INDUSTRIES BY FUEL

	1973	1978	1979	1980	1981	1985	1990
TOTAL (MTOE)	69.0	61.5	65.7	63.0	58.8	na	na
SOLID FUELS	9.0	5.2	5.8	6.1	6.7	na	na
OIL	41.6	34.9	36.8	32.8	28.0	na	na
GAS	8.3	10.2	11.4	12.1	12.5	na	na
ELECTRICITY	10.1	11.2	11.7	11.4	11.0	na	na
HEAT [1]	na	na	na	.6	.6	na	na
TOTAL (PERCENTS)	100.0	100.0	100.0	100.0	100.0	na	na
SOLID FUELS	13.0	8.5	8.8	9.8	11.4	na	na
OIL	60.3	56.7	56.1	52.0	47.6	na	na
GAS	12.1	16.6	17.3	19.3	21.3	na	na
ELECTRICITY	14.7	18.2	17.8	18.0	18.7	na	na
HEAT [1]	na	na	na	.9	1.1	na	na

Source: IEA/OECD Energy Balances and IEA Country Submissions (1982).

1. Includes only heat from combined heat and power plants.

GERMANY

5/D6 ENERGY USE IN INDUSTRY [1] BY FUEL

	1973	1978	1979	1980	1981	1985	1990
TOTAL (MTOE)	96.3	79.7	85.4	81.9	76.3	88.0	91.0
SOLID FUELS	24.1	16.3	18.0	18.6	19.0	16.0	17.0
OIL	46.1	37.2	39.2	34.2	28.7	32.0	31.0
GAS	14.0	13.0	14.4	15.0	14.9	24.0	25.0
ELECTRICITY	12.2	13.2	13.8	13.5	13.0	15.0	17.0
HEAT [2]	na	na	na	.6	.6	1.0	1.0
TOTAL (PERCENTS)	100.0	100.0	100.0	100.0	100.0	100.0	100.0
SOLID FUELS	25.0	20.5	21.1	22.7	24.9	18.2	18.7
OIL	47.8	46.7	45.9	41.8	37.6	36.4	34.1
GAS	14.5	16.3	16.8	18.3	19.6	27.3	27.5
ELECTRICITY	12.7	16.6	16.2	16.5	17.1	17.0	18.7
HEAT [2]	na	na	na	.7	.8	1.1	1.1

Source: IEA/OECD Energy Balances and IEA Country Submissions (1982).

1. Includes non-energy use of petroleum products.
2. Includes only heat from combined heat and power plants.

5/D7 ENERGY USE IN OTHER SECTORS [1] BY FUEL

	1973	1978	1979	1980	1981	1985	1990
TOTAL (MTOE)	78.1	81.2	85.4	77.7	75.3	79.0	85.0
SOLID FUELS	10.1	5.3	6.1	5.7	5.2	4.0	4.0
OIL	51.0	50.5	51.1	42.0	38.6	39.0	39.0
GAS	7.8	13.2	15.5	13.9	15.2	16.0	18.0
ELECTRICITY	9.1	12.3	12.7	13.0	13.4	16.0	19.0
HEAT [2]	na	na	na	3.0	2.9	4.0	5.0
TOTAL (PERCENTS)	100.0	100.0	100.0	100.0	100.0	100.0	100.0
SOLID FUELS	13.0	6.5	7.2	7.4	6.9	5.1	4.7
OIL	65.4	62.2	59.8	54.1	51.3	49.4	45.9
GAS	10.0	16.2	18.1	18.0	20.2	20.3	21.2
ELECTRICITY	11.7	15.1	14.9	16.7	17.7	20.3	22.4
HEAT [2]	na	na	na	3.8	3.9	5.1	5.9

Source: IEA/OECD Energy Balances and IEA Country Submissions (1982).

1. Includes use in agricultural, commercial, public services and residential sectors.
2. Includes only heat from combined heat and power plants.

GREECE

TOTAL ENERGY REQUIREMENTS BY FUEL

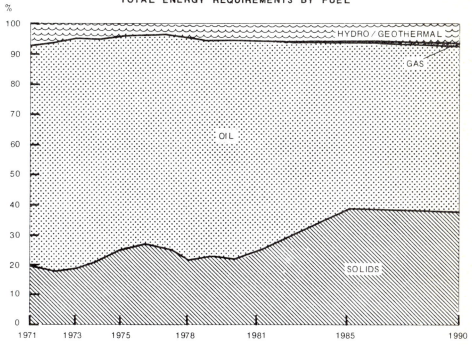

GREECE

5/A1 TOTAL PRIMARY ENERGY REQUIREMENTS (TPER) BY FUEL

	1973	1978	1979	1980	1981	1985	1990
TPER (MTOE) [1]	12.3	15.3	16.2	16.2	15.8	17.2	25.7
SOLID FUELS	2.3	3.3	3.7	3.5	3.9	6.7	10.4
OIL	9.4	11.3	11.6	11.7	11.0	9.4	13.9
GAS	-	-	-	-	-	.1	.1
NUCLEAR	-	-	-	-	-	-	-
HYDRO/GEOTHERMAL	.6	.8	.9	.9	.9	1.0	1.3
OTHER [2]	-	-	-	-	-	-	-
TPER (PERCENTS) [1]	100.0	100.0	100.0	100.0	100.0	100.0	100.0
SOLID FUELS	18.6	21.5	23.0	21.8	24.8	39.0	40.4
OIL	76.7	73.5	71.3	72.4	69.4	54.7	54.1
GAS	-	-	-	-	-	.6	.4
NUCLEAR	-	-	-	-	-	-	-
HYDRO/GEOTHERMAL	4.7	4.9	5.6	5.5	5.6	5.8	5.1
OTHER [2]	-	-	-	-	-	-	-

Source: IEA/OECD Energy Balances and IEA Country Submissions (1981 and 1982).

1. Net imports of electricity are only included in totals.
2. Includes other electricity generation (solar, wind, etc.) and traded synthetic fuels.

5/A2 TOTAL FINAL ENERGY CONSUMPTION (TFC) BY FUEL

	1973	1978	1979	1980	1981	1985	1990
TFC (MTOE)	9.0	10.9	11.3	11.3	10.9	12.0	18.0
SOLID FUELS	.7	.5	.6	.6	.5	1.4	1.9
OIL	7.2	8.8	9.1	9.1	8.8	8.6	12.7
GAS	.0	.0	.0	.0	.0	.1	.2
ELECTRICITY	1.1	1.5	1.6	1.6	1.6	1.9	3.2
HEAT [1]	-	-	-	-	-	-	-
TFC (PERCENTS)	100.0	100.0	100.0	100.0	100.0	100.0	100.0
SOLID FUELS	7.3	4.7	5.6	5.0	4.5	11.7	10.6
OIL	80.5	81.4	80.5	80.5	80.5	71.7	70.6
GAS	.0	.0	.0	.0	.0	.8	1.0
ELECTRICITY	12.1	13.9	13.8	14.5	15.0	15.8	17.8
HEAT [1]	-	-	-	-	-	-	-

Source: IEA/OECD Energy Balances and IEA Country Submissions (1981 and 1982).

1. Includes only heat production from combined heat and power plants.

GREECE

5/A3 TOTAL SOLID FUELS BALANCE
(MTOE)

	1973	1978	1979	1980	1981	1985	1990
PRODUCTION	1.8	3.0	3.3	3.2	3.8	5.6	8.8
IMPORTS	.5	.3	.5	.4	.2	1.1	1.6
EXPORTS	.0	.0	-	-	-	-	-
MARINE BUNKERS	-	-	-	-	-	-	-
STOCK CHANGE	-	.0	.0	-.1	-.1	-	-
PRIMARY REQUIREMENTS	2.3	3.3	3.7	3.5	3.9	6.7	10.4
ELECTRICITY GENERATION	-1.5	-2.7	-2.9	-2.8	-3.5	-5.3	-8.4
GAS MANUFACTURE	.0	.0	.0	.0	.0	-	-
LIQUEFACTION	-	-	-	-	-	-	-
ENERGY SECTOR OWN USE + LOSSES [1]	-.1	.0	-.2	-.1	.0	-	-.1
FINAL CONSUMPTION	.7	.5	.6	.6	.5	1.4	1.9
INDUSTRY SECTOR	.6	.5	.6	.5	.4	1.3	1.7
IRON AND STEEL	.4	.3	.3	.3	.2	na	na
CHEMICAL/ PETROCHEMICAL	.1	.1	.1	.1	.1	na	na
OTHER INDUSTRY	.1	.1	.1	.1	.1	na	na
TRANSPORTATION SECTOR	.0	.0	.0	.0	.0	-	-
OTHER SECTORS	.1	.0	.1	.1	.1	-	-
AGRICULTURE	-	-	-	-	-	-	-
COMMERCIAL AND PUBLIC SERVICE	-	-	-	-	.0	-	-
RESIDENTIAL	.1	.0	.1	.1	.0	-	-

Source: IEA/OECD Energy Balances and IEA Country Submissions (1981 and 1982).

1. Includes statistical difference.

GREECE

5/B1 SOLID FUELS INDIGENOUS PRODUCTION BY FUEL TYPE
(MTOE)

	1973	1978	1979	1980	1981	1985	1990
TOTAL SOLID FUELS	1.81	3.01	3.26	3.20	3.77	5.60	8.80
HARD COAL	-	-	-	-	-	-	-
COKING COAL	-	-	-	-	-	-	-
STEAM COAL	-	-	-	-	-	-	-
BROWN COAL/LIGNITE	1.81	3.01	3.26	3.20	3.77	5.60	8.80
OTHER SOLID FUELS	-	-	-	-	-	-	-

Source: IEA/OECD Coal Statistics, IEA/OECD Energy Balances, IEA Country Submissions (1981 and 1982).

5/B2 COAL PRODUCTION BY TYPE
(THOUSAND METRIC TONS)

	1973	1978	1979	1980	1981
TOTAL COAL	13118	21815	23621	23198	27315
HARD COAL	-	-	-	-	-
COKING COAL	-	-	-	-	-
STEAM COAL	-	-	-	-	-
BROWN COAL/LIGNITE	13118	21815	23621	23198	27315
DERIVED PRODUCTS:					
PATENT FUEL	-	-	-	-	-
COKE OVEN COKE	400	186	228	247	45
GAS COKE	10	12	11	14	13
BROWN COAL BRIQUETTES	105	73	95	97	198
COKE OVEN GAS(TCAL)	860	417	511	552	460
BLAST FURNACE GAS(TCAL)	1000	600	764	657	600

Source: IEA/OECD Coal Statistics.

GREECE

5/C1 SOLID FUELS TRADE BY TYPE
(MTOE)

	1973	1978	1979	1980	1981	1985	1990
IMPORTS							
TOTAL SOLID FUELS	.49	.29	.50	.40	.22	1.10	1.60
HARD COAL	.46	.24	.44	.37	.20	1.10e	1.60
COKING COAL	na	.15	.26	.27	.05	0.20e	0.40
STEAM COAL	na	.10	.18	.10	.15	0.90e	1.20
BROWN COAL/LIGNITE	-	-	-	-	-	-	-
OTHER SOLID FUELS [1]	.03	.05	.06	.03	.02	-	-
EXPORTS							
TOTAL SOLID FUELS	.02	.04	-	-	-	-	-
HARD COAL	-	-	-	-	-	-	-
COKING COAL	-	-	-	-	-	-	-
STEAM COAL	-	-	-	-	-	-	-
BROWN COAL/LIGNITE	-	-	-	-	-	-	-
OTHER SOLID FUELS [1]	.02	.04	-	-	-	-	-

Source: IEA/OECD Coal Statistics, IEA/OECD Energy Balances, IEA Country Submissions (1981 and 1982).

1.   Includes derived coal products, e.g. coke.

GREECE

5/C2 COAL IMPORTS BY ORIGIN
(THOUSAND METRIC TONS)

	1978	1979	1980	1981
HARD COAL IMPORTS	349	628	533	287
COKING COAL IMPORTS FROM:	213	378	384	67
AUSTRALIA	159	253	285	-
CANADA	-	-	-	-
GERMANY	-	-	-	-
UNITED KINGDOM	-	-	-	-
UNITED STATES	-	-	-	-
OTHER OECD	-	-	-	-
CZECHOSLOVAKIA	-	-	-	-
POLAND	54	125	99	67
USSR	-	-	-	-
CHINA	-	-	-	-
SOUTH AFRICA	-	-	-	-
OTHER NON-OECD	-	-	-	-
STEAM COAL IMPORTS FROM:	136	250	149	220
AUSTRALIA	48	155	123	-
CANADA	-	-	-	-
GERMANY	-	-	-	-
UNITED KINGDOM	-	-	-	-
UNITED STATES	-	-	-	213
OTHER OECD	-	-	-	-
CZECHOSLOVAKIA	1	1	1	2
POLAND	75	68	-	5
USSR	12	26	25	-
CHINA	-	-	-	-
SOUTH AFRICA	-	-	-	-
OTHER NON-OECD	-	-	-	-

Source: IEA/OECD Coal Statistics.

GREECE

5/D1 FUEL INPUT IN ELECTRICITY GENERATION

	1973	1978	1979	1980	1981	1985	1990
TOTAL (MTOE)	3.9	5.3	5.7	5.9	6.1	6.7	10.1
SOLID FUELS	1.5	2.7	2.9	2.8	3.5	5.3	8.4
OIL	1.8	1.8	1.9	2.1	1.7	.4	.4
GAS	-	-	-	-	-	-	-
NUCLEAR	-	-	-	-	-	-	-
HYDRO/GEOTHERMAL	.6	.8	.9	.9	.9	1.0	1.3
OTHER [1]	-	-	-	-	-	-	-
TOTAL (PERCENTS)	100.0	100.0	100.0	100.0	100.0	100.0	100.0
SOLID FUELS	39.3	51.6	50.6	48.6	57.3	79.1	83.2
OIL	45.7	34.2	33.2	36.4	28.2	6.0	4.0
GAS	-	-	-	-	-	-	-
NUCLEAR	-	-	-	-	-	-	-
HYDRO/GEOTHERMAL	15.0	14.2	16.1	15.0	14.5	14.9	12.8
OTHER [1]	-	-	-	-	-	-	-

Source: IEA/OECD Energy Balances and IEA Country Submissions (1981 and 1982).

1. Solar, wind, etc.

5/D2 ELECTRICTY PRODUCTION BY FUEL TYPE

	1973	1978	1979	1980	1981	1985	1990
TOTAL (TWH)	14.8	21.0	22.1	22.7	23.4	26.1	36.5
SOLID FUELS	5.3	10.5	10.7	10.3	12.5	20.5	30.0
OIL	7.3	7.6	7.9	9.0	7.6	1.7	1.5
GAS	-	-	-	-	-	-	-
NUCLEAR	-	-	-	-	-	-	-
HYDRO/GEOTHERMAL	2.2	3.0	3.6	3.4	3.4	3.9	5.0
OTHER [1]	-	-	-	-	-	-	-
TOTAL (PERCENTS)	100.0	100.0	100.0	100.0	100.0	100.0	100.0
SOLID FUELS	35.5	49.7	48.2	45.4	53.2	78.5	82.2
OIL	49.5	36.1	35.6	39.6	32.3	6.5	4.1
GAS	-	-	-	-	-	-	-
NUCLEAR	-	-	-	-	-	-	-
HYDRO/GEOTHERMAL	15.0	14.2	16.1	15.0	14.5	14.9	13.7
OTHER [1]	-	-	-	-	-	-	-

Source: IEA/OECD Energy Balances and IEA Country Submissions (1981 and 1982).

1. Solar, wind, etc.

GREECE

5/D3 PROJECTIONS OF ELECTRICITY GENERATION CAPACITY BY FUEL
(GW)

	Nuclear	Hydro/ Geoth.	Solid Fuels	Oil	Gas	Multi Firing	Total
I 1980							
Operating	-	1.7	2.4	1.8	-	-	6.0
II 1980-1985							
Capacity:							
Under Construction	-	0.9	0.6	0.2	-	-	1.7
Authorised	-	0.2	1.5	-	-	-	1.7
Other Planned	-	-	-	-	-	-	-
Conversion	-	-	-	-	-	-	-
Decommissioning	-	-	-0.2	-0.3	-	-	-0.5
Total Operating 1985	-	2.8	4.3	1.7	-	-	8.9
III 1986-1990							
Capacity:							
Under Construction	-	-	-	-	-	-	-
Authorised	-	1.8	-	-	-	-	1.8
Other Planned	-	-	1.2	0.2	-	-	1.4
Conversion	-	-	-	-	-	-	-
Decommissioning	-	-	-0.2	-0.3	-	-	-0.5
Total Operating 1990	-	4.6	5.3	1.6	-	-	11.6

Source: IEA/OECD Electricity Statistics and IEA Country Submissions (1981)

GREECE

5/D6 ENERGY USE IN INDUSTRY [1] BY FUEL

	1973	1978	1979	1980	1981	1985	1990
TOTAL (MTOE)	3.7	4.2	4.6	4.5	4.2	5.0	7.6
SOLID FUELS	.6	.5	.6	.5	.4	1.4	1.7
OIL	2.5	3.0	3.2	3.2	2.9	2.5	4.2
GAS	.0	.0	.0	.0	-	.1	.0
ELECTRICITY	.6	.8	.8	.8	.8	1.0	1.7
HEAT [2]	-	-	-	-	-	-	-
TOTAL (PERCENTS)	100.0	100.0	100.0	100.0	100.0	100.0	100.0
SOLID FUELS	15.8	10.9	12.3	11.2	10.5	28.0	22.4
OIL	67.0	70.7	69.9	70.4	70.4	50.0	55.3
GAS	.0	.0	.0	.0	-	2.0	.0
ELECTRICITY	17.2	18.4	17.7	18.4	19.1	20.0	22.4
HEAT [2]	-	-	-	-	-	-	-

Source: IEA/OECD Energy Balances and IEA Country Submissions (1981 and 1982).

1. Includes non-energy use of petroleum products.
2. Includes only heat from combined heat and power plants.

5/D7 ENERGY USE IN OTHER SECTORS [1] BY FUEL

	1973	1978	1979	1980	1981	1985	1990
TOTAL (MTOE)	2.7	2.9	2.9	2.8	2.7	2.9	4.0
SOLID FUELS	.1	.0	.1	.1	.1	-	-
OIL	2.1	2.1	2.0	1.9	1.8	2.0	2.4
GAS	.0	.0	.0	.0	.0	-	.1
ELECTRICITY	.5	.7	.7	.8	.8	.9	1.5
HEAT [2]	-	-	-	-	-	-	-
TOTAL (PERCENTS)	100.0	100.0	100.0	100.0	100.0	100.0	100.0
SOLID FUELS	2.1	1.5	2.4	2.0	2.0	-	-
OIL	80.7	73.5	71.3	69.0	66.7	69.0	60.0
GAS	.1	.1	.1	.1	.1	-	2.5
ELECTRICITY	17.1	24.9	26.3	29.0	31.2	31.0	37.5
HEAT [2]	-	-	-	-	-	-	-

Source: IEA/OECD Energy Balances and IEA Country Submissions (1981 and 1982).

1. Includes use in agricultural, commercial, public services and residential sectors.
2. Includes only heat from combined heat and power plants.

ICELAND

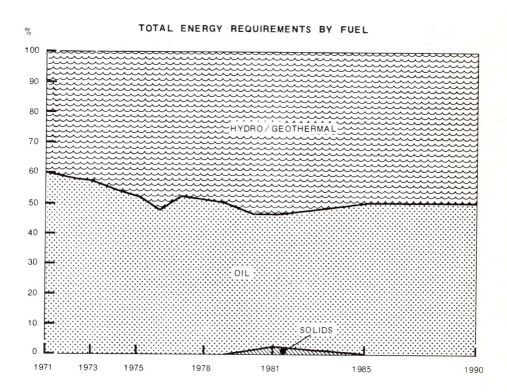

TOTAL ENERGY REQUIREMENTS BY FUEL

%

HYDRO / GEOTHERMAL

OIL

SOLIDS

100
90
80
70
60
50
40
30
20
10
0

1971   1973   1975   1978   1981   1985   1990

ICELAND

5/A1 TOTAL PRIMARY ENERGY REQUIREMENTS (TPER) BY FUEL

	1973	1978	1979	1980	1981	1985	1990
TPER (MTOE) [1]	1.2	1.3	1.4	1.4	1.4	1.5e	1.6e
SOLID FUELS	.0	-	-	.0	.0	-	-
OIL	.7	.7	.7	.6	.6	.7e	.8e
GAS	-	-	-	-	-	-	-
NUCLEAR	-	-	-	-	-	-	-
HYDRO/GEOTHERMAL	.5	.6	.7	.7	.8	.7e	.8e
OTHER [2]	-	-	-	-	-	-	-
TPER (PERCENTS) [1]	100.0	100.0	100.0	100.0	100.0	100.0	100.0
SOLID FUELS	.1	-	-	1.4	2.5	-	-
OIL	57.2	51.2	49.8	44.9	43.7	50.3e	50.3e
GAS	-	-	-	-	-	-	-
NUCLEAR	-	-	-	-	-	-	-
HYDRO/GEOTHERMAL	42.8	48.8	50.2	53.7	53.8	49.7e	49.7e
OTHER [2]	-	-	-	-	-	-	-

Source: IEA/OECD Energy Balances.
1. Net imports of electricity are only included in totals.
2. Includes other electricity generation (solar, wind, etc.) and traded synthetic fuels.

5/A2 TOTAL FINAL ENERGY CONSUMPTION (TFC) BY FUEL

	1973	1978	1979	1980	1981	1985	1990
TFC (MTOE)	.9	.8	.8	.8	.8	.9e	1.0e
SOLID FUELS	.0	-	-	.0	.0	-	-
OIL	.7	.6	.6	.5	.6	.7e	.8e
GAS	-	-	-	-	-	-	-
ELECTRICITY	.2	.2	.2	.2	.3	.2e	.3e
HEAT [1]	-	-	-	-	-	-	-
TFC (PERCENTS)	100.0	100.0	100.0	100.0	100.0	100.0	100.0
SOLID FUELS	.1	-	-	2.4	4.1	-	-
OIL	79.1	74.4	72.0	66.5	66.2	74.1e	74.1e
GAS	-	-	-	-	-	-	-
ELECTRICITY	20.8	25.6	28.0	31.1	29.7	25.9e	25.9e
HEAT [1]	-	-	-	-	-	-	-

Source: IEA/OECD Energy Balances.

1. Includes only heat production from combined heat and power plants.

ICELAND

5/D1 FUEL INPUT IN ELECTRICITY GENERATION

	1973	1978	1979	1980	1981	1985	1990
TOTAL (MTOE)	.5	.7	.7	.8	.8	.8e	.9e
SOLID FUELS	-	-	-	-	-	-	-
OIL	.0	.0	.1	.0	.1	.1e	.1e
GAS	-	-	-	-	-	-	-
NUCLEAR	-	-	-	-	-	-	-
HYDRO/GEOTHERMAL	.5	.6	.7	.7	.8	.7e	.8e
OTHER [1]	-	-	-	-	-	-	-
TOTAL (PERCENTS)	100.0	100.0	100.0	100.0	100.0	100.0	100.0
SOLID FUELS	-	-	-	-	-	-	-
OIL	4.4	6.5	7.3	6.2	7.0	5.9e	5.9e
GAS	-	-	-	-	-	-	-
NUCLEAR	-	-	-	-	-	-	-
HYDRO/GEOTHERMAL	95.6	93.5	92.7	93.8	93.0	94.1e	94.1e
OTHER [1]	-	-	-	-	-	-	-

Source: IEA/OECD Energy Balances.

1. Solar, wind, etc.

5/D2 ELECTRICTY PRODUCTION BY FUEL TYPE

	1973	1978	1979	1980	1981	1985	1990
TOTAL (TWH)	2.3	2.7	3.0	3.2	3.3	3.3e	3.4e
SOLID FUELS	-	-	-	-	-	-	-
OIL	.1	.1	.1	.0	.1	.1e	.1e
GAS	-	-	-	-	-	-	-
NUCLEAR	-	-	-	-	-	-	-
HYDRO/GEOTHERMAL	2.2	2.7	2.9	3.1	3.3	3.2e	3.3e
OTHER [1]	-	-	-	-	-	-	-
TOTAL (PERCENTS)	100.0	100.0	100.0	100.0	100.0	100.0	100.0
SOLID FUELS	-	-	-	-	-	-	-
OIL	3.7	2.1	2.0	1.5	1.7	3.0e	2.9e
GAS	-	-	-	-	-	-	-
NUCLEAR	-	-	-	-	-	-	-
HYDRO/GEOTHERMAL	96.3	97.9	98.0	98.5	98.3	97.0e	97.1e
OTHER [1]	-	-	-	-	-	-	-

Source: IEA/OECD Energy Balances.

1. Solar, wind, etc.

ICELAND

5/D3 PROJECTIONS OF ELECTRICITY GENERATION CAPACITY BY FUEL
(GW)

	Nuclear	Hydro/ Geoth.	Solid Fuels	Oil	Gas	Multi Firing	Total
I 1981							
Operating	-	0.6	-	0.2	-	-	0.8
II 1981-1985							
Capacity:							
Under Construction							
Authorised							
Other Planned							
Conversion		na					
Decommissioning							
Total Operating 1985							
III 1986-1990							
Capacity:							
Under Construction							
Authorised							
Other Planned							
Conversion		na					
Decommissioning							
Total Operating 1990							

Source: IEA/OECD Electricity Statistics.

ICELAND

5/D6 ENERGY USE IN INDUSTRY [1] BY FUEL

	1973	1978	1979	1980	1981	1985	1990
TOTAL (MTOE)	.2	.2	.3	.3	.4	.2e	.3e
SOLID FUELS	-	-	-	.0	.0	-	-
OIL	.1	.1	.1	.1	.2	.1e	.1e
GAS	-	-	-	-	-	-	-
ELECTRICITY	.1	.1	.2	.2	.2	.2e	.2e
HEAT [2]	-	-	-	-	-	-	-
TOTAL (PERCENTS)	100.0	100.0	100.0	100.0	100.0	100.0	100.0
SOLID FUELS	-	-	-	5.9	8.1	-	-
OIL	44.2	41.3	44.4	41.7	51.7	35.4e	35.4e
GAS	-	-	-	-	-	-	-
ELECTRICITY	55.8	58.7	55.6	52.4	40.1	64.6e	64.6e
HEAT [2]	-	-	-	-	-	-	-

Source: IEA/OECD Energy Balances.

1. Includes non-energy use of petroleum products.
2. Includes only heat from combined heat and power plants.

5/D7 ENERGY USE IN OTHER SECTORS [1] BY FUEL

	1973	1978	1979	1980	1981	1985	1990
TOTAL (MTOE)	.2	.2	.2	.2	.1	.2e	.3e
SOLID FUELS	.0	-	-	-	-	-	-
OIL	.2	.1	.1	.1	.1	.1e	.2e
GAS	-	-	-	-	-	-	-
ELECTRICITY	.0	.1	.1	.1	.1	.1e	.1e
HEAT [2]	-	-	-	-	-	-	-
TOTAL (PERCENTS)	100.0	100.0	100.0	100.0	100.0	100.0	100.0
SOLID FUELS	.3	-	-	-	-	-	-
OIL	78.5	65.7	58.8	53.3	39.8	65.2e	65.4e
GAS	-	-	-	-	-	-	-
ELECTRICITY	21.2	34.3	41.2	46.7	60.2	34.8e	34.6e
HEAT [2]	-	-	-	-	-	-	-

Source: IEA/OECD Energy Balances.

1. Includes use in agricultural, commercial, public services and residential sectors.
2. Includes only heat from combined heat and power plants.

IRELAND

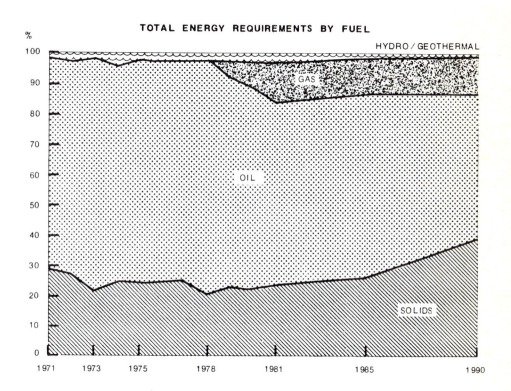

IRELAND

5/A1 TOTAL PRIMARY ENERGY REQUIREMENTS (TPER) BY FUEL

	1973	1978	1979	1980	1981	1985	1990
TPER (MTOE) [1]	7.4	7.7	9.0	8.6	8.7	10.9	12.9
SOLID FUELS	1.6	1.6	2.0	1.9	2.0	2.8	5.0
OIL	5.7	5.9	6.3	5.7	5.2	6.6	6.1
GAS	-	-	.5	.8	1.1	1.3	1.6
NUCLEAR	-	-	-	-	-	-	-
HYDRO/GEOTHERMAL	.2	.2	.3	.3	.3	.2	.2
OTHER [2]	-	-	-	-	-	-	-
TPER (PERCENTS) [1]	100.0	100.0	100.0	100.0	100.0	100.0	100.0
SOLID FUELS	21.4	20.2	22.1	22.1	23.3	25.8	39.1
OIL	76.4	76.6	69.4	65.9	60.0	60.5	47.3
GAS	-	-	5.2	8.7	13.2	12.0	12.2
NUCLEAR	-	-	-	-	-	-	-
HYDRO/GEOTHERMAL	2.2	3.2	3.3	3.3	3.5	1.7	1.5
OTHER [2]	-	-	-	-	-	-	-

Source: IEA/OECD Energy Balances and IEA Country Submissions (1982).

1. Net imports of electricity are only included in totals.
2. Includes other electricity generation (solar, wind, etc.) and traded synthetic fuels.

5/A2 TOTAL FINAL ENERGY CONSUMPTION (TFC) BY FUEL

	1973	1978	1979	1980	1981	1985	1990
TFC (MTOE)	5.4	5.9	6.9	6.5	6.7	8.3	9.7
SOLID FUELS	.9	1.0	1.4	1.3	1.6	1.7	2.2
OIL	3.9	4.2	4.4	4.1	4.0	4.9	5.3
GAS	.1	.1	.4	.4	.5	.7	.9
ELECTRICITY	.5	.7	.7	.7	.7	1.0	1.3
HEAT [1]	-	-	-	-	-	-	-
TFC (PERCENTS)	100.0	100.0	100.0	100.0	100.0	100.0	100.0
SOLID FUELS	17.0	16.6	20.1	19.6	23.3	20.9	22.3
OIL	71.4	70.6	63.9	62.4	59.1	59.4	55.2
GAS	1.9	1.4	5.2	6.7	6.7	8.2	9.7
ELECTRICITY	9.7	11.4	10.8	11.3	10.9	11.5	12.9
HEAT [1]	-	-	-	-	-	-	-

Source: IEA/OECD Energy Balances and IEA Country Submissions (1982).

1. Includes only heat from combined heat and power plants.

IRELAND

5/A3 TOTAL SOLID FUELS BALANCE
(MTOE)

	1973	1978	1979	1980	1981	1985	1990
PRODUCTION	1.1	1.1	.9	1.1	1.2	1.6	1.7
IMPORTS	.6	.4	.9	.8	.9	1.2	3.3
EXPORTS	-.1	-.1	-.1	.0	.0	-	-
MARINE BUNKERS	-	-	-	-	-	-	-
STOCK CHANGE	.0	.1	.3	.0	-.1	-	-
PRIMARY REQUIREMENTS	1.6	1.6	2.0	1.9	2.0	2.8	5.0
ELECTRICITY GENERATION	-.6	-.6	-.6	-.6	-.5	-1.1	-2.9
GAS MANUFACTURE	.0	.0	.0	.0	-	-	-
LIQUEFACTION	-	-	-	-	-	-	-
ENERGY SECTOR							
OWN USE + LOSSES [1]	-	.0	.0	.0	.1	-	-
FINAL CONSUMPTION	.9	1.0	1.4	1.3	1.6	1.7	2.2
INDUSTRY SECTOR	.0	.0	.1	.1	.1	.3	.5
IRON AND STEEL	.0	.0	.0	.0	.0	na	na
CHEMICAL/							
PETROCHEMICAL	-	-	-	-	-	na	na
OTHER INDUSTRY	.0	.0	.0	.1	.1	na	na
TRANSPORTATION SECTOR	-	-	-	-	-	-	-
OTHER SECTORS	.9	.9	1.3	1.2	1.4	1.4	1.7
AGRICULTURE	-	-	-	-	-	-	-
COMMERCIAL AND							
PUBLIC SERVICE	-	-	.1	.1	.0	-	-
RESIDENTIAL	.9	.9	1.2	1.1	1.4	-	-

Source: IEA/OECD Energy Balances and IEA Country Submissions (1982).

1. Includes statistical difference.

IRELAND

5/B1 SOLID FUELS INDIGENOUS PRODUCTION BY FUEL TYPE
(MTOE)

	1973	1978	1979	1980	1981	1985	1990
TOTAL SOLID FUELS	1.06	1.06	.93	1.08	1.23	1.61	1.73
HARD COAL	.04	.01	.04	.04	.05	.08	.08
COKING COAL	na	-	-	-	-	-	-
STEAM COAL	na	.01	.04	.04	.05	.08	.08
BROWN COAL/LIGNITE	-	-	-	-	-	-	-
OTHER SOLID FUELS	1.02	1.05	.89	1.04	1.18	1.53	1.65

Source: IEA/OECD Coal Statistics, IEA/OECD Energy Balances, IEA Country Submissions (1982).

5/B2 COAL PRODUCTION BY TYPE
(THOUSAND METRIC TONS)

	1973	1978	1979	1980	1981
TOTAL COAL	64	21	63	60	69
HARD COAL	64	21	63	60	69
COKING COAL	na	-	-	-	-
STEAM COAL	na	21	63	60	69
BROWN COAL/LIGNITE	-	-	-	-	-
DERIVED PRODUCTS:					
PATENT FUEL	-	-	-	-	-
COKE OVEN COKE	-	-	-	-	-
GAS COKE	37	30	27	-	-
BROWN COAL BRIQUETTES	-	337	340	308	341
COKE OVEN GAS(TCAL)	-	-	-	-	-
BLAST FURNACE GAS(TCAL)	-	-	-	-	-

Source: IEA/OECD Coal Statistics.

IRELAND

5/C1 SOLID FUELS TRADE BY TYPE
(MTOE)

	1973	1978	1979	1980	1981	1985	1990
IMPORTS							
TOTAL SOLID FUELS	.57	.40	.86	.84	.91	1.22	3.31
HARD COAL	.56	.40	.84	.83	.90	1.22	3.31
COKING COAL	-	-	-	-	-	-	-
STEAM COAL	-	.40	.84	.83	.90	1.22	3.31
BROWN COAL/LIGNITE	-	-	-	-	-	-	-
OTHER SOLID FUELS [1]	.01	-	.02	.01	.01	-	-
EXPORTS							
TOTAL SOLID FUELS	.07	.06	.07	.03	.01	-	-
HARD COAL	.05	.04	.05	.03	.01	-	-
COKING COAL	-	-	-	-	-	-	-
STEAM COAL	-	.04	.05	.03	.01	-	-
BROWN COAL/LIGNITE	-	-	-	-	-	-	-
OTHER SOLID FUELS [1]	.02	.02	.02	-	-	-	-

Source: IEA/OECD Coal Statistics, IEA/OECD Energy Balances, IEA Country Submissions (1982).

1. Includes derived coal products, e.g. coke.

5/C2 COAL IMPORTS BY ORIGIN
(THOUSAND METRIC TONS)

	1978	1979	1980	1981
HARD COAL IMPORTS	565	1206	1188	1289
COKING COAL IMPORTS	-	-	-	-
STEAM COAL IMPORTS FROM:	565	1206	1188	1289
AUSTRALIA	6	20	2	-
CANADA	-	-	-	-
GERMANY	-	8	9	29
UNITED KINGDOM	214	238	276	445
UNITED STATES	-	220	408	585
OTHER OECD	-	-	-	28
CZECHOSLOVAKIA	-	-	-	-
POLAND	331	706	483	184
USSR	-	1	-	-
CHINA	-	-	-	-
SOUTH AFRICA	11	13	10	12
OTHER NON-OECD	3	-	-	6

Source: IEA/OECD Coal Statistics.

IRELAND

5/D1 FUEL INPUT IN ELECTRICITY GENERATION

	1973	1978	1979	1980	1981	1985	1990
TOTAL (MTOE)	1.9	2.4	2.7	2.7	2.6	3.5	4.3
SOLID FUELS	.6	.6	.6	.6	.5	1.1	2.9
OIL	1.1	1.6	1.6	1.4	1.1	1.6	.6
GAS	-	-	.2	.4	.7	6	.6
NUCLEAR	-	-	-	-	-	-	-
HYDRO/GEOTHERMAL	.2	.2	.3	.3	.3	.2	.2
OTHER [1]	-	-	-	-	-	-	-
TOTAL (PERCENTS)	100.0	100.0	100.0	100.0	100.0	100.0	100.0
SOLID FUELS	34.3	24.4	21.6	21.8	20.1	31.3	66.2
OIL	56.9	65.3	60.2	53.0	40.2	45.1	14.9
GAS	-	-	7.3	14.7	28.2	18.1	14.5
NUCLEAR	-	-	-	-	-	-	-
HYDRO/GEOTHERMAL	8.8	10.3	10.9	10.6	11.4	5.5	4.4
OTHER [1]	-	-	-	-	-	-	-

Source: IEA/OECD Energy Balances and IEA Country Submissions (1982).

1. Solar, wind, etc.

5/D2 ELECTRICTY PRODUCTION BY FUEL TYPE

	1973	1978	1979	1980	1981	1985	1990
TOTAL (TWH)	7.3	10.0	11.0	10.9	10.9	13.7	17.4
SOLID FUELS	1.8	2.0	1.8	1.7	1.6	3.4	11.2
OIL	4.9	6.9	7.2	6.4	4.7	6.9	2.8
GAS	-	-	.8	1.6	3.4	2.7	2.7
NUCLEAR	-	-	-	-	-	-	-
HYDRO/GEOTHERMAL	.6	1.0	1.2	1.2	1.2	.7	.7
OTHER [1]	-	-	-	-	-	-	-
TOTAL (PERCENTS)	100.0	100.0	100.0	100.0	100.0	100.0	100.0
SOLID FUELS	24.9	20.1	16.2	15.6	14.5	25.0	64.2
OIL	66.3	69.6	65.8	59.0	42.9	50.3	16.4
GAS	-	-	7.1	14.8	31.2	19.5	15.3
NUCLEAR	-	-	-	-	-	-	-
HYDRO/GEOTHERMAL	8.8	10.3	10.9	10.6	11.4	5.2	4.1
OTHER [1]	-	-	-	-	-	-	-

Source: IEA/OECD Energy Balances and IEA Country Submissions (1982).

1. Solar, wind, etc.

IRELAND

5/D3 PROJECTIONS OF ELECTRICITY GENERATION CAPACITY BY FUEL
(GW)

	Nuclear	Hydro/ Geoth.	Solid Fuels	Oil	Gas	Multi Firing [1]	Total
I 1981							
Operating	-	0.5	0.4	1.7	0.4	0.2	3.2
II 1981-1985							
Capacity:							
Under Construction	-	-	0.3	-	-	0.4	0.7
Authorised	-	-	-	-	-	-	-
Other Planned	-	-	-	-	-	-	-
Conversion	-	-	-	-0.5	-	0.5	-
Decommissioning	-	-	-	-	-	-	-
Total Operating 1985	-	0.5	0.7	1.2	0.4	1.1	3.9
III 1986-1990							
Capacity:							
Under Construction	-	-	0.6	-	-	-	0.6
Authorised	-	-	-	-	-	-	-
Other Planned	-	-	-	-	-	-	-
Conversion	-	-	-	-	-0.3	0.3	-
Decommissioning	-	-	-	-	-	-	-
Total Operating 1990	-	0.5	1.3	1.2	0.1	1.4	4.5

Source: IEA/OECD Electricity Statistics and IEA Country Submissions (1982).

1. Includes "Other" capacity.

IRELAND

5/D6 ENERGY USE IN INDUSTRY [1] BY FUEL

	1973	1978	1979	1980	1981	1985	1990
TOTAL (MTOE)	1.5	1.7	2.2	2.1	2.1	3.3	4.0
SOLID FUELS	.0	.0	.1	.1	.1	.3	.5
OIL	1.3	1.4	1.5	1.4	1.3	2.0	2.2
GAS	.0	.0	.3	.4	.4	.5	.7
ELECTRICITY	.2	.3	.3	.3	.3	.4	.5
HEAT [2]	-	-	-	-	-	-	-
TOTAL (PERCENTS)	100.0	100.0	100.0	100.0	100.0	100.0	100.0
SOLID FUELS	2.8	2.0	2.9	4.8	6.7	10.1	12.6
OIL	83.3	82.2	70.6	64.7	62.1	62.4	55.4
GAS	1.6	1.1	13.5	17.5	18.7	16.2	18.9
ELECTRICITY	12.3	14.7	12.9	13.0	12.5	11.3	13.1
HEAT [2]	-	-	-	-	-	-	-

Source: IEA/OECD Energy Balances and IEA Country Submissions (1982).

1. Includes non-energy use of petroleum products.
2. Includes only heat from combined heat and power plants.

5/D7 ENERGY USE IN OTHER SECTORS [1] BY FUEL

	1973	1978	1979	1980	1981	1985	1990
TOTAL (MTOE)	2.5	2.4	2.9	2.6	2.8	3.0	3.4
SOLID FUELS	.9	.9	1.3	1.2	1.4	1.4	1.7
OIL	1.2	1.0	1.1	.9	.9	.8	.9
GAS	.1	.1	.1	.1	.1	.1	.2
ELECTRICITY	.3	.4	.5	.5	.5	.6	.7
HEAT [2]	-	-	-	-	-	-	-
TOTAL (PERCENTS)	100.0	100.0	100.0	100.0	100.0	100.0	100.0
SOLID FUELS	35.3	38.8	45.3	44.7	49.9	47.3	48.1
OIL	47.9	41.3	36.6	35.2	32.1	27.9	25.2
GAS	3.2	2.7	2.3	2.6	1.8	5.0	5.5
ELECTRICITY	13.6	17.2	15.8	17.5	16.3	19.8	21.2
HEAT [2]	-	-	-	-	-	-	-

Source: IEA/OECD Energy Balances and IEA Country Submissions (1982).

1. Includes use in agricultural, commercial, public services and residential sectors.
2. Includes only heat from combined heat and power plants.

ITALY

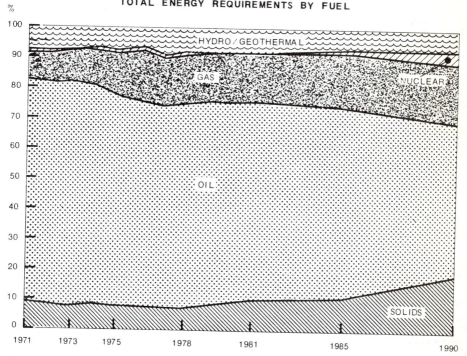

TOTAL ENERGY REQUIREMENTS BY FUEL

ITALY

5/A1 TOTAL PRIMARY ENERGY REQUIREMENTS (TPER) BY FUEL

	1973	1978	1979	1980	1981	1985	1990
TPER (MTOE) [1]	132.9	137.6	144.1	142.4	139.3	160.3	179.9
SOLID FUELS	10.3	10.3	11.8	13.6	14.4	17.7	34.0
OIL	97.8	92.3	97.0	93.4	90.2	100.7	89.3
GAS	14.5	22.8	23.1	23.1	22.3	28.9	35.0
NUCLEAR	.7	1.0	.6	.5	.6	2.0	8.0
HYDRO/GEOTHERMAL	9.5	11.1	11.2	11.2	10.9	10.5	11.6
OTHER [2]	-	-	-	-	-	.5	2.0
TPER (PERCENTS) [1]	100.0	100.0	100.0	100.0	100.0	100.0	100.0
SOLID FUELS	7.7	7.5	8.2	9.6	10.3	11.0	18.9
OIL	73.6	67.1	67.3	65.6	64.8	62.8	49.6
GAS	10.9	16.6	16.0	16.2	16.0	18.0	19.5
NUCLEAR	.5	.7	.4	.3	.4	1.2	4.4
HYDRO/GEOTHERMAL	7.1	8.0	7.8	7.9	7.8	6.6	6.4
OTHER [2]	-	-	-	-	-	.3	1.1

Source: IEA/OECD Energy Balances and IEA Country Submissions (1982).

1. Net imports of electricity are only included in totals.
2. Includes other electricity generation (solar, wind, etc.) and traded synthetic fuels.

5/A2 TOTAL FINAL ENERGY CONSUMPTION (TFC) BY FUEL

	1973	1978	1979	1980	1981	1985	1990
TFC (MTOE)	103.0	104.0	109.4	105.9	102.7	114.4	123.7
SOLID FUELS	6.3	5.3	5.4	5.8	6.0	8.5	9.0
OIL	72.8	65.7	70.2	65.5	62.8	61.6	58.6
GAS	13.3	20.2	20.3	20.8	20.2	26.0	31.6
ELECTRICITY	10.6	12.8	13.5	13.8	13.7	17.8	22.5
HEAT [1]	-	-	-	-	-	.5	2.0
TFC (PERCENTS)	100.0	100.0	100.0	100.0	100.0	100.0	100.0
SOLID FUELS	6.1	5.1	5.0	5.4	5.8	7.4	7.3
OIL	70.7	63.2	64.1	61.9	61.2	53.8	47.4
GAS	12.9	19.5	18.6	19.7	19.6	22.7	25.5
ELECTRICITY	10.3	12.3	12.3	13.0	13.4	15.6	18.2
HEAT [1]	-	-	-	-	-	.4	1.6

Source: IEA/OECD Energy Balances and IEA Country Submissions (1982).

1. Includes only heat production from combined heat and power plants.

ITALY

5/A3 TOTAL SOLID FUELS BALANCE
(MTOE)

	1973	1978	1979	1980	1981	1985	1990
PRODUCTION	1.8	1.0	1.1	.9	.9	1.3	2.0
IMPORTS	8.8	9.4	10.7	13.4	14.6	16.4	32.0
EXPORTS	-.4	-.5	-.5	-.5	-.5	-	-
MARINE BUNKERS	-	-	-	-	-	-	-
STOCK CHANGE	.1	.3	.5	-.2	-.6	-	-
PRIMARY REQUIREMENTS	10.3	10.3	11.8	13.6	14.4	17.7	34.0
ELECTRICITY GENERATION	-1.6	-2.7	-4.0	-5.0	-5.7	-6.7	-22.4
GAS MANUFACTURE	.0	.0	.0	.0	.0	-	-
LIQUEFACTION	-	-	-	-	-	-	-
ENERGY SECTOR							
OWN USE + LOSSES [1]	-2.4	-2.3	-2.4	-2.6	-2.5	-2.5	-2.6
FINAL CONSUMPTION	6.3	5.3	5.4	5.8	6.0	8.5	9.0
INDUSTRY SECTOR	4.6	4.5	4.5	5.1	5.4	6.7	7.4
IRON AND STEEL	3.5	3.7	3.5	3.6	3.6	na	na
CHEMICAL/							
PETROCHEMICAL	.2	.1	.2	.5	.4	na	na
OTHER INDUSTRY	.9	.8	.8	1.0	1.3	na	na
TRANSPORTATION SECTOR	.2	.0	.0	.0	.0	.1	.1
OTHER SECTORS	1.6	.7	.9	.7	.6	1.3	1.3
AGRICULTURE	-	-	-	-	-	-	-
COMMERCIAL AND							
PUBLIC SERVICE	-	-	-	-	.0	-	-
RESIDENTIAL	1.6	.7	.9	.7	.6	-	-

Source: IEA/OECD Energy Balances and IEA Country Submissions (1982).

1. Includes statistical difference.

ITALY

5/B1 SOLID FUELS INDIGENOUS PRODUCTION BY FUEL TYPE
(MTOE)

	1973	1978	1979	1980	1981	1985	1990
TOTAL SOLID FUELS	1.85	1.02	1.12	.93	.91	1.30	2.00
HARD COAL	-	-	-	-	-	-	-
COKING COAL							
STEAM COAL	-	-	-	-	-	-	-
BROWN COAL/LIGNITE	.30	.30	.35	.48	.46	0.85e	1.50
OTHER SOLID FUELS	1.55	.72	.77	.45	.45	0.45e	.50

Source: IEA/OECD Coal Statistics, IEA/OECD Energy Balances, IEA Country Submissions (1982).

5/B2 COAL PRODUCTION BY TYPE
(THOUSAND METRIC TONS)

	1973	1978	1979	1980	1981
TOTAL COAL	1190	1208	1412	1933	1850
HARD COAL	-	6	-	-	-
COKING COAL	-	-	-	-	-
STEAM COAL	-	6	-	-	-
BROWN COAL/LIGNITE	1190	1202	1412	1933	1850
DERIVED PRODUCTS:					
PATENT FUEL	47	12	22	10	10
COKE OVEN COKE	7665	7315	7501	8264	8054
GAS COKE	-	-	-	-	-
BROWN COAL BRIQUETTES	-	-	-	-	-
COKE OVEN GAS(TCAL)	15117	13099	13558	14511	14556
BLAST FURNACE GAS(TCAL)	13671	13895	14113	15190	15304

Source: IEA/OECD Coal Statistics.

ITALY

5/C1 SOLID FUELS TRADE BY TYPE
(MTOE)

	1973	1978	1979	1980	1981	1985	1990
IMPORTS							
TOTAL SOLID FUELS	8.82	9.41	10.70	13.41	14.60	16.40	32.00
HARD COAL	8.49	9.22	10.45	12.63	13.68	16.40	32.00
COKING COAL	na	7.41	7.32	8.32	7.97	8.00	8.00
STEAM COAL	na	1.81	3.14	4.31	5.72	8.40e	24.00e
BROWN COAL/LIGNITE	.03	.02	.01	.01	-	-	-
OTHER SOLID FUELS [1]	.30	.17	.24	.77	.92	-	-
EXPORTS							
TOTAL SOLID FUELS	.44	.49	.47	.54	.52	-	-
HARD COAL	-	-	-	-	-	-	-
COKING COAL	-	-	-	-	-	-	-
STEAM COAL	-	-	-	-	-	-	-
BROWN COAL/LIGNITE	-	-	-	-	-	-	-
OTHER SOLID FUELS [1]	.44	.49	.47	.54	.52	-	-

Source: IEA/OECD Coal Statistics, IEA/OECD Energy Balances, IEA Country Submissions (1982).

1.   Includes derived coal products, e.g. coke.

ITALY

5/C2 COAL IMPORTS BY ORIGIN
(THOUSAND METRIC TONS)

	1978	1979	1980	1981
HARD COAL IMPORTS	12458	14128	17061	18492
COKING COAL IMPORTS FROM:	10007	9888	11241	10764
AUSTRALIA	1347	892	1206	1546
CANADA	-	-	-	67
GERMANY	2537	1941	2513	2646
UNITED KINGDOM	-	-	-	48
UNITED STATES	3027	4232	5420	5725
OTHER OECD	-	-	438	-
CZECHOSLOVAKIA	-	-	-	-
POLAND	1525	1214	678	570
USSR	1036	925	986	162
CHINA	-	-	-	-
SOUTH AFRICA	-	-	-	-
OTHER NON-OECD	535	684	-	-
STEAM COAL IMPORTS FROM:	2451	4240	5820	7728
AUSTRALIA	-	-	-	-
CANADA	-	-	10	-
GERMANY	-	-	88	100
UNITED KINGDOM	52	10	-	19
UNITED STATES	-	-	1327	3482
OTHER OECD	-	-	-	17
CZECHOSLOVAKIA	-	-	-	-
POLAND	1437	2080	1210	417
USSR	-	-	-	55
CHINA	-	-	53	-
SOUTH AFRICA	960	2130	3132	3638
OTHER NON-OECD	2	10	-	-

Source: IEA/OECD Coal Statistics.

ITALY

5/D1 FUEL INPUT IN ELECTRICITY GENERATION

	1973	1978	1979	1980	1981	1985	1990
TOTAL (MTOE)	33.0	38.8	39.9	41.6	41.0	53.6	70.7
SOLID FUELS	1.6	2.7	4.0	5.0	5.7	6.7	22.4
OIL	20.2	21.6	21.7	22.8	21.8	31.4	23.8
GAS	1.0	2.3	2.4	2.0	1.9	2.5	2.9
NUCLEAR	.7	1.0	.6	.5	.6	2.0	8.0
HYDRO/GEOTHERMAL	9.5	11.1	11.2	11.2	10.9	10.5	11.6
OTHER [1]	-	-	-	-	-	.5	2.0
TOTAL (PERCENTS)	100.0	100.0	100.0	100.0	100.0	100.0	100.0
SOLID FUELS	4.8	7.0	10.0	12.1	13.9	12.5	31.7
OIL	61.1	55.9	54.4	54.9	53.3	58.6	33.7
GAS	3.2	6.1	6.1	4.9	4.6	4.7	4.1
NUCLEAR	2.2	2.5	1.4	1.2	1.5	3.7	11.3
HYDRO/GEOTHERMAL	28.7	28.5	28.0	27.0	26.6	19.6	16.4
OTHER [1]	-	-	-	-	-	.9	2.8

Source: IEA/OECD Energy Balances and IEA Country Submissions (1982).

1. Solar, wind, etc.

5/D2 ELECTRICTY PRODUCTION BY FUEL TYPE

	1973	1978	1979	1980	1981	1985	1990
TOTAL (TWH)	144.8	175.0	181.3	185.7	181.7	231.0	301.0
SOLID FUELS	6.5	11.6	16.3	19.5	22.5	25.6	95.7
OIL	89.0	98.1	100.5	104.6	99.3	134.0	101.0
GAS	4.5	11.0	11.1	9.2	8.7	10.2	12.0
NUCLEAR	3.1	4.4	2.6	2.2	2.7	9.0	35.0
HYDRO/GEOTHERMAL	41.6	49.9	50.7	50.2	48.4	52.2	57.3
OTHER [1]	-	-	-	-	-	-	-
TOTAL (PERCENTS)	100.0	100.0	100.0	100.0	100.0	100.0	100.0
SOLID FUELS	4.5	6.6	9.0	10.5	12.4	11.1	31.8
OIL	61.5	56.1	55.4	56.3	54.7	58.0	33.6
GAS	3.1	6.3	6.1	5.0	4.8	4.4	4.0
NUCLEAR	2.2	2.5	1.4	1.2	1.5	3.9	11.6
HYDRO/GEOTHERMAL	28.7	28.5	28.0	27.0	26.6	22.6	19.0
OTHER [1]	-	-	-	-	-	-	-

Source: IEA/OECD Energy Balances and IEA Country Submissions (1982).

1. Solar, wind, etc.

ITALY

5/D3 PROJECTIONS OF ELECTRICITY GENERATION CAPACITY BY FUEL
(GW)

	Nuclear	Hydro/ Geoth.	Coal	Oil	Gas	Multi Firing [1]	Total
I 1981							
Operating	1.3	16.2	5.0	16.6	0.4	8.2	47.7
II 1981-1985							
Capacity:							
Under Construction	0.4	3.1	-	-	7.8	-	11.3
Authorised	-	-	-	-	-	-	-
Other Planned	-	-	-	-	-	1.5	-
Conversion	-	-	2.3	-2.3	-	-	1.5
Decommissioning	-	-	-	-	-	-	-
Total Operating 1985	1.7	19.3	7.3	14.3	8.2	9.7	60.5
III 1986-1990							
Capacity:							
Under Construction	2.0	2.2	-	-	-	-	4.2
Authorised	2.0	-	11.5	-	-	-	13.5
Other Planned	-	2.0	-	-	0.5	1.1	3.6
Conversion	-	-	0.9	-0.9	-	-	-
Decommissioning	-	-	-	-	-	-	-
Total Operating 1990	5.7	23.5	19.7	13.4	8.7	10.8	81.8

Source: IEA/OECD Electricity Statistics and IEA Country Submissions (1982).

1.   Includes "Other" capacity.

ITALY

5/D4 ENERGY USE IN IRON AND STEEL INDUSTRY BY FUEL

	1973	1978	1979	1980	1981	1985	1990
TOTAL (MTOE)	7.7	8.5	8.5	8.4	7.9	na	na
SOLID FUELS	3.5	3.7	3.5	3.6	3.6	na	na
OIL	1.1	1.0	1.1	1.1	.9	na	na
GAS	1.9	2.2	2.2	2.1	1.9	na	na
ELECTRICITY	1.2	1.6	1.7	1.7	1.6	na	na
HEAT [1]	-	-	-	-	-	na	na
TOTAL (PERCENTS)	100.0	100.0	100.0	100.0	100.0	na	na
SOLID FUELS	45.7	43.0	41.2	42.6	45.5	na	na
OIL	14.2	11.7	13.4	12.8	10.8	na	na
GAS	24.7	26.4	25.4	24.3	23.8	na	na
ELECTRICITY	15.4	18.9	20.0	20.2	19.8	na	na
HEAT [1]	-	-	-	-	-	na	na

Source: IEA/OECD Energy Balances and IEA Country Submissions (1982).

1. Includes only heat from combined heat and power plants.

5/D5 ENERGY USE IN OTHER INDUSTRIES BY FUEL

	1973	1978	1979	1980	1981	1985	1990
TOTAL (MTOE)	43.6	38.4	40.9	38.2	35.9	na	na
SOLID FUELS	1.1	.9	1.0	1.5	1.8	na	na
OIL	29.9	21.9	24.1	20.9	19.2	na	na
GAS	7.2	9.6	9.5	9.4	8.7	na	na
ELECTRICITY	5.5	6.0	6.3	6.4	6.2	na	na
HEAT [1]	-	-	-	-	-	na	na
TOTAL (PERCENTS)	100.0	100.0	100.0	100.0	100.0	na	na
SOLID FUELS	2.5	2.3	2.5	3.8	4.9	na	na
OIL	68.6	57.1	59.0	54.8	53.5	na	na
GAS	16.4	25.0	23.1	24.7	24.2	na	na
ELECTRICITY	12.5	15.6	15.3	16.7	17.3	na	na
HEAT [1]	-	-	-	-	-	na	na

Source: IEA/OECD Energy Balances and IEA Country Submissions (1982).

1. Includes only heat from combined heat and power plants.

ITALY

5/D6 ENERGY USE IN INDUSTRY [1] BY FUEL

	1973	1978	1979	1980	1981	1985	1990
TOTAL (MTOE)	51.3	46.9	49.4	46.6	43.8	50.7	55.0
SOLID FUELS	4.6	4.5	4.5	5.1	5.4	7.1	7.6
OIL	31.0	22.9	25.3	22.0	20.1	17.8	17.5
GAS	9.1	11.8	11.6	11.5	10.6	15.6	17.7
ELECTRICITY	6.6	7.6	8.0	8.1	7.8	10.1	11.8
HEAT [2]	-	-	-	-	-	.1	.4
TOTAL (PERCENTS)	100.0	100.0	100.0	100.0	100.0	100.0	100.0
SOLID FUELS	9.0	9.7	9.1	10.8	12.2	14.0	13.8
OIL	60.4	48.9	51.2	47.2	45.8	35.1	31.8
GAS	17.7	25.2	23.5	24.6	24.1	30.8	32.2
ELECTRICITY	13.0	16.2	16.1	17.3	17.8	19.9	21.5
HEAT [2]	-	-	-	-	-	.2	.7

Source: IEA/OECD Energy Balances and IEA Country Submissions (1982).

1. Includes non-energy use of petroleum products.
2. Includes only heat from combined heat and power plants.

5/D7 ENERGY USE IN OTHER SECTORS [1] BY FUEL

	1973	1978	1979	1980	1981	1985	1990
TOTAL (MTOE)	32.1	34.7	35.5	34.9	34.5	36.6	39.6
SOLID FUELS	1.6	.7	.9	.7	.6	1.3	1.3
OIL	22.9	21.1	21.1	19.9	19.0	17.5	13.1
GAS	4.1	8.2	8.5	9.1	9.3	10.1	13.5
ELECTRICITY	3.6	4.7	5.0	5.2	5.5	7.3	10.1
HEAT [2]	-	-	-	-	-	.4	1.6
TOTAL (PERCENTS)	100.0	100.0	100.0	100.0	100.0	100.0	100.0
SOLID FUELS	4.9	2.0	2.5	2.0	1.8	3.6	3.3
OIL	71.2	60.9	59.4	56.9	55.3	47.8	33.1
GAS	12.7	23.6	23.9	26.1	27.0	27.6	34.1
ELECTRICITY	11.3	13.6	14.2	15.0	15.8	19.9	25.5
HEAT [2]	-	-	-	-	-	1.1	4.0

Source: IEA/OECD Energy Balances and IEA Country Submissions (1982).

1. Includes use in agricultural, commercial, public services and residential sectors.
2. Includes only heat from combined heat and power plants.

LUXEMBOURG

TOTAL ENERGY REQUIREMENTS BY FUEL

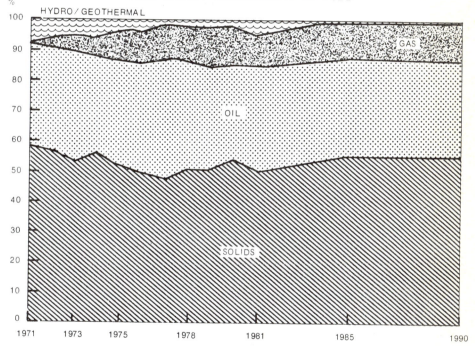

LUXEMBOURG

5/A1 TOTAL PRIMARY ENERGY REQUIREMENTS (TPER) BY FUEL

	1973	1978	1979	1980	1981	1985	1990
TPER (MTOE) [1]	4.8	4.2	3.9	3.7	3.3	4.0	4.2
SOLID FUELS	2.4	2.0	1.8	1.8	1.5	2.0	2.1
OIL	1.7	1.4	1.3	1.1	1.0	1.2	1.3
GAS	.2	.5	.5	.4	.3	.4	.5
NUCLEAR	-	-	-	-	-	-	-
HYDRO/GEOTHERMAL [2]	.3	.1	.1	.1	.2	.0	.0
OTHER [3]	-	-	-	-	-	-	-
TPER (PERCENTS) [1]	100.0	100.0	100.0	100.0	100.0	100.0	100.0
SOLID FUELS	50.5	47.3	46.7	49.7	45.8	50.3	50.0
OIL	35.4	33.8	32.4	29.5	31.5	30.2	29.8
GAS	4.7	11.1	12.2	11.8	10.0	11.3	11.9
NUCLEAR	-	-	-	-	-	-	-
HYDRO/GEOTHERMAL [2]	5.7	2.4	2.7	2.4	5.1	.8	.7
OTHER [3]	-	-	-	-	-	-	-

Source: IEA/OECD Energy Balances and IEA Country Submissions (1982).

1. Net imports of electricity are only included in totals.
2. Includes Vianden pumped storage plant, not connected to Luxembourg grid.
3. Includes other electricity generation (solar, wind, etc.) and traded synthetic fuels.

5/A2 TOTAL FINAL ENERGY CONSUMPTION (TFC) BY FUEL

	1973	1978	1979	1980	1981	1985	1990
TFC (MTOE)	4.1	3.7	3.5	3.3	3.0	3.7	4.0
SOLID FUELS	2.1	1.7	1.6	1.6	1.3	1.8	1.9
OIL	1.5	1.4	1.3	1.1	1.0	1.2	1.3
GAS	.2	.3	.3	.4	.3	.4	.4
ELECTRICITY	.3	.3	.3	.3	.3	.3	.4
HEAT [1]	-	-	-	-	-	-	-
TFC (PERCENTS)	100.0	100.0	100.0	100.0	100.0	100.0	100.0
SOLID FUELS	51.7	46.8	46.1	47.4	45.6	49.3	47.7
OIL	37.7	36.6	35.7	32.3	34.3	32.7	31.3
GAS	4.4	8.7	9.7	11.1	10.2	9.8	10.5
ELECTRICITY	6.2	7.8	8.5	9.2	9.9	8.2	10.5
HEAT [1]	-	-	-	-	-	-	-

Source: IEA/OECD Energy Balances and IEA Country Submissions (1982).

1. Includes only heat production from combined heat and power plants.

LUXEMBOURG

5/A3 TOTAL SOLID FUELS BALANCE
(MTOE)

	1973	1978	1979	1980	1981	1985	1990
PRODUCTION	-	.0	.0	.0	.0	-	-
IMPORTS	2.4	1.9	1.9	1.8	1.4	2.0	2.1
EXPORTS	-	-	-	-	-	-	-
MARINE BUNKERS	-	-	-	-	-	-	-
STOCK CHANGE	.0	.0	.0	.0	.0	-	-
PRIMARY REQUIREMENTS	2.4	2.0	1.8	1.8	1.5	2.0	2.1
ELECTRICITY GENERATION	-.3	-.1	-.2	-.2	-.1	-.2	-.2
GAS MANUFACTURE	-	-	-	-	-	-	-
LIQUEFACTION	-	-	-	-	-	-	-
ENERGY SECTOR							
OWN USE + LOSSES [1]	.0	-.1	-.1	-.1	.0	-	-
FINAL CONSUMPTION	2.1	1.7	1.6	1.6	1.3	1.8	1.9
INDUSTRY SECTOR	2.1	1.7	1.6	1.5	1.3	1.8	1.9
IRON AND STEEL	2.1	1.7	1.6	1.5	1.3	na	na
CHEMICAL/							
PETROCHEMICAL	-	-	-	-	-	na	na
OTHER INDUSTRY	.0	.0	.0	.1	.1	na	na
TRANSPORTATION SECTOR	-	-	-	-	-	-	-
OTHER SECTORS	.0	.0	.0	.0	.0	-	-
AGRICULTURE	-	-	-	-	-	-	-
COMMERCIAL AND							
PUBLIC SERVICE	-	-	-	-	-	-	-
RESIDENTIAL	.0	.0	.0	.0	.0	-	-

Source: IEA/OECD Energy Balances and IEA Country Submissions (1982).

1. Includes statistical difference.

LUXEMBOURG

5/C1 SOLID FUELS TRADE BY TYPE
(MTOE)

	1973	1978	1979	1980	1981	1985	1990
IMPORTS							
TOTAL SOLID FUELS	2.41	1.94	1.86	1.82	1.44	2.00	2.10
HARD COAL	.21	.35	.24	.25	.21	.30e	.30e
COKING COAL	na	-	-	-	-	-	-
STEAM COAL	na	.35	.24	.25	.21	.30e	.30e
BROWN COAL/LIGNITE	-	.01	.01	.01	-	-	-
OTHER SOLID FUELS [1]	2.20	1.58	1.61	1.56	1.23	1.70e	1.80e

Source: IEA/OECD Coal Statistics, IEA/OECD Energy Balances, IEA Country Submissions (1982).

1.   Includes derived coal products, e.g. coke.

5/C2 COAL IMPORTS BY ORIGIN
(THOUSAND METRIC TONS)

	1978	1979	1980	1981
HARD COAL IMPORTS	495	349	364	297
COKING COAL	-	-	-	-
STEAM COAL IMPORTS FROM:	495	349	364	297
AUSTRALIA	-	-	-	-
CANADA	-	-	-	-
GERMANY	299	146	75	12
UNITED KINGDOM	12	15	50	55
UNITED STATES	1	16	171	100
OTHER OECD	6	16	23	6
CZECHOSLOVAKIA	-	-	-	-
POLAND	-	-	-	-
USSR	-	-	-	18
CHINA	-	-	-	-
SOUTH AFRICA	125	139	45	106
OTHER NON-OECD	52	17	-	-

Source: IEA/OECD Coal Statistics.

LUXEMBOURG

5/D1 FUEL INPUT IN ELECTRICITY GENERATION

	1973	1978	1979	1980	1981	1985	1990
TOTAL (MTOE)	.7	.4	.4	.3	.3	.3	.3
SOLID FUELS	.3	.1	.2	.2	.1	.2	.2
OIL	.1	.0	.0	.0	.0	-	-
GAS	.0	.1	.1	.1	.0	.1	.1
NUCLEAR	-	-	-	-	-	-	-
HYDRO/GEOTHERMAL [1]	.3	.1	.1	.1	.2	.0	.0
OTHER [2]	-	-	-	-	-	-	-
TOTAL (PERCENTS)	100.0	100.0	100.0	100.0	100.0	100.0	100.0
SOLID FUELS	40.4	35.0	36.5	49.4	38.8	61.3	63.3
OIL	15.3	9.4	6.9	4.4	4.9	-	-
GAS	5.9	32.4	31.8	20.0	8.4	29.0	26.7
NUCLEAR	-	-	-	-	-	-	-
HYDRO/GEOTHERMAL [1]	38.4	23.2	24.8	26.1	47.9	9.7	10.0
OTHER [2]	-	-	-	-	-	-	-

Source: IEA/OECD Energy Balances and IEA Country Submissions (1982).

1. Includes Vianden pumped storage plant, not connected to Luxembourg grid.
2. Solar, wind, etc.

5/D2 ELECTRICTY PRODUCTION BY FUEL TYPE

	1973	1978	1979	1980	1981	1985	1990
TOTAL (TWH)	2.2	1.4	1.3	1.1	1.2	1.2	1.2
SOLID FUELS	.8	.4	.4	.5	.5	.7	.7
OIL	.4	.2	.2	.1	.1	-	-
GAS	.1	.4	.4	.2	.1	.3	.3
NUCLEAR	-	-	-	-	-	-	-
HYDRO/GEOTHERMAL [1]	.8	.3	.3	.3	.6	.1	.1
OTHER [2]	-	-	-	-	-	-	-
TOTAL (PERCENTS)	100.0	100.0	100.0	100.0	100.0	100.0	100.0
SOLID FUELS	37.5	28.0	27.9	45.4	38.2	61.3	63.4
OIL	17.6	17.7	17.3	9.0	6.3	-	-
GAS	6.5	31.2	30.0	19.5	7.5	29.1	26.7
NUCLEAR	-	-	-	-	-	-	-
HYDRO/GEOTHERMAL [1]	38.4	23.2	24.8	26.1	47.9	9.6	10.0
OTHER [2]	-	-	-	-	-	-	-

Source: IEA/OECD Energy Balances and IEA Country Submissions (1982).

1. Includes Vianden pumped storage plant, not connected to Luxembourg grid.
2. Solar, wind, etc.

LUXEMBOURG

5/D3 PROJECTIONS OF ELECTRICITY GENERATION CAPACITY BY FUEL
(GW)

	Nuclear	Hydro/ Geoth.	Solid Fuels	Oil	Gas	Multi Firing	Total
I 1981							
Operating	-	1.1	-	-	-	0.2	1.3
II 1981-1985							
Capacity:							
Under Construction							
Authorised							
Other Planned			na				
Conversion							
Decommissioning							
Total Operating 1985							
III 1986-1990							
Capacity:							
Under Construction							
Authorised							
Other Planned			na				
Conversion							
Decommissioning							
Total Operating 1990							

Source: IEA/OECD Electricity Statistics and IEA Country Submissions (1982).

LUXEMBOURG

5/D4 ENERGY USE IN IRON AND STEEL INDUSTRY BY FUEL

	1973	1978	1979	1980	1981	1985	1990
TOTAL (MTOE)	3.0	2.4	2.2	1.9	1.6	na	na
SOLID FUELS	2.1	1.7	1.6	1.5	1.3	na	na
OIL	.6	.3	.2	.1	.0	na	na
GAS	.1	.2	.2	.2	.2	na	na
ELECTRICITY	.2	.1	.1	.1	.1	na	na
HEAT [1]	-	-	-	-	-	-	-
TOTAL (PERCENTS)	100.0	100.0	100.0	100.0	100.0	na	na
SOLID FUELS	69.1	70.7	72.6	76.3	79.1	na	na
OIL	21.2	14.3	10.0	4.2	2.9	na	na
GAS	4.3	9.4	11.0	12.5	10.3	na	na
ELECTRICITY	5.4	5.6	6.3	7.1	7.7	na	na
HEAT [1]	-	-	-	-	-	-	-

Source: IEA/OECD Energy Balances and IEA Country Submissions (1982).

1. Includes only heat from combined heat and power plants.

5/D5 ENERGY USE IN OTHER INDUSTRIES BY FUEL

	1973	1978	1979	1980	1981	1985	1990
TOTAL (MTOE)	.3	.3	.3	.3	.3	na	na
SOLID FUELS	.0	.0	.0	.1	.1	na	na
OIL	.2	.2	.2	.1	.1	na	na
GAS	.0	.0	.0	.0	.0	na	na
ELECTRICITY	.0	.1	.1	.1	.1	na	na
HEAT [1]	-	-	-	-	-	-	-
TOTAL (PERCENTS)	100.0	100.0	100.0	100.0	100.0	na	na
SOLID FUELS	7.2	4.2	13.1	25.5	19.7	na	na
OIL	70.8	67.5	58.9	46.9	50.2	na	na
GAS	5.8	1.9	1.5	4.0	4.4	na	na
ELECTRICITY	16.2	26.4	26.5	23.5	25.7	na	na
HEAT [1]	-	-	-	-	-	-	-

Source: IEA/OECD Energy Balances and IEA Country Submissions (1982).

1. Includes only heat from combined heat and power plants.

LUXEMBOURG

5/D6 ENERGY USE IN INDUSTRY [1] BY FUEL

	1973	1978	1979	1980	1981	1985	1990
TOTAL (MTOE)	3.2	2.7	2.4	2.2	1.9	2.5	2.8
SOLID FUELS	2.1	1.7	1.6	1.5	1.3	1.8	1.9
OIL	.8	.5	.4	.2	.2	.3	.5
GAS	.1	.2	.2	.3	.2	.2	.2
ELECTRICITY	.2	.2	.2	.2	.2	.2	.3
HEAT [2]	-	-	-	-	-	-	-
TOTAL (PERCENTS)	100.0	100.0	100.0	100.0	100.0	100.0	100.0
SOLID FUELS	64.3	64.0	65.9	69.4	69.9	73.3	68.2
OIL	25.1	19.7	15.5	9.9	10.2	11.3	16.8
GAS	4.4	8.6	9.9	11.3	9.4	7.3	6.1
ELECTRICITY	6.3	7.7	8.6	9.3	10.5	8.1	8.9
HEAT [2]	-	-	-	-	-	-	-

Source: IEA/OECD Energy Balances and IEA Country Submissions (1982). Disaggregated demand figures for 1985 and 1990 are Secretariat estimates.

1. Includes non-energy use of petroleum products.
2. Includes only heat from combined heat and power plants.

5/D7 ENERGY USE IN OTHER SECTORS [1] BY FUEL

	1973	1978	1979	1980	1981	1985	1990
TOTAL (MTOE)	.6	.6	.6	.6	.6	.6	.6
SOLID FUELS	.0	.0	.0	.0	.0	-	-
OIL	.4	.4	.4	.4	.3	.3	.2
GAS	.0	.1	.1	.1	.1	.2	.3
ELECTRICITY	.0	.1	.1	.1	.1	.1	.2
HEAT [2]	-	-	-	-	-	-	-
TOTAL (PERCENTS)	100.0	100.0	100.0	100.0	100.0	100.0	100.0
SOLID FUELS	6.0	3.6	3.9	4.1	6.1	-	-
OIL	78.5	68.6	67.0	61.5	54.8	53.3	30.0
GAS	6.8	14.9	15.7	19.1	22.6	30.0	41.7
ELECTRICITY	8.7	12.8	13.4	15.4	16.4	16.7	28.3
HEAT [2]	-	-	-	-	-	-	-

Source: IEA/OECD Energy Balances and IEA Country Submissions (1982). Disaggregated demand figures for 1985 and 1990 are Secretariat estimates.

1. Includes use in agricultural, commercial, public services and residential sectors.
2. Includes only heat from combined heat and power plants.

NETHERLANDS

TOTAL ENERGY REQUIREMENTS BY FUEL

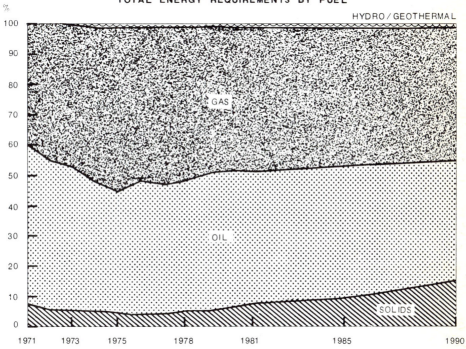

NETHERLANDS

5/A1 TOTAL PRIMARY ENERGY REQUIREMENTS (TPER) BY FUEL

	1973	1978	1979	1980	1981	1985	1990
TPER (MTOE) [1]	61.9	65.9	69.5	65.7	61.8	64.3	70.0
SOLID FUELS	3.2	3.4	3.6	4.2	4.6	5.9	10.5
OIL	29.5	28.5	31.7	29.5	27.0	28.1	27.4
GAS	29.0	33.2	33.4	31.0	29.5	29.3	30.7
NUCLEAR	.3	.9	.7	.9	.8	1.0	1.0
HYDRO/GEOTHERMAL	-	-	-	-	-	-	-
OTHER [2]	-	-	-	-	-	-	-
TPER (PERCENTS) [1]	100.0	100.0	100.0	100.0	100.0	100.0	100.0
SOLID FUELS	5.2	5.1	5.2	6.4	7.4	9.2	15.0
OIL	47.7	43.2	45.6	45.0	43.7	43.7	39.1
GAS	46.9	50.3	48.1	47.2	47.7	45.6	43.9
NUCLEAR	.4	1.3	1.1	1.4	1.2	1.6	1.4
HYDRO/GEOTHERMAL	-	-	-	-	-	-	-
OTHER [2]	-	-	-	-	-	-	-

Source: IEA/OECD Energy Balances and IEA Country Submissions (1982).

1. Net imports of electricity are only included in totals.
2. Includes other electricity generation (solar, wind, etc.) and traded synthetic fuels.

5/A2 TOTAL FINAL ENERGY CONSUMPTION (TFC) BY FUEL

	1973	1978	1979	1980	1981	1985	1990
TFC (MTOE)	50.2	55.5	59.3	51.4	48.7	51.8	58.2
SOLID FUELS	2.1	1.6	1.7	1.6	1.8	2.3	3.5
OIL	24.6	24.4	26.5	20.2	18.3	21.2	23.5
GAS	19.7	24.7	26.2	24.7	23.7	22.8	25.0
ELECTRICITY	3.8	4.7	4.9	4.9	4.9	4.8	5.0
HEAT [1]	-	-	-	-	-	.7	1.2
TFC (PERCENTS)	100.0	100.0	100.0	100.0	100.0	100.0	100.0
SOLID FUELS	4.3	2.9	2.9	3.1	3.6	4.4	6.0
OIL	49.0	44.0	44.6	39.3	37.5	40.9	40.4
GAS	39.1	44.6	44.1	48.0	48.8	44.0	43.0
ELECTRICITY	7.6	8.5	8.3	9.6	10.0	9.3	8.6
HEAT [1]	-	-	-	-	-	1.4	2.1

Source: IEA/OECD Energy Balances and IEA Country Submissions (1982).

1. Includes only heat production from combined heat and power plants.

NETHERLANDS

5/A3 TOTAL SOLID FUELS BALANCE
(MTOE)

	1973	1978	1979	1980	1981	1985	1990
PRODUCTION	1.3	-	-	-	-	.2	.2
IMPORTS	3.2	3.9	5.0	5.7	6.3	6.2	10.8
EXPORTS	-1.5	-.8	-1.2	-1.6	-1.2	-.5	-.5
MARINE BUNKERS	-	-	-	-	-	-	-
STOCK CHANGE	.2	.3	-.2	.1	-.5	-	-
PRIMARY REQUIREMENTS	3.2	3.4	3.6	4.2	4.6	5.9	10.5
ELECTRICITY GENERATION	-.9	-1.5	-1.2	-2.0	-2.4	-2.9	-6.2
GAS MANUFACTURE	-	-	-	-	-	-	-
LIQUEFACTION	-	-	-	-	-	-	-
ENERGY SECTOR OWN USE + LOSSES [1]	-.2	-.3	-.7	-.6	-.4	-.7	-.8
FINAL CONSUMPTION	2.1	1.6	1.7	1.6	1.8	2.3	3.5
INDUSTRY SECTOR	1.8	1.5	1.6	1.5	1.7	2.3	3.3
IRON AND STEEL	1.5	1.3	1.4	1.2	1.4	1.7	2.5
CHEMICAL/ PETROCHEMICAL	.2	.1	.2	.1	.1	.6	.8
OTHER INDUSTRY	.1	.1	.1	.1	.1	-	-
TRANSPORTATION SECTOR	-	-	-	-	-	-	-
OTHER SECTORS	.4	.1	.1	.1	.1	-	.2
AGRICULTURE	-	-	-	-	-	-	-
COMMERCIAL AND PUBLIC SERVICE	-	-	-	-	-	-	.2
RESIDENTIAL	.3	.1	.0	.1	.0	-	-

Source: IEA/OECD Energy Balances and IEA Country Submissions (1982).

1. Includes statistical difference.

NETHERLANDS

5/B1 SOLID FUELS INDIGENOUS PRODUCTION BY FUEL TYPE
(MTOE)

	1973	1978	1979	1980	1981	1985	1990
TOTAL SOLID FUELS	1.28	-	-	-	-	.20	.20
HARD COAL	1.28	-	-	-	-	-	-
COKING COAL	na	-	-	-	-	-	-
STEAM COAL	na	-	-	-	-	-	-
BROWN COAL/LIGNITE	-	-	-	-	-	-	-
OTHER SOLID FUELS	-	-	-	-	-	.20	.20

Source: IEA/OECD Coal Statistics, IEA/OECD Energy Balances, IEA Country Submissions (1982).

5/B2 COAL PRODUCTION BY TYPE
(THOUSAND METRIC TONS)

	1973	1978	1979	1980	1981
TOTAL COAL	1829	-	-	-	-
HARD COAL	1829	-	-	-	-
COKING COAL	na	-	-	-	-
STEAM COAL	na	-	-	-	-
BROWN COAL/LIGNITE	-	-	-	-	-
DERIVED PRODUCTS:					
PATENT FUEL	250	-	-	-	-
COKE OVEN COKE	2655	2401	2529	2455	2242
GAS COKE	-	-	-	-	-
BROWN COAL BRIQUETTES	-	-	-	-	-
COKE OVEN GAS(TCAL)	5474	4605	5037	4912	4538
BLAST FURNACE GAS(TCAL)	10655	5869	6677	5776	6076

Source: IEA/OECD Coal Statistics.

NETHERLANDS

5/C1 SOLID FUELS TRADE BY TYPE
(MTOE)

	1973	1978	1979	1980	1981	1985	1990
IMPORTS							
TOTAL SOLID FUELS	3.24	3.90	5.02	5.68	6.32	6.20	10.80
HARD COAL	2.78	3.53	4.35	5.01	5.56	6.00	10.80
COKING COAL	na	2.00	2.55	2.38	2.24	2.30e	4.10e
STEAM COAL	na	1.53	1.80	2.63	3.32	3.70e	6.70e
BROWN COAL/LIGNITE	-	.03	.07	.07	.07	-	-
OTHER SOLID FUELS [1]	.46	.34	.60	.60	.69	.20	-
EXPORTS							
TOTAL SOLID FUELS	1.51	.83	1.23	1.55	1.18	.50	.50
HARD COAL	.87	.34	.73	1.08	.64	.50	.50
COKING COAL	na	.01	.02	.03	.02	-	-
STEAM COAL	na	.33	.71	1.06	.61	.50e	.50e
BROWN COAL/LIGNITE	-	-	-	-	-	-	-
OTHER SOLID FUELS [1]	.64	.49	.50	.47	.54	-	-

Source: IEA/OECD Coal Statistics, IEA/OECD Energy Balances, IEA Country Submissions (1982).

1. Includes derived coal products, e.g. coke.

NETHERLANDS

5/C2 COAL IMPORTS BY ORIGIN
(THOUSAND METRIC TONS)

	1978	1979	1980	1981
HARD COAL IMPORTS	5038	6217	7156	7949
COKING COAL IMPORTS FROM:	2853	3642	3399	3207
AUSTRALIA	961	693	382	518
CANADA	-	-	-	76
GERMANY	782	644	670	513
UNITED KINGDOM	-	108	-	47
UNITED STATES	644	1565	1749	1709
OTHER OECD	1	-	1	8
CZECHOSLOVAKIA	89	81	114	40
POLAND	376	487	482	-
USSR	-	7	-	-
CHINA	-	-	-	-
SOUTH AFRICA	-	57	-	7
OTHER NON-OECD	-	-	1	289
STEAM COAL IMPORTS FROM:	2185	2575	3757	4742
AUSTRALIA	529	673	792	839
CANADA	55	-	-	2
GERMANY	703	825	706	676
UNITED KINGDOM	166	81	156	429
UNITED STATES	14	20	1053	2340
OTHER OECD	10	31	63	27
CZECHOSLOVAKIA	62	53	68	14
POLAND	264	218	460	154
USSR	57	-	-	29
CHINA	-	-	-	7
SOUTH AFRICA	309	645	459	224
OTHER NON-OECD	16	29	-	1

Source: IEA/OECD Coal Statistics.

NETHERLANDS

5/C3 HARD COAL EXPORTS BY DESTINATION
(THOUSAND METRIC TONS)

	1978	1979	1980	1981
TOTAL EXPORTS	488	1042	1547	910
EXPORTS TO:				
AUSTRALIA	-	-	-	-
AUSTRIA	-	-	-	-
BELGIUM	244	466	584	377
CANADA	-	-	-	-
DENMARK	-	61	59	14
FINLAND	-	-	173	15
FRANCE	9	28	87	27
GERMANY	228	396	409	1
GREECE	-	-	-	-
ICELAND	-	-	-	-
IRELAND	-	3	-	4
ITALY	-	-	1	-
JAPAN	-	-	-	306
LUXEMBOURG	-	13	-	-
NETHERLANDS	-	-	-	-
NEW ZEALAND	-	-	-	-
NORWAY	-	-	-	3
PORTUGAL	-	1	4	-
SPAIN	-	-	-	-
SWEDEN	-	-	24	16
SWITZERLAND	5	36	121	72
TURKEY	-	-	-	15
UNITED KINGDOM	2	9	70	3
UNITED STATES	-	-	-	-
TOTAL NON-OECD	-	29	15	57

Source: IEA/OECD Coal Statistics.

NETHERLANDS

5/D1 FUEL INPUT IN ELECTRICITY GENERATION

	1973	1978	1979	1980	1981	1985	1990
TOTAL (MTOE)	12.2	13.0	13.8	13.8	13.4	13.6	13.4
SOLID FUELS	.9	1.5	1.2	2.0	2.4	2.9	6.2
OIL	1.6	1.9	4.7	5.1	5.0	3.4	.6
GAS	9.5	8.7	7.2	5.8	5.2	6.3	5.6
NUCLEAR	.3	.9	.7	.9	.8	1.0	1.0
HYDRO/GEOTHERMAL	-	-	-	-	-	-	-
OTHER [1]	-	-	-	-	-	-	-
TOTAL (PERCENTS)	100.0	100.0	100.0	100.0	100.0	100.0	100.0
SOLID FUELS	7.0	11.7	8.6	14.6	17.8	21.3	46.3
OIL	12.9	14.7	33.8	36.8	37.4	25.0	4.5
GAS	78.0	67.1	52.2	42.1	39.1	46.3	41.8
NUCLEAR	2.1	6.6	5.4	6.5	5.7	7.4	7.5
HYDRO/GEOTHERMAL	-	-	-	-	-	-	-
OTHER [1]	-	-	-	-	-	-	-

Source: IEA/OECD Energy Balances and IEA Country Submissions (1982).

1. Solar, wind, etc.

5/D2 ELECTRICTY PRODUCTION BY FUEL TYPE

	1973	1978	1979	1980	1981	1985	1990
TOTAL (TWH)	52.6	61.6	64.3	64.8	64.1	56.8	54.6
SOLID FUELS	3.2	8.3	6.5	10.3	11.9	11.4	24.7
OIL	6.5	10.4	21.8	24.5	25.1	14.4	2.6
GAS	41.9	38.9	32.5	25.8	23.3	26.8	23.0
NUCLEAR	1.1	4.1	3.5	4.2	3.7	4.2	4.3
HYDRO/GEOTHERMAL	-	-	-	-	-	-	-
OTHER [1]	-	-	-	-	-	-	-
TOTAL (PERCENTS)	100.0	100.0	100.0	100.0	100.0	100.0	100.0
SOLID FUELS	6.0	13.4	10.1	15.9	18.7	20.1	45.2
OIL	12.3	16.9	34.0	37.8	39.2	25.4	4.8
GAS	79.5	63.1	50.5	39.8	36.4	47.2	42.1
NUCLEAR	2.1	6.6	5.4	6.5	5.7	7.4	7.9
HYDRO/GEOTHERMAL	-	-	-	-	-	-	-
OTHER [1]	-	-	-	-	-	-	-

Source: IEA/OECD Energy Balances and IEA Country Submissions (1982).

1. Solar, wind, etc.

NETHERLANDS

5/D3 PROJECTIONS OF ELECTRICITY GENERATION CAPACITY BY FUEL
(GW)

	Nuclear	Hydro/ Geoth.	Solid Fuels [2]	Oil	Gas	Multi Firing [1]	Total
I 1981							
Operating	0.5	-	2.6	0.7	2.1	9.9	15.8
II 1981-1985							
Capacity:							
Under Construction	-	-	-	-	-	0.3	0.3
Authorised	-	-	-	-	-	0.1 [3]	0.1
Other Planned	-	-	-	-	-	0.7	0.7
Conversion		-	-	-	-	-	-
Decommissioning	-	-	-	-0.5	-0.2	-0.5	-1.2
Total Operating 1990	0.5	-	2.6	0.2	1.9	10.5	15.7
III 1986-1990							
Capacity:							
Under Construction	-	-	-	-	-	0.5	0.5
Authorised	-	-	0.6	-	-	-	0.6
Other Planned	-	-	0.5	-	-	0.2	0.7
Conversion	-	-	1.3	-	-	-1.3	-
Decommissioning	-	-	-0.7	-0.1	-	-1.1	-1.9
Total Operating 1990	0.5	-	4.3	0.1	1.9	8.8	15.6

Source: IEA/OECD Electricity Statistics and IEA Country Submissions (1982).

1.   Includes « Other » capacity.
2.   Coal/Oil and Coal/Gas included.
3.   Steam and Gas Turb. unit on coal-gasification base.

NETHERLANDS

5/D4 ENERGY USE IN IRON AND STEEL INDUSTRY BY FUEL

	1973	1978	1979	1980	1981	1985	1990
TOTAL (MTOE)	2.8	2.3	2.4	2.1	2.2	3.0	4.0
SOLID FUELS	1.5	1.3	1.4	1.2	1.4	1.7	2.5
OIL	.5	.4	.5	.4	.2	.4	.4
GAS	.5	.4	.4	.4	.4	.2	.3
ELECTRICITY	.2	.2	.2	.2	.2	.7	.8
HEAT [1]	-	-	-	-	-	-	-
TOTAL (PERCENTS)	100.0	100.0	100.0	100.0	100.0	100.0	100.0
SOLID FUELS	54.7	56.0	56.5	57.6	66.2	56.7	62.5
OIL	19.1	18.8	19.0	16.9	8.6	13.3	10.0
GAS	19.5	18.0	17.5	18.1	17.9	6.7	7.5
ELECTRICITY	6.7	7.2	7.1	7.4	7.3	23.3	20.0
HEAT [1]	-	-	-	-	-	-	-

Source: IEA/OECD Energy Balances and IEA Country Submissions (1982).

1. Includes only heat from combined heat and power plants.

5/D5 ENERGY USE IN OTHER INDUSTRIES BY FUEL

	1973	1978	1979	1980	1981	1985	1990
TOTAL (MTOE)	19.5	23.0	25.8	19.1	18.2	19.7	23.0
SOLID FUELS	.3	.2	.3	.3	.2	.6	.8
OIL	9.7	11.7	13.9	8.4	7.3	9.9	11.2
GAS	7.8	8.8	9.3	8.2	8.4	7.5	9.4
ELECTRICITY	1.8	2.2	2.3	2.2	2.2	1.5	1.6
HEAT [1]	-	-	-	-	-	.2	-
TOTAL (PERCENTS)	100.0	100.0	100.0	100.0	100.0	100.0	100.0
SOLID FUELS	1.4	.9	1.1	1.5	1.3	3.0	3.5
OIL	49.7	51.0	54.0	44.0	40.3	50.3	48.7
GAS	39.8	38.5	35.9	42.8	46.1	38.1	40.9
ELECTRICITY	9.1	9.7	9.0	11.8	12.2	7.6	7.0
HEAT [1]	-	-	-	-	-	1.0	-

Source: IEA/OECD Energy Balances and IEA Country Submissions (1982).

1. Includes only heat from combined heat and power plants.

NETHERLANDS

5/D6 ENERGY USE IN INDUSTRY [1] BY FUEL

	1973	1978	1979	1980	1981	1985	1990
TOTAL (MTOE)	22.3	25.3	28.2	21.3	20.4	22.7	27.0
SOLID FUELS	1.8	1.5	1.6	1.5	1.7	2.3	3.3
OIL	10.2	12.2	14.4	8.8	7.5	10.3	11.6
GAS	8.3	9.3	9.7	8.6	8.8	7.7	9.7
ELECTRICITY	2.0	2.4	2.5	2.4	2.4	2.2	2.4
HEAT [2]	-	-	-	-	-	.2	-
TOTAL (PERCENTS)	100.0	100.0	100.0	100.0	100.0	100.0	100.0
SOLID FUELS	8.0	5.9	5.8	7.1	8.2	10.1	12.2
OIL	46.0	48.0	51.0	41.2	37.0	45.4	43.0
GAS	37.3	36.6	34.3	40.3	43.1	33.9	35.9
ELECTRICITY	8.8	9.4	8.9	11.3	11.7	9.7	8.9
HEAT [2]	-	-	-	-	-	.9	-

Source: IEA/OECD Energy Balances and IEA Country Submissions (1982).

1. Includes non-energy use of petroleum products.
2. Includes only heat from combined heat and power plants.

5/D7 ENERGY USE IN OTHER SECTORS [1] BY FUEL

	1973	1978	1979	1980	1981	1985	1990
TOTAL (MTOE)	20.5	21.8	22.7	21.6	19.7	20.4	22.0
SOLID FUELS	.4	.1	.1	.1	.1	-	.2
OIL	7.0	3.9	3.7	2.9	2.2	2.4	2.9
GAS	11.4	15.5	16.5	16.1	15.0	15.0	15.2
ELECTRICITY	1.8	2.3	2.4	2.4	2.4	2.5	2.5
HEAT [2]	-	-	-	-	-	.5	1.2
TOTAL (PERCENTS)	100.0	100.0	100.0	100.0	100.0	100.0	100.0
SOLID FUELS	1.7	.4	.4	.5	.5	-	.9
OIL	34.1	18.0	16.4	13.4	11.2	11.8	13.2
GAS	55.4	71.1	72.8	74.8	76.1	73.5	69.1
ELECTRICITY	8.7	10.4	10.4	11.3	12.3	12.3	11.4
HEAT [2]	-	-	-	-	-	2.5	5.5

Source: IEA/OECD Energy Balances and IEA Country Submissions (1982).

1. Includes use in agricultural, commercial, public services and residential sectors.
2. Includes only heat from combined heat and power plants.

NORWAY

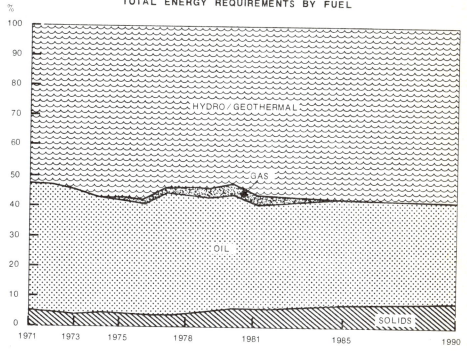

TOTAL ENERGY REQUIREMENTS BY FUEL

NORWAY

5/A1 TOTAL PRIMARY ENERGY REQUIREMENTS (TPER) BY FUEL

	1973	1978	1979	1980	1981	1985	1990
TPER (MTOE)[1]	19.5	22.3	24.2	23.9	24.2	24.7	26.7
SOLID FUELS	.8	.9	1.3	1.4	1.4	1.8	2.1
OIL	8.3	8.9	9.2	9.0	8.6	8.7	8.9
GAS	-	.6	.7	.9	.8	-	-
NUCLEAR	-	-	-	-	-	-	-
HYDRO/GEOTHERMAL	10.9	12.1	13.4	12.6	13.9	14.2	15.7
OTHER [2]	-	-	-	-	-	-	-
TPER (PERCENTS) [1]	100.0	100.0	100.0	100.0	100.0	100.0	100.0
SOLID FUELS	3.9	4.2	5.5	6.0	5.8	7.3	7.9
OIL	42.4	39.9	38.0	37.7	35.6	35.2	33.3
GAS	-	2.9	3.0	3.7	3.1	-	-
NUCLEAR	-	-	-	-	-	-	-
HYDRO/GEOTHERMAL	56.0	54.4	55.2	52.7	57.4	57.5	58.8
OTHER [2]	-	-	-	-	-	-	-

Source: IEA/OECD Energy Balances and IEA Country Submissions (1982).

1. Net imports of electricity are only included in totals.
2. Includes other electricity generation (solar, wind, etc.) and traded synthetic fuels.

5/A2 TOTAL FINAL ENERGY CONSUMPTION (TFC) BY FUEL

	1973	1978	1979	1980	1981	1985	1990
TFC (MTOE)	13.8	15.1	17.0	16.3	16.0	16.9	18.3
SOLID FUELS	.7	.8	1.3	1.4	1.4	1.7	2.0
OIL	7.8	8.3	9.2	8.5	8.0	8.1	8.3
GAS	.0	.0	.0	.0	.0	-	-
ELECTRICITY	5.2	5.9	6.5	6.4	6.6	7.1	8.0
HEAT [1]	-	-	-	-	-	-	-
TFC (PERCENTS)	100.0	100.0	100.0	100.0	100.0	100.0	100.0
SOLID FUELS	5.4	5.5	7.8	8.3	8.6	10.1	10.9
OIL	56.7	55.1	54.2	52.3	49.8	47.9	45.4
GAS	.1	.0	.0	.0	.0	-	-
ELECTRICITY	37.9	39.3	38.0	39.4	41.5	42.0	43.7
HEAT [1]	-	-	-	-	-	-	-

Source: IEA/OECD Energy Balances and IEA Country Submissions (1982).

1. Includes only heat production from combined heat and power plants.

NORWAY

5/A3 TOTAL SOLID FUELS BALANCE

	1973	1978	1979	1980	1981	1985	1990
PRODUCTION	.3	.4	.5	.6	.7	.7	.8
IMPORTS	.5	.5	.9	.9	.8	1.2	1.4
EXPORTS	-.1	-.1	-.1	-.1	-.1	-.1	-.1
MARINE BUNKERS	-	-	-	-	-	-	-
STOCK CHANGE	.0	.1	.0	.0	-	-	-
PRIMARY REQUIREMENTS	.8	.9	1.3	1.4	1.4	1.8	2.1
ELECTRICITY GENERATION	.0	.0	.0	.0	.0	-	-
GAS MANUFACTURE	-	-	-	-	-	-	-
LIQUEFACTION	-	-	-	-	-	-	-
ENERGY SECTOR OWN USE + LOSSES [1]	.0	-.1	.0	-.1	.0	-.1	-.1
FINAL CONSUMPTION	.7	.8	1.3	1.4	1.4	1.7	2.0
INDUSTRY SECTOR	.7	.7	1.0	.9	.9	1.2	1.4
IRON AND STEEL	.4	.5	.7	.7	.6	na	na
CHEMICAL/ PETROCHEMICAL	.0	.1	.1	.1	.1	na	na
OTHER INDUSTRY	.2	.1	.1	.2	.2	na	na
TRANSPORTATION SECTOR	-	-	-	-	-	-	-
OTHER SECTORS	.1	.1	.3	.4	.4	.5	.6
AGRICULTURE	-	-	-	-	-	-	-
COMMERCIAL AND PUBLIC SERVICE	-	-	-	-	-	-	-
RESIDENTIAL	.1	.1	.3	.4	.4	-	-

Source: IEA/OECD Energy Balances and IEA Country Submissions (1982).

1.   Includes statistical difference.

NORWAY

5/B1 SOLID FUELS INDIGENOUS PRODUCTION BY FUEL TYPE
(MTOE)

	1973	1978	1979	1980	1981	1985	1990
TOTAL SOLID FUELS	.29	.43	.54	.64	.71	.70	.80
HARD COAL	.29	.28	.20	.20	.28	.30e	.30e
COKING COAL	na	.27	.19	.19	.27	.30e	.30e
STEAM COAL	na	.01	.01	.01	.01	-	-
BROWN COAL/LIGNITE	-	-	-	-	-	-	-
OTHER SOLID FUELS	-	.15	.34	.44	.43	.40e	.50e

Source: IEA/OECD Coal Statistics, IEA/OECD Energy Balances, IEA Country Submissions (1982).

5/B2 COAL PRODUCTION BY TYPE
(THOUSAND METRIC TONS)

	1973	1978	1979	1980	1981
TOTAL COAL	415	402	282	288	394
HARD COAL	415	402	282	288	394
COKING COAL	na	383	266	277	384
STEAM COAL	na	19	16	11	10
BROWN COAL/LIGNITE	-	-	-	-	-
DERIVED PRODUCTS:					
PATENT FUEL	-	-	-	-	-
COKE OVEN COKE	320	320	341	349	359
GAS COKE	-	-	-	-	-
BROWN COAL BRIQUETTES	-	-	-	-	-
COKE OVEN GAS(TCAL)	700	688	729	729	707
BLAST FURNACE GAS(TCAL)	514	478	513	493	476

Source: IEA/OECD Coal Statistics.

NORWAY

5/C1 SOLID FUELS TRADE BY TYPE
(MTOE)

	1973	1978	1979	1980	1981	1985	1990
IMPORTS							
TOTAL SOLID FUELS	.55	.54	.89	.87	.82	1.20	1.40
HARD COAL	.28	.32	.47	.52	.48	.60e	.70e
COKING COAL	na	.31	.45	.44	.35	.50e	.60e
STEAM COAL	na	.01	.02	.07	.14	.10e	.10e
BROWN COAL/LIGNITE	-	-	-	-	-	-	-
OTHER SOLID FUELS [1]	.27	.22	.42	.35	.34	.60e	.70e
EXPORTS							
TOTAL SOLID FUELS	.09	.11	.08	.10	.13	.10e	.10
HARD COAL	.06	.05	.04	.07	.06	.10e	.10e
COKING COAL	na	.05	.04	.06	.06	.10	.10e
STEAM COAL	na	-	-	.01	-	-	-
BROWN COAL/LIGNITE	-	-	-	-	-	-	-
OTHER SOLID FUELS [1]	.03	.06	.04	.03	.07	-	-

Source: IEA/OECD Coal Statistics, IEA/OECD Energy Balances, IEA Country Submissions (1982).

1. Includes derived coal products, e.g. coke.

NORWAY

5/C2 COAL IMPORTS BY ORIGIN
(THOUSAND METRIC TONS)

	1978	1979	1980	1981
HARD COAL IMPORTS	450	673	738	692
COKING COAL IMPORTS FROM:	440	639	635	493
AUSTRALIA	-	-	-	-
CANADA	-	-	-	-
GERMANY	38	13	1	-
UNITED KINGDOM	28	55	3	17
UNITED STATES	108	126	335	240
OTHER OECD	58	86	60	115
CZECHOSLOVAKIA	-	-	-	-
POLAND	204	346	236	118
USSR	-	-	-	-
CHINA	-	-	-	-
SOUTH AFRICA	4	13	-	3
OTHER NON-OECD	-	-	-	-
STEAM COAL	10	34	103	199

Source: IEA/OECD Coal Statistics.

NORWAY

5/D1 FUEL INPUT IN ELECTRICITY GENERATION

	1973	1978	1979	1980	1981	1985	1990
TOTAL (MTOE)	11.0	12.2	13.4	12.6	13.9	14.2	15.7
SOLID FUELS	.0	.0	.0	.0	.0	-	-
OIL	.0	.0	.0	.0	.0	-	-
GAS	-	-	-	-	-	-	-
NUCLEAR	-	-	-	-	-	-	-
HYDRO/GEOTHERMAL	10.9	12.1	13.4	12.6	13.9	14.2	15.7
OTHER [1]	-	-	-	-	-	-	-
TOTAL (PERCENTS)	100.0	100.0	100.0	100.0	100.0	100.0	100.0
SOLID FUELS	.1	.1	.1	.1	.1	-	-
OIL	.1	.1	.1	.2	.2	-	-
GAS	-	-	-	-	-	-	-
NUCLEAR	-	-	-	-	-	-	-
HYDRO/GEOTHERMAL	99.8	99.8	99.8	99.8	99.8	100.0	100.0
OTHER [1]	-	-	-	-	-	-	-

Source: IEA/OECD Energy Balances and IEA Country Submissions (1982).

1.   Solar, wind, etc.

5/D2 ELECTRICTY PRODUCTION BY FUEL TYPE

	1973	1978	1979	1980	1981	1985	1990
TOTAL (TWH)	73.1	81.0	89.1	84.1	92.8	94.0	104.7
SOLID FUELS	.0	.0	.0	.0	.0	-	-
OIL	.1	.1	.1	.1	.1	-	-
GAS	-	-	-	-	-	-	-
NUCLEAR	-	-	-	-	-	-	-
HYDRO/GEOTHERMAL	72.9	80.9	89.0	84.0	92.7	94.0	104.7
OTHER [1]	-	-	-	-	-	-	-
TOTAL (PERCENTS)	100.0	100.0	100.0	100.0	100.0	100.0	100.0
SOLID FUELS	.0	.0	.0	.0	.0	-	-
OIL	.2	.1	.1	.1	.1	-	-
GAS	-	-	-	-	-	-	-
NUCLEAR	-	-	-	-	-	-	-
HYDRO/GEOTHERMAL	99.8	99.8	99.8	99.8	99.9	100.0	100.0
OTHER [1]	-	-	-	-	-	-	-

Source: IEA/OECD Energy Balances and IEA Country Submissions (1982).

1.   Solar, wind, etc.

NORWAY

5/D3 PROJECTIONS OF ELECTRICITY GENERATION CAPACITY BY FUEL
(GW)

	Nuclear	Hydro/ Geoth.	Solid Fuels	Oil	Gas	Multi Firing	Total
I 1981							
Operating	-	21.4	0.3	-	-	-	21.7
II 1981-1985							
Capacity:							
Under Construction	-	2.5	-	-	-	-	2.5
Authorised	-	-	-	-	-	-	-
Other Planned	-	0.2	-	-	-	-	0.2
Conversion	-	-	-	-	-	-	-
Decommissioning	-	-	-	-	-	-	-
Total Operating 1985	-	24.1	-	-	-	-	24.4
III 1986-1990							
Capacity:							
Under Construction	-	1.4	-	-	-	-	1.4
Authorised	-	-	-	-	-	-	-
Other Planned	-	2.6	-	-	-	-	2.6
Conversion	-	-	-	-	-	-	-
Decommissioning	-	-	-	-	-	-	-
Total Operating 1990	-	27.1	0.3	-	-	-	27.4

Source: IEA/OECD Electricity Statistics and IEA Country Submissions (1982).

NORWAY

5/D4 ENERGY USE IN IRON AND STEEL INDUSTRY BY FUEL

	1973	1978	1979	1980	1981	1985	1990
TOTAL (MTOE)	1.3	1.2	1.6	1.4	1.3	na	na
SOLID FUELS	.4	.5	.7	.7	.6	na	na
OIL	.1	-	.0	.0	.0	na	na
GAS	-	-	-	-	-	na	na
ELECTRICITY	.7	.7	.8	.7	.7	na	na
HEAT [1]	-	-	-	-	-	-	-
TOTAL (PERCENTS)	100.0	100.0	100.0	100.0	100.0	na	na
SOLID FUELS	35.1	42.5	47.5	46.4	47.6	na	na
OIL	8.4	-	2.6	2.5	2.4	na	na
GAS	-	-	-	-	-	na	na
ELECTRICITY	56.5	57.5	50.0	51.1	50.0	na	na
HEAT [1]	-	-	-	-	-	-	-

Source: IEA/OECD Energy Balances and IEA Country Submissions (1982).

1. Includes only heat from combined heat and power plants.

5/D5 ENERGY USE IN OTHER INDUSTRIES BY FUEL

	1973	1978	1979	1980	1981	1985	1990
TOTAL (MTOE)	5.7	6.1	6.6	6.5	6.2	na	na
SOLID FUELS	.2	.2	.2	.3	.3	na	na
OIL	3.0	3.4	3.7	3.5	3.1	na	na
GAS	.0	.0	.0	.0	.0	na	na
ELECTRICITY	2.5	2.5	2.7	2.7	2.8	na	na
HEAT [1]	-	-	-	-	-	-	-
TOTAL (PERCENTS)	100.0	100.0	100.0	100.0	100.0	na	na
SOLID FUELS	3.7	3.5	3.6	4.1	5.1	na	na
OIL	52.4	55.5	56.0	54.3	50.3	na	na
GAS	.0	.0	.0	.0	.0	na	na
ELECTRICITY	43.9	40.9	40.4	41.6	44.6	na	na
HEAT [1]	-	-	-	-	-	-	-

Source: IEA/OECD Energy Balances and IEA Country Submissions (1982).

1. Includes only heat from combined heat and power plants.

NORWAY

5/D6 ENERGY USE IN INDUSTRY [1] BY FUEL

	1973	1978	1979	1980	1981	1985	1990
TOTAL (MTOE)	6.9	7.3	8.2	7.9	7.6	7.3	7.7
SOLID FUELS	.7	.7	1.0	.9	.9	1.2	1.4
OIL	3.1	3.4	3.8	3.5	3.2	2.5	2.5
GAS	.0	.0	.0	.0	.0	-	-
ELECTRICITY	3.2	3.2	3.5	3.4	3.4	3.6	3.8
HEAT [2]	-	-	-	-	-	-	-
TOTAL (PERCENTS)	100.0	100.0	100.0	100.0	100.0	100.0	100.0
SOLID FUELS	9.4	9.8	12.0	11.7	12.5	16.4	18.2
OIL	44.4	46.6	45.8	45.0	42.0	34.2	32.5
GAS	.0	.0	.0	.0	.0	-	-
ELECTRICITY	46.2	43.6	42.2	43.3	45.5	49.3	49.4
HEAT [2]	-	-	-	-	-	-	-

Source: IEA/OECD Energy Balances and IEA Country Submissions (1982).

1. Includes non-energy use of petroleum products.
2. Includes only heat from combined heat and power plants.

5/D7 ENERGY USE IN OTHER SECTORS [1] BY FUEL

	1973	1978	1979	1980	1981	1985	1990
TOTAL (MTOE)	4.4	4.7	5.4	5.1	5.1	6.4	7.3
SOLID FUELS	.1	.1	.3	.4	.4	.5	.6
OIL	2.3	1.8	2.1	1.7	1.5	2.5	2.6
GAS	.0	.0	.0	.0	.0	-	-
ELECTRICITY	2.0	2.7	3.0	3.0	3.1	3.4	4.1
HEAT [2]	-	-	-	-	-	-	-
TOTAL (PERCENTS)	100.0	100.0	100.0	100.0	100.0	100.0	100.0
SOLID FUELS	2.0	2.6	6.5	8.5	8.4	7.8	8.2
OIL	52.3	39.3	38.8	33.1	30.0	39.1	35.6
GAS	.2	.1	.1	.1	.1	-	-
ELECTRICITY	45.5	58.0	54.6	58.3	61.5	53.1	56.2
HEAT [2]	-	-	-	-	-	-	-

Source: IEA/OECD Energy Balances and IEA Country Submissions (1982).

1. Includes use in agricultural, commercial, public services and residential sectors.
2. Includes only heat from combined heat and power plants.

PORTUGAL

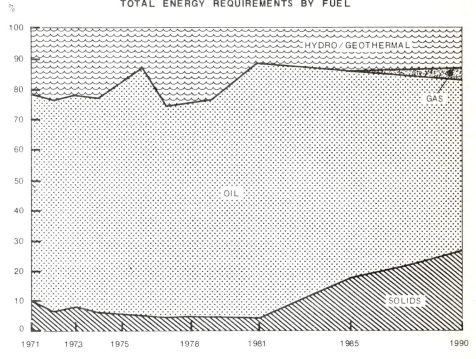

TOTAL ENERGY REQUIREMENTS BY FUEL

PORTUGAL

5/A1 TOTAL PRIMARY ENERGY REQUIREMENTS (TPER) BY FUEL

	1973	1978	1979	1980	1981	1985	1990
TPER (MTOE) [1]	7.7	10.2	11.0	10.7	10.4	15.7	19.1
SOLID FUELS	.6	.5	.5	.5	.4	2.6	4.9
OIL	5.4	7.3	7.9	8.2	8.5	10.7	10.7
GAS	-	-	-	-	-	-	.8
NUCLEAR	-	-	-	-	-	-	-
HYDRO/GEOTHERMAL	1.7	2.6	2.6	1.9	1.2	2.3	2.7
OTHER [2]	-	-	-	-	-	-	.0
TPER (PERCENTS) [1]	100.0	100.0	100.0	100.0	100.0	100.0	100.0
SOLID FUELS	7.7	4.5	4.3	4.3	4.0	16.6	25.8
OIL	69.8	70.7	71.8	76.5	81.8	68.6	55.9
GAS	-	-	-	-	-	-	4.2
NUCLEAR	-	-	-	-	-	-	-
HYDRO/GEOTHERMAL	22.5	24.9	24.0	17.7	11.7	14.9	14.0
OTHER [2]	-	-	-	-	-	-	.0

Source: IEA/OECD Energy Balances and IEA Country Submissions (1982).

1. Net imports of electricity are only included in totals.
2. Includes other electricity generation (solar, wind, etc.) and traded synthetic fuels.

5/A2 TOTAL FINAL ENERGY CONSUMPTION (TFC) BY FUEL

	1973	1978	1979	1980	1981	1985	1990
TFC (MTOE)	5.7	7.0	7.6	7.8	7.7	11.6	14.0
SOLID FUELS	.3	.3	.3	.3	.2	1.9	2.7
OIL	4.7	5.7	6.2	6.3	6.2	8.1	8.5
GAS	.0	.0	.0	.1	.1	.1	.8
ELECTRICITY	.7	1.0	1.2	1.2	1.2	1.5	1.9
HEAT [1]	na	na	na	.0	.0	.0	.0
TFC (PERCENTS)	100.0	100.0	100.0	100.0	100.0	100.0	100.0
SOLID FUELS	4.7	3.8	3.4	3.3	2.6	16.4	19.2
OIL	82.0	80.6	80.9	80.0	80.7	69.6	60.8
GAS	.9	.7	.6	.6	.7	.6	6.0
ELECTRICITY	12.4	14.9	15.0	15.7	15.8	13.4	13.9
HEAT [1]	na	na	na	.3	.3	.0	.1

Source: IEA/OECD Energy Balances and IEA Country Submissions (1982).

1. Includes only heat production from combined heat and power plants.

PORTUGAL

5/A3 TOTAL SOLID FUELS BALANCE
(MTOE)

	1973	1978	1979	1980	1981	1985	1990
PRODUCTION	.2	.1	.1	.1	.1	1.3	1.7
IMPORTS	.3	.3	.3	.3	.2	1.5	3.3
EXPORTS	.0	-	-	-	-	-	-
MARINE BUNKERS	-	-	-	-	-	-	-
STOCK CHANGE	.1	.0	.0	.0	.1	-.2	-.1
PRIMARY REQUIREMENTS	.6	.5	.5	.5	.4	2.6	4.9
ELECTRICITY GENERATION	-.2	-.1	-.2	-.2	-.2	-.6	-1.9
GAS MANUFACTURE	-	-	-	-	-	.0	-.1
LIQUEFACTION	-	-	-	-	-	-	.0
ENERGY SECTOR OWN USE + LOSSES [1]	-.1	-.1	.0	.0	.0	-.1	-.2
FINAL CONSUMPTION	.3	.3	.3	.3	.2	1.9	2.7
INDUSTRY SECTOR	.2	.3	.3	.3	.2	1.2	2.0
IRON AND STEEL	.1	.2	.2	.2	.2	.2	.5
CHEMICAL/ PETROCHEMICAL	-	.0	.0	.0	.0	.0	.1
OTHER INDUSTRY	.1	.0	.0	.0	.0	.9	1.4
TRANSPORTATION SECTOR	.0	.0	.0	.0	.0	-	-
OTHER SECTORS	.0	.0	.0	.0	.0	.7	.6
AGRICULTURE	-	-	-	-	-	-	-
COMMERCIAL AND PUBLIC SERVICE	-	.0	.0	.0	.0	.0	.0
RESIDENTIAL	.0	.0	.0	.0	-	.7	.6

Source: IEA/OECD Energy Balances and IEA Country Submissions (1982).

1. Includes statistical difference.

PORTUGAL

5/B1 SOLID FUELS INDIGENOUS PRODUCTION BY FUEL TYPE
(MTOE)

	1973	1978	1979	1980	1981	1985	1990
TOTAL SOLID FUELS	.15	.12	.12	.12	.12	1.29	1.71
HARD COAL	.15	.12	.12	.12	.12	.10	.10
COKING COAL	na	-	-	-	-	-	-
STEAM COAL	na	.12	.12	.12	.12	.10	.10
BROWN COAL/LIGNITE	-	-	-	-	-	-	.30
OTHER SOLID FUELS	-	-	-	-	-	1.19	1.31

Source: IEA/OECD Coal Statistics, IEA/OECD Energy Balances, IEA Country Submissions (1982).

5/B2 COAL PRODUCTION BY TYPE
(THOUSAND METRIC TONS)

	1973	1978	1979	1980	1981
TOTAL COAL	221	180	179	177	183
HARD COAL	221	180	179	177	183
COKING COAL	na	-	-	-	-
STEAM COAL	na	180	179	177	183
BROWN COAL/LIGNITE	-	-	-	-	-
DERIVED PRODUCTS:					
PATENT FUEL	34	-	-	-	-
COKE OVEN COKE	269	244	237	216	198
GAS COKE	-	-	-	-	-
BROWN COAL BRIQUETTES	-	-	-	-	-
COKE OVEN GAS(TCAL)	460	364	432	439	388
BLAST FURNACE GAS(TCAL)	583	399	430	482	393

Source: IEA/OECD Coal Statistics.

PORTUGAL

5/C1 SOLID FUELS TRADE BY TYPE
(MTOE)

	1973	1978	1979	1980	1981	1985	1990
IMPORTS							
TOTAL SOLID FUELS	.33	.34	.32	.33	.24	1.54	3.34
HARD COAL	.30	.27	.25	.26	.20	1.54	3.34
COKING COAL	-	.26	.24	.26	.20	.34	.70
STEAM COAL	-	.01	-	-	-	1.20	2.64
BROWN COAL/LIGNITE	-	-	-	-	-	-	-
OTHER SOLID FUELS [1]	.03	.07	.07	.07	.04	-	-

Source: IEA/OECD Coal Statistics, IEA/OECD Energy Balances, IEA Country Submissions (1982).

1. Includes derived coal products, e.g. coke.

5/C2 COAL IMPORTS BY ORIGIN
(THOUSAND METRIC TONS)

	1978	1979	1980	1981
HARD COAL IMPORTS	412	380	398	313
COKING COAL	404	375	393	311
IMPORTS FROM:				
AUSTRALIA	-	-	-	-
CANADA	-	-	-	-
GERMANY	8	2	1	3
UNITED KINGDOM	3	2	2	31
UNITED STATES	257	258	255	254
OTHER OECD	14	4	10	8
CZECHOSLOVAKIA	-	-	-	-
POLAND	122	104	120	4
USSR	-	-	-	-
CHINA	-	-	-	-
SOUTH AFRICA	-	-	-	-
OTHER NON-OECD	-	5	5	11
STEAM COAL IMPORTS	8	5	5	2

Source: IEA/OECD Coal Statistics.

PORTUGAL

5/D1 FUEL INPUT IN ELECTRICITY GENERATION

	1973	1978	1979	1980	1981	1985	1990
TOTAL (MTOE)	2.3	3.4	3.9	3.5	3.1	4.9	6.2
SOLID FUELS	.2	.1	.2	.2	.2	.6	1.9
OIL	.3	.7	1.0	1.4	1.7	2.0	1.5
GAS	-	-	-	-	-	-	-
NUCLEAR	-	-	-	-	-	-	-
HYDRO/GEOTHERMAL	1.7	2.6	2.6	1.9	1.2	2.3	2.7
OTHER [1]	-	-	-	-	-	-	.0
TOTAL (PERCENTS)	100.0	100.0	100.0	100.0	100.0	100.0	100.0
SOLID FUELS	10.7	4.0	4.3	4.7	5.3	11.9	31.5
OIL	14.5	20.3	27.1	40.8	55.2	40.3	25.0
GAS	-	-	-	-	-	-	-
NUCLEAR	-	-	-	-	-	-	-
HYDRO/GEOTHERMAL	74.8	75.7	68.6	54.4	39.5	47.8	43.4
OTHER [1]	-	-	-	-	-	-	.0

Source: IEA/OECD Energy Balances and IEA Country Submissions (1982).

1. Solar, wind, etc.

5/D2 ELECTRICTY PRODUCTION BY FUEL TYPE

	1973	1978	1979	1980	1981	1985	1990
TOTAL (TWH)	9.8	14.7	16.2	15.3	13.9	20.6	25.8
SOLID FUELS	.6	.5	.6	.7	.7	2.3	7.9
OIL	1.9	3.3	4.3	6.5	8.1	8.4	6.6
GAS	-	-	-	-	-	-	-
NUCLEAR	-	-	-	-	-	-	-
HYDRO/GEOTHERMAL	7.4	10.9	11.3	8.1	5.2	9.9	11.4
OTHER [1]	-	-	-	-	-	-	.0
TOTAL (PERCENTS)	100.0	100.0	100.0	100.0	100.0	100.0	100.0
SOLID FUELS	6.0	3.2	3.9	4.4	4.7	11.2	30.5
OIL	19.2	22.7	26.5	42.7	58.1	40.8	25.4
GAS	-	-	-	-	-	-	-
NUCLEAR	-	-	-	-	-	-	-
HYDRO/GEOTHERMAL	74.9	74.1	69.7	52.9	37.2	48.1	44.1
OTHER [1]	-	-	-	-	-	-	.0

Source: IEA/OECD Energy Balances and IEA Country Submissions (1982).

2. Solar, wind, etc.

PORTUGAL

5/D3 PROJECTIONS OF ELECTRICITY GENERATION CAPACITY BY FUEL
(GW)

	Nuclear	Hydro/ Geoth.	Coal	Oil	Gas	Multi Firing [1]	Total
I 1981							
Operating	-	2.6	0.1	1.8	-	-	4.5
II 1981-1985							
Capacity:							
Under Construction	-	0.5	0.3	0.7	-	-	1.5
Authorised	-	-	-	-	-	-	-
Other Planned	-	-	-	0.1	-	0.2	0.3
Conversion	-	-	-	-	-	-	-
Decommissioning	-	-	-	-	-	-	-
Total Operating 1985	-	3.1	0.4	2.6	-	0.2	6.3
III 1986-1990							
Capacity:							
Under Construction	-	0.9	0.3	-	-	-	1.2
Authorised	-	0.1	0.3	-	-	-	0.4
Other Planned	-	0.2	0.3	0.1	-	-	0.6
Conversion	-	-	-	-	-	-	-
Decommissioning	-	-	-	-	-	-	-
Total Operating 1990	-	4.3	1.3	2.7	-	0.2	8.5

Source: IEA/OECD Electricity Statistics and IEA Country Submissions (1982).

1. Includes "Other" capacity.

PORTUGAL

5/D6 ENERGY USE IN INDUSTRY [1] BY FUEL

	1973	1978	1979	1980	1981	1985	1990
TOTAL (MTOE)	2.5	3.2	3.6	3.7	3.4	6.2	7.9
SOLID FUELS	.2	.3	.3	.3	.2	1.2	2.0
OIL	1.9	2.3	2.7	2.7	2.5	4.0	4.0
GAS	.0	.0	.0	.0	.0	.0	.6
ELECTRICITY	.4	.6	.7	.7	.7	1.0	1.3
HEAT [2]	-	-	-	-	-	-	-
TOTAL (PERCENTS)	100.0	100.0	100.0	100.0	100.0	100.0	100.0
SOLID FUELS	8.3	8.2	7.0	6.9	5.8	19.1	26.0
OIL	74.1	72.7	74.3	74.0	73.6	65.0	50.4
GAS	.1	.1	.1	.1	.1	.0	7.3
ELECTRICITY	17.4	19.1	18.6	19.1	20.5	15.8	16.3
HEAT [2]	-	-	-	-	-	-	-

Source: IEA/OECD Energy Balances and IEA Country Submissions (1982).

1. Includes non-energy use of petroleum products.
2. Includes only heat from combined heat and power plants.

5/D7 ENERGY USE IN OTHER SECTORS [1] BY FUEL

	1973	1978	1979	1980	1981	1985	1990
TOTAL (MTOE)	1.0	1.3	1.4	1.5	1.6	2.3	2.5
SOLID FUELS	.0	.0	.0	.0	.0	.7	.6
OIL	.7	.8	.8	.9	1.0	1.0	.9
GAS	.0	.0	.0	.0	.0	.1	.3
ELECTRICITY	.2	.4	.5	.5	.5	.5	.6
HEAT [2]	na	na	na	.0	.0	.0	.0
TOTAL (PERCENTS)	100.0	100.0	100.0	100.0	100.0	100.0	100.0
SOLID FUELS	3.8	.5	.6	.4	.2	30.9	25.9
OIL	66.5	62.6	61.8	61.6	63.8	42.1	36.7
GAS	4.8	3.7	3.5	3.2	3.0	3.1	11.0
ELECTRICITY	25.0	33.2	34.1	33.5	31.5	23.6	26.2
HEAT [2]	na	na	na	1.4	1.5	.2	.3

Source: IEA/OECD Energy Balances and IEA Country Submissions (1982).

1. Includes use in agricultural, commercial, public services and residential sectors.
2. Includes only heat from combined heat and power plants.

SPAIN

TOTAL ENERGY REQUIREMENTS BY FUEL

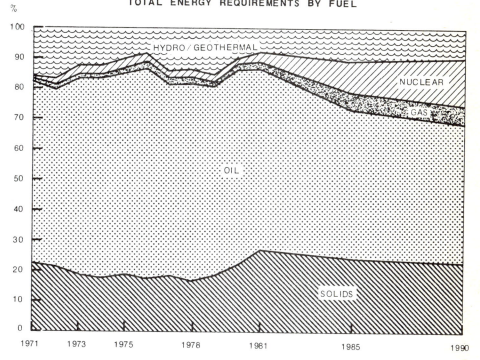

SPAIN

5/A1 TOTAL PRIMARY ENERGY REQUIREMENTS (TPER) BY FUEL

	1973	1978	1979	1980	1981	1985	1990
TPER (MTOE) [1]	57.8	73.1	75.8	76.6	76.2	87.0	106.4
SOLID FUELS	10.9	12.2	14.1	16.7	20.7	21.1	24.3
OIL	37.6	48.1	47.5	49.9	45.9	42.9	48.2
GAS	1.0	1.4	1.6	1.6	2.0	4.7	6.5
NUCLEAR	1.5	1.8	1.6	1.2	2.4	9.2	16.0
HYDRO/GEOTHERMAL	6.9	9.7	11.2	7.2	5.5	8.8	9.7
OTHER [2]	-	-	-	-	-	.3	1.8
TPER (PERCENTS) [1]	100.0	100.0	100.0	100.0	100.0	100.0	100.0
SOLID FUELS	18.9	16.7	18.6	21.8	27.1	24.3	22.8
OIL	65.1	65.7	62.7	65.1	60.2	49.3	45.2
GAS	1.7	2.0	2.1	2.1	2.6	5.4	6.1
NUCLEAR	2.7	2.5	2.1	1.6	3.1	10.5	15.1
HYDRO/GEOTHERMAL	12.0	13.3	14.7	9.5	7.2	10.1	9.1
OTHER [2]	-	-	-	-	-	.4	1.6

Source: IEA/OECD Energy Balances and IEA Country Submissions (1982).

1. Net imports of electricity are only included in totals.
2. Includes other electricity generation (solar, wind, etc.), traded synthetic fuels and direct use of solar, wind and other non-solid renewable energy sources.

5/A2 TOTAL FINAL ENERGY CONSUMPTION (TFC) BY FUEL

	1973	1978	1979	1980	1981	1985	1990
TFC (MTOE)	42.5	50.5	52.3	51.3	48.5	58.9	71.2
SOLID FUELS	5.7	3.8	4.2	3.6	3.8	8.1	9.3
OIL	31.0	38.7	39.6	38.8	35.7	36.6	42.1
GAS	.7	1.2	1.3	1.2	1.1	4.2	6.0
ELECTRICITY	5.1	6.8	7.3	7.7	7.8	9.9	13.0
OTHER [1]	-	-	-	-	-	.1	.9
TFC (PERCENTS)	100.0	100.0	100.0	100.0	100.0	100.0	100.0
SOLID FUELS	13.5	7.5	7.9	7.1	7.9	13.7	13.0
OIL	72.9	76.7	75.7	75.5	73.7	62.1	59.1
GAS	1.7	2.3	2.5	2.3	2.3	7.1	8.4
ELECTRICITY	11.9	13.5	13.9	15.0	16.2	16.9	18.2
OTHER [1]	-	-	-	-	-	.2	1.3

Source: IEA/OECD Energy Balances and IEA Country Submissions (1982).

1. Includes heat production from combined heat and power plants and direct use of solar, wind and other non-solid renewable energy sources.

SPAIN

5/A3 TOTAL SOLID FUELS BALANCE
(MTOE)

	1973	1978	1979	1980	1981	1985	1990
PRODUCTION	8.0	10.9	12.0	14.4	17.2	14.0	14.8
IMPORTS	2.5	2.6	3.3	4.3	5.3	7.1	9.5
EXPORTS	.0	.0	.0	.0	.0	-	-
MARINE BUNKERS	.0	-	-	-	-	-	-
STOCK CHANGE	.4	-1.2	-1.2	-2.0	-1.8	-	-
PRIMARY REQUIREMENTS	10.9	12.2	14.1	16.7	20.7	21.1	24.3
ELECTRICITY GENERATION	-3.9	-7.7	-9.0	-11.9	-14.8	-13.0	-15.0
GAS MANUFACTURE	-	-	-	.0	.0	-	-
LIQUEFACTION	-	-	-	-	-	-	-
ENERGY SECTOR OWN USE + LOSSES [1]	-1.3	-.7	-.9	-1.2	-2.0	-	-
FINAL CONSUMPTION	5.7	3.8	4.2	3.6	3.8	8.1	9.3
INDUSTRY SECTOR	5.4	3.5	3.5	3.3	3.4	7.3	8.4
IRON AND STEEL	3.8	2.9	2.8	2.6	2.6	3.9	4.0
CHEMICAL/ PETROCHEMICAL	.3	.2	.2	.2	.3	-	-
OTHER INDUSTRY	1.2	.3	.4	.4	.5	3.4	4.4
TRANSPORTATION SECTOR	-	.0	.0	.0	.0	-	-
OTHER SECTORS	.4	.3	.6	.3	.4	.8	.8
AGRICULTURE	-	-	-	-	-	-	-
COMMERCIAL AND PUBLIC SERVICE	-	.0	-	-	-	-	-
RESIDENTIAL	.4	.3	.6	.3	.4	.8	.8

Source: IEA/OECD Energy Balances and IEA Country Submissions (1982).

1. Includes statistical difference.

Note: Later information indicates that the conversion factor for solid fuels in 1980 and 1981 should be adjusted. New information on the increased production of lower grade coals, especially lignite, indicates that the indigenous production of solid fuels should be reduced by about 2 Mtoe in 1980 and 4 Mtoe in 1981.

SPAIN

5/B1 SOLID FUELS INDIGENOUS PRODUCTION BY FUEL TYPE
(MTOE)

TOTAL SOLID FUELS	1973	1978	1979	1980	1981	1985	1990
TOTAL SOLID FUELS	8.04	10.86	12.04	14.40	17.22	14.03	14.76
HARD COAL	6.99	7.97	8.30	8.99	9.91	8.53	9.16
COKING COAL	na	1.26	1.24	.93	1.03	1.65	1.80
STEAM COAL	na	6.71	7.05	8.05	8.88	6.88	7.36
BROWN COAL/LIGNITE	1.05	2.89	3.74	5.41	7.31	5.50	5.60
OTHER SOLID FUELS	-	-	-	-	-	-	-

Source: IEA/OECD Coal Statistics, IEA/OECD Energy Balances, IEA Country Submissions (1982).

Note: Later information indicates that the conversion factor for solid fuels in 1980 and 1981 should be adjusted. New information on the increased production of lower grade coals, especially lignite, indicates that the indigenous production of solid fuels should be reduced by about 2 Mtoe in 1980 and 4 Mtoe in 1981.

5/B2 COAL PRODUCTION BY TYPE
(THOUSAND METRIC TONS)

TOTAL SOLID FUELS	1973	1978	1979	1980	1981
TOTAL COAL	12994	19648	22548	28292	35048
HARD COAL	9991	11387	11852	12838	14162
COKING COAL	na	1800	1776	1333	1471
STEAM COAL	na	9587	10076	11505	12691
BROWN COAL/LIGNITE	3003	8261	10696	15454	20886
DERIVED PRODUCTS:					
PATENT FUEL	147	62	40	60	60
COKE OVEN COKE	4475	3887	3843	3900	3800
GAS COKE	-	-	-	-	-
BROWN COAL BRIQUETTES	-	-	-	-	-
COKE OVEN GAS(TCAL)	7684	7750	6939	7078	7100
BLAST FURNACE GAS(TCAL)	11616	10050	10395	8319	8500

Source: IEA/OECD Coal Statistics.

SPAIN

5/C1 SOLID FUELS TRADE BY TYPE
(MTOE)

	1973	1978	1979	1980	1981	1985	1990
IMPORTS							
TOTAL SOLID FUELS	2.49	2.58	3.30	4.35	5.27	7.06	9.50
HARD COAL	2.18	2.41	2.94	3.97	4.93	7.06	9.50
COKING COAL	na	2.14	2.62	2.85	2.70	2.60	3.10
STEAM COAL	na	.27	.32	1.12	2.23	4.46	6.40
BROWN COAL/LIGNITE	.01	-	-	-	-	-	-
OTHER SOLID FUELS [1]	.30	.17	.36	.38	.34	-	-
EXPORTS							
TOTAL SOLID FUELS	.01	.01	.01	.01	.01	-	-
HARD COAL	.01	.01	.01	.01	.01	-	-
COKING COAL	na	-	-	-	-	-	-
STEAM COAL	na	.01	.01	.01	.01	-	-
BROWN COAL/LIGNITE	-	-	-	-	-	-	-
OTHER SOLID FUELS [1]	-	-	-	-	-	-	-

Source: IEA/OECD Coal Statistics, IEA/OECD Energy Balances, IEA Country Submissions (1982).

1. Includes derived coal products, e.g. coke.

SPAIN

5/C2 COAL IMPORTS BY ORIGIN
(THOUSAND METRIC TONS)

	1978	1979	1980	1981
HARD COAL IMPORTS	3443	4198	5678	7045
COKING COAL IMPORTS FROM:	3063	3746	4074	3854
AUSTRALIA	579	709	636	562
CANADA	135	163	46	-
GERMANY	110	134	54	31
UNITED KINGDOM	-	-	-	70
UNITED STATES	1072	1312	2340	2765
OTHER OECD	-	-	-	-
CZECHOSLOVAKIA	-	-	-	-
POLAND	1167	1428	998	426
USSR	-	-	-	-
CHINA	-	-	-	-
SOUTH AFRICA	-	-	-	-
OTHER NON-OECD	-	-	-	-
STEAM COAL IMPORTS FROM:	380	452	1604	3191
AUSTRALIA	-	-	-	-
CANADA	-	-	-	54
GERMANY	-	-	-	-
UNITED KINGDOM	-	-	-	44
UNITED STATES	-	-	723	2710
OTHER OECD	26	31	15	5
CZECHOSLOVAKIA	-	-	-	-
POLAND	27	32	89	-
USSR	181	215	-	8
CHINA	-	-	72	-
SOUTH AFRICA	146	174	705	370
OTHER NON-OECD	-	-	-	-

Source: IEA/OECD Coal Statistics.

SPAIN

5/D1 FUEL INPUT IN ELECTRICITY GENERATION

	1973	1978	1979	1980	1981	1985	1990
TOTAL (MTOE)	17.9	25.7	28.4	29.8	32.4	34.0	43.6
SOLID FUELS	3.9	7.7	9.0	11.9	14.8	13.0	15.0
OIL	5.4	6.2	6.4	9.2	8.9	2.8	2.0
GAS	.1	.2	.2	.3	.8	-	-
NUCLEAR	1.5	1.8	1.6	1.2	2.4	9.2	16.0
HYDRO/GEOTHERMAL	6.9	9.7	11.2	7.2	5.5	8.8	9.7
OTHER [1]	-	-	-	-	-	.3	.8
TOTAL (PERCENTS)	100.0	100.0	100.0	100.0	100.0	100.0	100.0
SOLID FUELS	21.5	30.1	31.8	40.0	45.8	38.2	34.4
OIL	30.4	24.3	22.6	30.7	27.5	8.3	4.5
GAS	.7	.6	.7	.9	2.5	-	-
NUCLEAR	8.6	7.0	5.5	4.1	7.3	27.0	36.8
HYDRO/GEOTHERMAL	38.7	38.0	39.3	24.3	16.9	25.7	22.3
OTHER [1]	-	-	-	-	-	.7	1.9

Source: IEA/OECD Energy Balances and IEA Country Submissions (1982).

1.   Solar, wind, etc.

5/D2 ELECTRICTY PRODUCTION BY FUEL TYPE

	1973	1978	1979	1980	1981	1985	1990
TOTAL (TWH)	76.3	99.4	105.8	110.5	111.2	133.2	171.8
SOLID FUELS	14.3	20.3	23.3	32.4	39.5	47.4	56.1
OIL	25.1	28.8	26.6	38.5	33.8	10.3	6.7
GAS	.8	1.1	1.8	3.6	4.6	-	-
NUCLEAR	6.5	7.6	6.7	5.2	10.1	37.2	65.0
HYDRO/GEOTHERMAL	29.5	41.5	47.5	30.8	23.2	36.2	41.4
OTHER [1]	-	-	-	-	-	2.1	2.5
TOTAL (PERCENTS)	100.0	100.0	100.0	100.0	100.0	100.0	100.0
SOLID FUELS	18.8	20.4	22.0	29.3	35.5	35.6	32.7
OIL	32.9	29.0	25.1	34.9	30.4	7.7	3.9
GAS	1.0	1.1	1.7	3.3	4.1	-	-
NUCLEAR	8.6	7.7	6.3	4.7	9.1	27.9	37.9
HYDRO/GEOTHERMAL	38.7	41.8	44.9	27.9	20.9	27.2	24.1
OTHER [1]	-	-	-	-	-	1.5	1.5

Source: IEA/OECD Energy Balances and IEA Country Submissions (1982).

1.   Solar, wind, etc.

SPAIN

5/D3 PROJECTIONS OF ELECTRICITY GENERATION CAPACITY BY FUEL
(GW)

	Nuclear	Hydro/ Geoth.	Solid Fuels	Oil	Gas	Multi Firing [1]	Total
I 1981							
Operating	2.1	13.6	4.9	9.3	-	3.0	32.9
II 1981-1985							
Capacity:							
Under Construction	5.6	1.0	4.8	0.3	-	-	11.7
Authorised	-	-	-	-	-	-	-
Other Planned	-	-	-	-	-	-	-
Conversion	-	-	0.2	-0.2	-	-	-
Decommissioning	-	-	-	-	-	-	-
Total Operating 1985	7.7	14.6	9.9	9.4	-	3.0	44.6
III 1986-1990							
Capacity:							
Under Construction	-	-	-	-	-	-	-
Authorised	5.0	3.7	0.2	0.2	-	-	9.1
Other Planned	-	-	-	-	-	-	-
Conversion	-	-	1.7	-1.7	-	-	-
Decommissioning	-	-	-	-	-	-	-
Total Operating 1990	12.7	18.3	11.8	7.9	-	3.0	53.7

Source: IEA/OECD Electricity Statistics and IEA Country Submissions (1982).

1.   Includes "Other" capacity.

SPAIN

5/D4 ENERGY USE IN IRON AND STEEL INDUSTRY BY FUEL

	1973	1978	1979	1980	1981	1985	1990
TOTAL (MTOE)	5.7	4.6	4.4	4.1	4.9	4.9	5.8
SOLID FUELS	3.8	2.9	2.8	2.6	2.6	3.9	4.0
OIL	1.2	.9	.7	.6	1.4	-	-
GAS	.0	.0	.0	.0	.0	-	.5
ELECTRICITY	.6	.8	.9	.9	.8	1.0	1.3
OTHER [1]	-	-	-	-	-	-	-
TOTAL (PERCENTS)	100.0	100.0	100.0	100.0	100.0	100.0	100.0
SOLID FUELS	67.0	63.6	64.6	63.3	54.2	79.3	69.0
OIL	21.8	19.3	15.5	14.7	28.1	-	-
GAS	.3	.4	.4	.5	.4	-	8.6
ELECTRICITY	10.9	16.7	19.4	21.5	17.3	20.7	22.4
OTHER [1]	-	-	-	-	-	-	-

Source: IEA/OECD Energy Balances and IEA Country Submissions (1982).

1. Includes heat from combined heat and power plants and direct use of solar, wind and other non-solid renewable energy sources.

5/D5 ENERGY USE IN OTHER INDUSTRIES BY FUEL

	1973	1978	1979	1980	1981	1985	1990
TOTAL (MTOE)	18.0	20.2	21.4	20.7	17.6	24.4	29.6
SOLID FUELS	1.6	.6	.7	.7	.8	3.4	4.4
OIL	13.4	15.4	16.1	15.2	12.0	13.4	15.0
GAS	.4	.8	.8	.7	.6	2.9	3.9
ELECTRICITY	2.6	3.5	3.7	4.1	4.2	4.7	5.8
OTHER [1]	-	-	-	-	-	-	.5
TOTAL (PERCENTS)	100.0	100.0	100.0	100.0	100.0	100.0	100.0
SOLID FUELS	8.7	2.8	3.1	3.2	4.4	13.8	15.0
OIL	74.4	76.1	75.5	73.4	67.9	54.9	50.8
GAS	2.2	3.7	3.9	3.5	3.7	11.9	13.1
ELECTRICITY	14.7	17.4	17.5	19.8	24.0	19.4	19.5
OTHER [1]	-	-	-	-	-	-	1.6

Source: IEA/OECD Energy Balances and IEA Country Submissions (1982).

1. Includes heat from combined heat and power plants and direct use of solar, wind and other non-solid renewable energy sources.

SPAIN

5/D6 ENERGY USE IN INDUSTRY [1] BY FUEL

	1973	1978	1979	1980	1981	1985	1990
TOTAL (MTOE)	23.7	24.8	25.8	24.8	22.5	29.3	35.4
SOLID FUELS	5.4	3.5	3.5	3.3	3.4	7.3	8.4
OIL	14.6	16.3	16.8	15.8	13.3	13.4	15.0
GAS	.4	.8	.9	.7	.7	2.9	4.4
ELECTRICITY	3.3	4.3	4.6	5.0	5.1	5.8	7.1
OTHER [2]	-	-	-	-	-	-	.5
TOTAL (PERCENTS)	100.0	100.0	100.0	100.0	100.0	100.0	100.0
SOLID FUELS	22.6	14.1	13.6	13.2	15.1	24.8	23.8
OIL	61.8	65.5	65.2	63.7	59.3	45.7	42.5
GAS	1.8	3.1	3.3	3.0	3.0	9.9	12.4
ELECTRICITY	13.8	17.3	17.8	20.1	22.6	19.6	20.0
OTHER [2]	-	-	-	-	-	-	1.4

Source: IEA/OECD Energy Balances and IEA Country Submissions (1982).

1. Includes non-energy use of petroleum products.
2. Includes heat from combined heat and power plants and direct use of solar, wind and other non-solid renewable energy sources.

5/D7 ENERGY USE IN OTHER SECTORS [1] BY FUEL

	1973	1978	1979	1980	1981	1985	1990
TOTAL (MTOE)	7.6	10.8	11.7	11.4	10.9	13.7	17.4
SOLID FUELS	.4	.3	.6	.3	.4	.8	.8
OIL	5.2	7.8	8.1	8.0	7.5	7.5	8.9
GAS	.3	.4	.4	.5	.4	1.3	1.6
ELECTRICITY	1.7	2.4	2.5	2.6	2.6	4.0	5.6
OTHER [2]	-	-	-	-	-	.1	.4
TOTAL (PERCENTS)	100.0	100.0	100.0	100.0	100.0	100.0	100.0
SOLID FUELS	4.9	2.6	5.4	3.0	3.8	5.9	4.7
OIL	68.8	71.6	69.1	70.5	68.5	54.8	51.3
GAS	4.0	3.7	3.7	3.9	3.9	9.5	9.2
ELECTRICITY	22.3	22.1	21.7	22.5	23.8	29.1	32.4
OTHER [2]	-	-	-	-	-	.7	2.4

Source: IEA/OECD Energy Balances and IEA Country Submissions (1982).

1. Includes use in agricultural, commercial, public services and residential sectors.
2. Includes heat from combined heat and power plants and direct use of solar, wind and other non-solid renewable energy sources.

SWEDEN

TOTAL ENERGY REQUIREMENTS BY FUEL

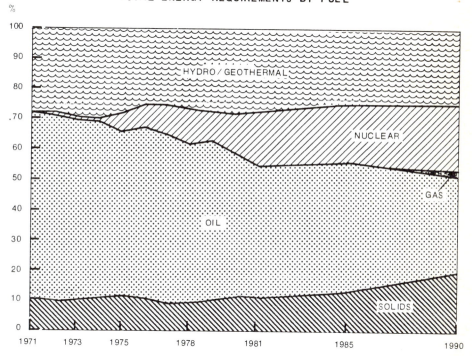

SWEDEN

5/A1 TOTAL PRIMARY ENERGY REQUIREMENTS (TPER) BY FUEL

	1973	1978	1979	1980	1981	1985	1990
TPER (MTOE) [1]	47.7	49.9	51.6	48.7	50.2	56.8	58.9
SOLID FUELS	4.8	4.7	5.0	5.6	5.4	7.3	11.5
OIL	28.3	26.1	27.1	22.9	22.2	24.0	17.9
GAS	-	-	-	-	-	-	.3
NUCLEAR	.5	5.6	4.9	6.2	8.9	10.8	13.5
HYDRO/GEOTHERMAL	14.1	13.6	14.4	13.9	13.9	14.2	14.6
OTHER [2]	-	-	-	-	-	.5	1.0
TPER (PERCENTS) [1]	100.0	100.0	100.0	100.0	100.0	100.0	100.0
SOLID FUELS	10.0	9.3	9.8	11.5	10.8	12.9	19.5
OIL	59.3	52.4	52.6	47.0	44.2	42.2	30.4
GAS	-	-	-	-	-	-	.6
NUCLEAR	1.0	11.2	9.6	12.8	17.7	19.0	22.9
HYDRO/GEOTHERMAL	29.5	27.2	27.9	28.6	27.8	25.0	24.8
OTHER [2]	-	-	-	-	-	.9	1.7

Source: IEA/OECD Energy Balances and IEA Country Submissions (1982).

1. Net imports of electricity are only included in totals.
2. Includes other electricity generation (solar, wind, etc.) and traded synthetic fuels.

5/A2 TOTAL FINAL ENERGY CONSUMPTION (TFC) BY FUEL

	1973	1978	1979	1980	1981	1985	1990
TFC (MTOE)	36.1	33.3	34.7	34.5	33.1	37.1	37.1
SOLID FUELS	4.4	4.4	4.6	5.1	4.8	6.0	7.7
OIL	25.6	21.9	22.7	20.5	19.4	18.2	14.3
GAS	.1	.1	.1	.1	.1	.3	.5
ELECTRICITY	6.0	7.0	7.4	7.4	7.6	9.4	10.5
HEAT [1]	na	na	na	1.4	1.3	3.2	4.0
TFC (PERCENTS)	100.0	100.0	100.0	100.0	100.0	100.0	100.0
SOLID FUELS	12.3	13.2	13.3	14.7	14.4	16.3	20.8
OIL	71.0	65.7	65.3	59.5	58.6	49.1	38.7
GAS	.3	.2	.2	.2	.2	.8	1.5
ELECTRICITY	16.5	20.9	21.2	21.6	23.0	25.3	28.3
HEAT [1]	na	na	na	4.0	3.8	8.6	10.7

Source: IEA/OECD Energy Balances and IEA Country Submissions (1982).

1. Includes only heat production from combined heat and power plants.

SWEDEN

5/A3 TOTAL SOLID FUELS BALANCE

	1973	1978	1979	1980	1981	1985	1990
PRODUCTION	3.2	3.2	3.3	4.0	4.0	4.8	7.1
IMPORTS	1.7	1.3	1.8	1.7	1.5	2.5	4.4
EXPORTS	.0	-.1	.0	-.1	-.2	-	-
MARINE BUNKERS	-	-	-	-	-	-	-
STOCK CHANGE	-.1	.3	.0	.0	.1	-	-
PRIMARY REQUIREMENTS	4.8	4.7	5.0	5.6	5.4	7.3	11.5
ELECTRICITY GENERATION	-.2	-.1	-.1	-.3	-.4	-1.0	-3.5
GAS MANUFACTURE	-	-	-	-	-	-.2	-.2
LIQUEFACTION	-	-	-	-	-	-	-
ENERGY SECTOR							
OWN USE + LOSSES [1]	-.2	-.2	-.3	-.3	-.2	-.1	-.1
FINAL CONSUMPTION	4.4	4.4	4.6	5.1	4.8	6.0	7.7
INDUSTRY SECTOR	4.0	4.3	4.5	4.3	4.0	4.8	5.6
IRON AND STEEL	.9	.9	1.1	.9	.7	na	na
CHEMICAL/							
PETROCHEMICAL	.0	.0	.1	.1	.1	na	na
OTHER INDUSTRY	3.1	3.3	3.4	3.3	3.2	na	na
TRANSPORTATION SECTOR	-	-	.0	-	-	-	-
OTHER SECTORS	.4	.1	.1	.7	.8	1.2	2.1
AGRICULTURE	-	-	-	-	-	-	-
COMMERCIAL AND							
PUBLIC SERVICE	-	-	-	-	-	-	-
RESIDENTIAL	.4	.1	.1	.7	.8	-	-

Source: IEA/OECD Energy Balances and IEA Country Submissions (1982).

1. Includes statistical difference.

SWEDEN

5/B1 SOLID FUELS INDIGENOUS PRODUCTION BY FUEL TYPE
(MTOE)

	1973	1978	1979	1980	1981	1985	1990
TOTAL SOLID FUELS	3.22	3.17	3.29	3.96	3.98	4.82	7.14
HARD COAL	-	.01	.01	.01	.02	.02	.02
COKING COAL	-	-	-	-	-	-	-
STEAM COAL	-	.01	.01	.01	.02	.02	.02
BROWN COAL/LIGNITE	-	-	-	-	-	-	-
OTHER SOLID FUELS	3.22	3.16	3.28	3.95	3.96	4.80	7.12

Source: IEA/OECD Coal Statistics, IEA/OECD Energy Balances, IEA Country Submissions (1982).

5/B2 COAL PRODUCTION BY TYPE
(THOUSAND METRIC TONS)

	1973	1978	1979	1980	1981
HARD COAL	-	16	14	18	28
COKING COAL	-	-	-	-	-
STEAM COAL	-	16	14	18	28
BROWN COAL/LIGNITE	-	-	-	-	-
DERIVED PRODUCTS:					
PATENT FUEL	-	-	-	-	-
COKE OVEN COKE	533	933	1153	1189	1094
GAS COKE	-	-	-	-	-
BROWN COAL BRIQUETTES	-	-	-	-	-
COKE OVEN GAS(TCAL)	865	1600	1988	2048	1896
BLAST FURNACE GAS(TCAL)	3500	3084	3891	3169	2238

Source: IEA/OECD Coal Statistics.

SWEDEN

5/C1 SOLID FUELS TRADE BY TYPE
(MTOE)

	1973	1978	1979	1980	1981	1985	1990
IMPORTS							
TOTAL SOLID FUELS	1.67	1.29	1.80	1.72	1.48	2.49	4.37
HARD COAL	.67	1.00	1.38	1.42	1.32	2.49	4.37
COKING COAL	na	.81	1.09	1.16	.94	1.20	1.20
STEAM COAL	na	.20	.29	.25	.38	1.29	3.17
BROWN COAL/LIGNITE	-	-	-	-	-	-	-
OTHER SOLID FUELS [1]	1.00	.29	.42	.30	.16	-	-
EXPORTS							
TOTAL SOLID FUELS	.02	.08	.03	.08	.16	-	-
HARD COAL	.01	.02	.02	-	-	-	-
COKING COAL	na	-	-	-	-	-	-
STEAM COAL	na	.02	.02	-	-	-	-
BROWN COAL/LIGNITE	-	-	-	-	-	-	-
OTHER SOLID FUELS [1]	.01	.06	.01	.08	.16	-	-

Source: IEA/OECD Coal Statistics, IEA/OECD Energy Balances, IEA Country Submissions (1982).

1.   Includes derived coal products, e.g. coke.

SWEDEN

5/C2 COAL IMPORTS BY ORIGIN
(THOUSAND METRIC TONS)

	1978	1979	1980	1981
HARD COAL IMPORTS	1543	2119	2182	2028
COKING COAL IMPORTS FROM:	1239	1679	1790	1444
AUSTRALIA	-	-	-	53
CANADA	78	164	159	261
GERMANY	231	161	-	-
UNITED KINGDOM	5	71	-	33
UNITED STATES	338	755	1106	1017
OTHER OECD	-	-	-	-
CZECHOSLOVAKIA	99	107	122	45
POLAND	71	-	1	-
USSR	417	421	402	35
CHINA	-	-	-	-
SOUTH AFRICA	-	-	-	-
OTHER NON-OECD	-	-	-	-
STEAM COAL IMPORTS FROM:	304	440	392	584
AUSTRALIA	-	3	6	-
CANADA	-	-	28	-
GERMANY	42	2	-	2
UNITED KINGDOM	30	39	31	140
UNITED STATES	-	-	54	191
OTHER OECD	-	10	18	64
CZECHOSLOVAKIA	8	13	11	8
POLAND	172	282	171	57
USSR	52	91	73	122
CHINA	-	-	-	-
SOUTH AFRICA	-	-	-	-
OTHER NON-OECD	-	-	-	-

Source: IEA/OECD Coal Statistics.

SWEDEN

5/D1 FUEL INPUT IN ELECTRICITY GENERATION

	1973	1978	1979	1980	1981	1985	1990
TOTAL (MTOE)	17.1	22.0	22.7	23.1	25.0	30.0	34.3
SOLID FUELS	.2	.1	.1	.3	.4	1.0	3.5
OIL	2.4	2.8	3.2	2.7	1.8	3.6	1.7
GAS	-	.0	.0	.0	.0	-	-
NUCLEAR	.5	5.6	4.9	6.2	8.9	10.8	13.5
HYDRO/GEOTHERMAL	14.1	13.6	14.4	13.9	13.9	14.2	14.6
OTHER [1]	-	-	-	-	-	.5	1.0
TOTAL (PERCENTS)	100.0	100.0	100.0	100.0	100.0	100.0	100.0
SOLID FUELS	.9	.5	.6	1.2	1.5	3.2	10.2
OIL	14.0	12.6	14.2	11.6	7.0	11.9	4.9
GAS	-	.0	.0	.0	.0	-	-
NUCLEAR	2.9	25.3	21.8	26.9	35.6	35.9	39.3
HYDRO/GEOTHERMAL	82.2	61.6	63.4	60.3	55.8	47.2	42.6
OTHER [1]	-	-	-	-	-	1.7	3.0

Source: IEA/OECD Energy Balances and IEA Country Submissions (1982).

1.   Solar, wind, etc.

5/D2 ELECTRICTY PRODUCTION BY FUEL TYPE

	1973	1978	1979	1980	1981	1985	1990
TOTAL (TWH)	78.1	92.9	95.2	97.0	102.7	120.0	134.0
SOLID FUELS	.9	.3	.3	.8	.7	2.0	6.0
OIL	15.2	11.1	12.6	10.5	4.8	7.0	3.0
GAS	-	-	-	-	-	-	-
NUCLEAR	2.1	23.8	21.0	26.5	37.8	47.0	58.0
HYDRO/GEOTHERMAL	59.9	57.8	61.2	59.2	59.4	63.0	65.0
OTHER [1]	-	-	-	-	-	1.0	2.0
TOTAL (PERCENTS)	100.0	100.0	100.0	100.0	100.0	100.0	100.0
SOLID FUELS	1.2	.3	.3	.8	.7	1.7	4.5
OIL	19.4	11.9	13.3	10.8	4.7	5.8	2.2
GAS	-	-	-	-	-	-	-
NUCLEAR	2.7	25.6	22.1	27.3	36.8	39.2	43.3
HYDRO/GEOTHERMAL	76.7	62.2	64.3	61.1	57.8	52.5	48.5
OTHER [1]	-	-	-	-	-	.8	1.5

Source: IEA/OECD Energy Balances and IEA Country Submissions (1982).

1.   Solar, wind, etc.

SWEDEN

5/D3 PROJECTIONS OF ELECTRICITY GENERATION CAPACITY BY FUEL
(GW)

	Nuclear	Hydro/ Geoth.	Solid Fuels	Oil	Gas	Multi Firing [1]	Total
I 1981							
Operating	6.4	14.9	-	7.2	-	0.8	29.3
II 1981-1985							
Capacity:							
Under Construction	2.0	0.9	-	-	-	0.1	3.0
Authorised	-	-	-	-	-	-	-
Other Planned	-	-	-	-	-	-	-
Conversion	-	-	0.3	-0.3	-	-	-
Decommissioning	-	-	-	-0.2	-	-	-0.2
Total Operating 1985	8.4	15.8	0.3	6.7	-	0.9	32.1
III 1986-1990							
Capacity:							
Under Construction	1.0	0.4	-	-	-	-	1.4
Authorised	-	-	-	-	-	-	-
Other Planned	-	0.3	0.4	-	-	0.8	1.5
Conversion	-	-	1.3	-1.3	-	-	-
Decommissioning	-	-	-	-0.4	-	-	-0.4
Total Operating 1990	9.4	16.5	2.0	5.0	-	1.7	34.6

Source: IEA/OECD Electricity Statistics and IEA Country Submissions (1982).

1.   Includes "Other" capacity.

SWEDEN

5/D4 ENERGY USE IN IRON AND STEEL INDUSTRY BY FUEL

	1973	1978	1979	1980	1981	1985	1990
TOTAL (MTOE)	2.4	2.0	2.2	1.9	1.6	na	na
SOLID FUELS	.9	.9	1.1	.9	.7	na	na
OIL	1.0	.7	.7	.6	.5	na	na
GAS	-	-	-	-	-	na	na
ELECTRICITY	.5	.4	.4	.4	.4	na	na
HEAT [1]	-	-	-	-	-	na	na
TOTAL (PERCENTS)	100.0	100.0	100.0	100.0	100.0	na	na
SOLID FUELS	36.9	45.1	47.9	46.8	44.0	na	na
OIL	42.1	34.1	31.7	31.0	29.3	na	na
GAS	-	-	-	-	-	na	na
ELECTRICITY	20.9	20.8	20.4	22.2	26.7	na	na
HEAT [1]	-	-	-	-	-	na	na

Source: IEA/OECD Energy Balances and IEA Country Submissions (1982).

1. Includes only heat from combined heat and power plants.

5/D5 ENERGY USE IN OTHER INDUSTRIES BY FUEL

	1973	1978	1979	1980	1981	1985	1990
TOTAL (MTOE)	14.4	12.2	13.0	12.2	11.3	na	na
SOLID FUELS	3.2	3.4	3.5	3.4	3.3	na	na
OIL	8.3	5.9	6.5	5.7	5.0	na	na
GAS	.0	.0	.0	.0	.0	na	na
ELECTRICITY	2.9	3.0	3.1	3.1	3.1	na	na
HEAT [1]	-	-	-	-	-	na	na
TOTAL (PERCENTS)	100.0	100.0	100.0	100.0	100.0	na	na
SOLID FUELS	21.9	27.4	26.6	27.9	29.1	na	na
OIL	57.8	48.3	49.5	46.7	43.8	na	na
GAS	.1	.1	.1	.1	.1	na	na
ELECTRICITY	20.2	24.2	23.9	25.3	27.0	na	na
HEAT [1]	-	-	-	-	-	na	na

Source: IEA/OECD Energy Balances and IEA Country Submissions (1982).

1. Includes only heat from combined heat and power plants.

SWEDEN

5/D6 ENERGY USE IN INDUSTRY [1] BY FUEL

	1973	1978	1979	1980	1981	1985	1990
TOTAL (MTOE)	16.8	14.3	15.2	14.1	13.0	16.9	17.4
SOLID FUELS	4.0	4.3	4.5	4.3	4.0	4.8	5.6
OIL	9.3	6.6	7.2	6.3	5.4	6.9	5.8
GAS	.0	.0	.0	.0	.0	.2	.4
ELECTRICITY	3.4	3.4	3.6	3.5	3.5	4.7	5.1
HEAT [2]	-	-	-	-	-	.3	.4
TOTAL (PERCENTS)	100.0	100.0	100.0	100.0	100.0	100.0	100.0
SOLID FUELS	24.1	30.0	29.7	30.5	31.0	28.5	32.1
OIL	55.6	46.3	46.9	44.5	42.0	40.8	33.7
GAS	.1	.0	.0	.0	.0	1.2	2.6
ELECTRICITY	20.3	23.7	23.4	24.9	27.0	27.9	29.2
HEAT [2]	-	-	-	-	-	1.5	2.5

Source: IEA/OECD Energy Balances and IEA Country Submissions (1982).

1. Includes non-energy use of petroleum products.
2. Includes only heat from combined heat and power plants.

5/D7 ENERGY USE IN OTHER SECTORS [1] BY FUEL

	1973	1978	1979	1980	1981	1985	1990
TOTAL (MTOE)	14.1	12.8	13.3	14.4	14.3	13.8	13.3
SOLID FUELS	.4	.1	.1	.7	.8	1.2	2.1
OIL	11.3	9.2	9.5	8.5	8.3	5.1	2.4
GAS	.1	.1	.1	.1	.1	.1	.1
ELECTRICITY	2.4	3.4	3.6	3.7	3.9	4.5	5.2
HEAT [2]	na	na	na	1.4	1.3	2.9	3.5
TOTAL (PERCENTS)	100.0	100.0	100.0	100.0	100.0	100.0	100.0
SOLID FUELS	2.8	.9	.8	5.1	5.3	8.8	16.1
OIL	79.8	72.1	71.6	58.9	58.0	36.9	17.9
GAS	.6	.6	.5	.4	.4	.6	.8
ELECTRICITY	16.8	26.5	27.0	25.9	27.5	32.5	38.7
HEAT [2]	na	na	na	9.5	8.8	21.2	26.5

Source: IEA/OECD Energy Balances and IEA Country Submissions (1982).

1. Includes use in agricultural, commercial, public services and residential sectors.
2. Includes only heat from combined heat and power plants.

SWITZERLAND

TOTAL ENERGY REQUIREMENTS BY FUEL

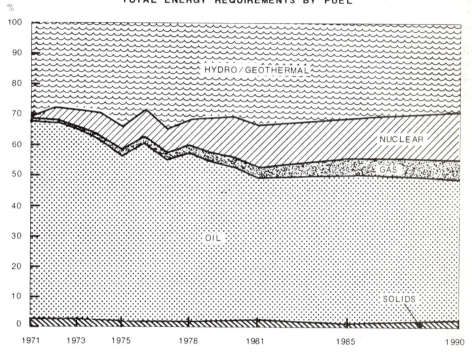

SWITZERLAND

5/A1 TOTAL PRIMARY ENERGY REQUIREMENTS (TPER) BY FUEL

	1973	1978	1979	1980	1981	1985	1990
TPER (MTOE) [1]	23.9	23.9	24.0	25.0	24.8	25.3	27.1
SOLID FUELS	.6	.5	.4	.6	.8	.4	.8
OIL	15.2	13.5	12.9	12.9	11.9	12.3	12.4
GAS	.2	.7	.8	.9	1.0	1.4	1.8
NUCLEAR	1.5	2.0	2.8	3.4	3.6	3.3	4.2
HYDRO/GEOTHERMAL	6.8	7.7	7.7	8.0	8.6	7.9	7.9
OTHER [2]	-	-	-	-	-	-	-
TPER (PERCENTS) [1]	100.0	100.0	100.0	100.0	100.0	100.0	100.0
SOLID FUELS	2.4	2.1	1.9	2.3	3.0	1.6	3.0
OIL	63.4	56.4	53.8	51.5	47.9	48.6	45.8
GAS	.6	2.9	3.2	3.6	3.8	5.5	6.6
NUCLEAR	6.2	8.3	11.6	13.5	14.4	13.0	15.5
HYDRO/GEOTHERMAL	28.6	32.3	32.0	31.9	34.6	31.2	29.2
OTHER [2]	-	-	-	-	-	-	-

Source: IEA/OECD Energy Balances and IEA Country Submissions (1982).

1. Net imports of electricity are only included in totals.
2. Includes other electricity generation (solar, wind, etc.) and traded synthetic fuels.

5/A2 TOTAL FINAL ENERGY CONSUMPTION (TFC) BY FUEL

	1973	1978	1979	1980	1981	1985	1990
TFC (MTOE)	17.6	16.8	16.3	16.9	16.2	17.7	18.9
SOLID FUELS	.5	.5	.4	.6	.7	.4	.5
OIL	14.4	12.9	12.3	12.6	11.6	12.3	12.4
GAS	.2	.6	.7	.8	.8	1.4	1.7
ELECTRICITY	2.4	2.8	2.9	3.0	3.1	3.6	4.1
HEAT [1]	-	-	-	-	-	-	.2
TFC (PERCENTS)	100.0	100.0	100.0	100.0	100.0	100.0	100.0
SOLID FUELS	3.0	2.9	2.7	3.3	4.6	2.3	2.6
OIL	81.8	76.8	75.5	74.2	71.3	69.5	65.6
GAS	1.3	3.6	4.1	4.7	5.0	7.9	9.0
ELECTRICITY	13.9	16.6	17.7	17.8	19.1	20.3	21.7
HEAT [1]	-	-	-	-	-	-	1.1

Source: IEA/OECD Energy Balances and IEA Country Submissions (1982).

1. Includes only heat production from combined heat and power plants.

SWITZERLAND

5/A3 TOTAL SOLID FUELS BALANCE
(MTOE)

	1973	1978	1979	1980	1981	1985	1990
PRODUCTION	.2	.3	.2	.2	.2	.2	.4
IMPORTS	.2	.2	.4	.5	.7	.2	.4
EXPORTS	.0	-	.0	-	.0	-	-
MARINE BUNKERS	-	-	-	-	-	-	-
STOCK CHANGE	.1	.0	-.1	-.2	-.2	-	-
PRIMARY REQUIREMENTS	.6	.5	.4	.6	.8	.4	.8
ELECTRICITY GENERATION	-	.0	.0	.0	.0	-	-.3
GAS MANUFACTURE	.0	-	-	-	-	-	-
LIQUEFACTION	-	-	-	-	-	-	-
ENERGY SECTOR OWN USE + LOSSES [1]	.0	-	-	-	-	-	-
FINAL CONSUMPTION	.5	.5	.4	.6	.7	.4	.5
INDUSTRY SECTOR	.1	.1	.1	.2	.4	.2	.3
IRON AND STEEL	.0	.0	.0	.0	.0	-	-
CHEMICAL/ PETROCHEMICAL	.0	.0	.0	.0	.0	-	-
OTHER INDUSTRY	.1	.1	.1	.2	.4	.2	.3
TRANSPORTATION SECTOR	-	-	-	-	-	-	-
OTHER SECTORS	.5	.4	.3	.3	.3	.2	.2
AGRICULTURE	-	-	-	-	-	-	-
COMMERCIAL AND PUBLIC SERVICE	-	-	-	-	-	-	-
RESIDENTIAL	.5	.4	.3	.3	.3	-	-

Source: IEA/OECD Energy Balances and IEA Country Submissions (1982).

1. Includes statistical difference.

SWITZERLAND

5/B1 SOLID FUELS INDIGENOUS PRODUCTION BY FUEL TYPE
(MTOE)

	1973	1978	1979	1980	1981	1985	1990
TOTAL SOLID FUELS	.24	.27	.22	.23	.25	.20	.40
HARD COAL	-	-	-	-	-	-	-
COKING COAL	-	-	-	-	-	-	-
STEAM COAL	-	-	-	-	-	-	-
BROWN COAL/LIGNITE	-	-	-	-	-	-	-
OTHER SOLID FUELS	.24	.27	.22	.23	.25	.20	.40

Source: IEA/OECD Coal Statistics, IEA/OECD Energy Balances, IEA Country Submissions (1982).

5/C1 SOLID FUELS TRADE BY TYPE
(MTOE)

	1973	1978	1979	1980	1981	1985	1990
IMPORTS							
TOTAL SOLID FUELS	.24	.21	.35	.53	.71	.20	.40
HARD COAL	.09	.10	.22	.40	.60	.20e	.40e
COKING COAL	na	-	-	-	-	-	-
STEAM COAL	na	.10	.22	.40	.60	.20e	.40e
BROWN COAL/LIGNITE	-	-	-	-	-	-	-
OTHER SOLID FUELS [1]	.15	.11	.13	.13	.11	-	-
EXPORTS							
TOTAL SOLID FUELS	.02	-	-	-	-	-	-
HARD COAL	-	-	-	-	-	-	-
COKING COAL	-	-	-	-	-	-	-
STEAM COAL	-	-	-	-	-	-	-
BROWN COAL/LIGNITE	-	-	-	-	-	-	-
OTHER SOLID FUELS [1]	.02	-	-	-	-	-	-

Source: IEA/OECD Coal Statistics, IEA/OECD Energy Balances, IEA Country Submissions (1982).

1. Includes derived coal products, e.g. coke.

SWITZERLAND

5/C2 COAL IMPORTS BY ORIGIN
(THOUSAND METRIC TONS)

	1978	1979	1980	1981
HARD COAL IMPORTS	141	317	574	863
COKING COAL	-	-	-	-
STEAM COAL IMPORTS FROM:	141	317	574	863
AUSTRALIA	-	-	-	-
CANADA	-	-	-	-
GERMANY	79	115	124	135
UNITED KINGDOM	-	-	-	18
UNITED STATES	-	6	17	448
OTHER OECD	17	26	31	92
CZECHOSLOVAKIA	14	-	11	12
POLAND	8	5	53	9
USSR	-	-	-	-
CHINA	-	-	-	-
SOUTH AFRICA	23	165	337	149
OTHER NON-OECD	-	-	1	-

Source: IEA/OECD Coal Statistics.

SWITZERLAND

5/D1 FUEL INPUT IN ELECTRICITY GENERATION

	1973	1978	1979	1980	1981	1985	1990
TOTAL (MTOE)	8.7	10.0	10.8	11.4	12.2	11.2	12.5
SOLID FUELS	-	.0	.0	.0	.0	-	.3
OIL	.4	.3	.3	.1	.1	-	-
GAS	-	.0	.0	.0	.0	-	.1
NUCLEAR	1.5	2.0	2.8	3.4	3.6	3.3	4.2
HYDRO/GEOTHERMAL	6.8	7.7	7.7	8.0	8.6	7.9	7.9
OTHER [1]	-	-	-	-	-	-	-
TOTAL (PERCENTS)	100.0	100.0	100.0	100.0	100.0	100.0	100.0
SOLID FUELS	-	.1	.1	.0	.1	-	2.4
OIL	4.8	3.4	2.9	.5	.5	-	-
GAS	-	.2	.3	.3	.2	-	.8
NUCLEAR	17.0	19.6	25.7	29.5	29.2	29.5	33.6
HYDRO/GEOTHERMAL	78.2	76.8	71.0	69.7	70.0	70.5	63.2
OTHER [1]	-	-	-	-	-	-	-

Source: IEA/OECD Energy Balances and IEA Country Submissions (1982).

1. Solar, wind, etc.

5/D2 ELECTRICTY PRODUCTION BY FUEL TYPE

	1973	1978	1979	1980	1981	1985	1990
TOTAL (TWH)	38.0	43.1	46.3	49.1	52.2	46.8	53.4
SOLID FUELS	-	.0	.0	.0	.0	-	.6
OIL	2.6	1.7	1.6	.6	.4	.5	-
GAS	-	.2	.2	.3	.1	.1	.6
NUCLEAR	6.3	8.4	11.8	14.3	15.2	13.9	18.6
HYDRO/GEOTHERMAL	29.1	32.8	32.7	33.9	36.5	32.4	33.6
OTHER [1]	-	-	-	-	-	-	-
TOTAL (PERCENTS)	100.0	100.0	100.0	100.0	100.0	100.0	100.0
SOLID FUELS	-	.0	.0	.1	.1	-	1.1
OIL	6.8	3.9	3.4	1.1	.8	1.1	-
GAS	-	.4	.4	.5	.2	-	1.1
NUCLEAR	16.6	19.5	25.5	29.2	29.1	29.7	34.8
HYDRO/GEOTHERMAL	76.6	76.2	70.6	69.0	69.8	69.2	62.9
OTHER [1]	-	-	-	-	-	-	-

Source: IEA/OECD Energy Balances and IEA Country Submissions (1982).

1. Solar, wind, etc.

SWITZERLAND

5/D3 PROJECTIONS OF ELECTRICITY GENERATION CAPACITY BY FUEL
(GW)

	Nuclear	Hydro/ Geoth.	Solid Fuels	Oil	Gas	Multi Firing [1]	Total
I 1981							
Operating	1.9	11.5	-	0.7		-	14.1
II 1981-1985							
Capacity:							
Under Construction	0.9	0.1	-	-		-	1.0
Authorised	-	-	-	-		-	-
Other Planned	-	-	-	-		-	-
Conversion	-	-	-	-		-	-
Decommissioning	-	-	-	-		-	-
Total Operating 1985	2.8	11.6	-	0.7		-	15.1
III 1986-1990							
Capacity:							
Under Construction	-	-	-	-		-	-
Authorised	-	0.1	-	-		-	3.3
Other Planned	3.2	-	-	-		-	-
Conversion	-	-	-	-		-	-
Decommissioning	-	-	-	-		-	-
Total Operating 1990	5.0	11.7	-	0.7		-	18.4

Source: IEA/OECD Electricity Statistics and IEA Country Submissions (1982).

1. Includes "Other" capacity.

SWITZERLAND

5/D6 ENERGY USE IN INDUSTRY [1] BY FUEL

	1973	1978	1979	1980	1981	1985	1990
TOTAL (MTOE)	4.8	4.3	4.2	4.5	4.5	4.1	4.1
SOLID FUELS	.1	.1	.1	.2	.4	.2	.3
OIL	3.7	2.9	2.8	2.8	2.7	2.1	1.9
GAS	.0	.3	.4	.4	.4	.6	.6
ELECTRICITY	.9	.9	1.0	1.0	1.0	1.2	1.3
HEAT [2]	-	-	-	-	-	-	-
TOTAL (PERCENTS)	100.0	100.0	100.0	100.0	100.0	100.0	100.0
SOLID FUELS	1.6	2.9	3.1	5.4	9.2	4.9	7.3
OIL	77.7	68.1	65.2	63.0	59.7	51.2	46.3
GAS	1.0	7.1	8.5	9.1	8.5	14.6	14.6
ELECTRICITY	19.6	22.0	23.2	22.4	22.6	29.3	31.7
HEAT [2]	-	-	-	-	-	-	-

Source: IEA/OECD Energy Balances and IEA Country Submissions (1982).

1. Includes non-energy use of petroleum products.
2. Includes only heat from combined heat and power plants.

5/D7 ENERGY USE IN OTHER SECTORS[1] BY FUEL

	1973	1978	1979	1980	1981	1985	1990
TOTAL (MTOE)	8.5	8.4	8.1	8.2	7.4	8.1	8.0
SOLID FUELS	.5	.4	.3	.3	.3	.2	.2
OIL	6.5	6.1	5.7	5.7	4.7	4.9	3.9
GAS	.2	.3	.3	.4	.4	.8	1.1
ELECTRICITY	1.3	1.7	1.7	1.8	1.9	2.2	2.6
HEAT [2]	-	-	-	-	-	-	.2
TOTAL (PERCENTS)	100.0	100.0	100.0	100.0	100.0	100.0	100.0
SOLID FUELS	5.3	4.3	3.8	3.9	4.4	2.5	2.5
OIL	76.9	72.3	71.0	69.2	64.2	60.5	48.8
GAS	2.1	3.6	3.7	4.6	5.8	9.9	13.7
ELECTRICITY	15.7	19.7	21.5	22.3	25.7	27.2	32.5
HEAT [2]	-	-	-	-	-	-	2.5

Source: IEA/OECD Energy Balances and IEA Country Submissions (1982).

1. Includes use in agricultural, commercial, public services and residential sectors.
2. Includes only heat from combined heat and power plants.

TURKEY

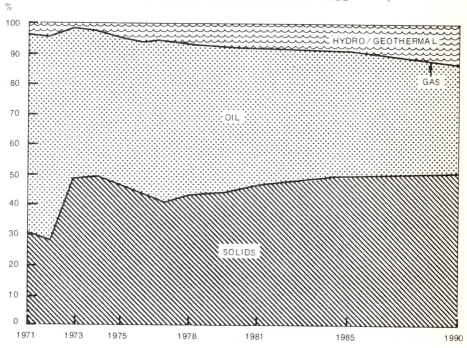

TOTAL ENERGY REQUIREMENTS BY FUEL

TURKEY

5/A1 TOTAL PRIMARY ENERGY REQUIREMENTS (TPER) BY FUEL

	1973	1978	1979	1980	1981	1985	1990
TPER (MTOE) [1]	24.9	31.2	30.9	32.5	34.2	54.8	75.0
SOLID FUELS	12.0	13.3	13.5	14.3	15.7	26.4	36.3
OIL	12.3	15.7	14.9	15.4	15.4	21.9	25.5
GAS	-	-	-	-	-	.1	.2
NUCLEAR	-	-	-	-	-	-	-
HYDRO/GEOTHERMAL	.6	2.2	2.4	2.7	3.0	4.9	9.8
OTHER [2]	-	-	-	-	-	1.4	3.1
TPER (PERCENTS) [1]	100.0	100.0	100.0	100.0	100.0	100.0	100.0
SOLID FUELS	48.2	42.5	43.6	44.1	46.0	48.2	48.5
OIL	49.3	50.2	48.2	47.4	44.9	39.9	34.0
GAS	-	-	-	-	-	.3	.3
NUCLEAR	-	-	-	-	-	-	-
HYDRO/GEOTHERMAL	2.5	7.0	7.8	8.2	8.7	9.0	13.1
OTHER [2]	-	-	-	-	-	2.6	4.1

Source: IEA/OECD Energy Balances and IEA Country Submissions (1982).

1. Net imports of electricity are only included in totals.
2. Includes other electricity generation (solar, wind, etc.), traded synthetic fuels and other non-specified fuels.

5/A2 TOTAL FINAL ENERGY CONSUMPTION (TFC) BY FUEL

	1973	1978	1979	1980	1981	1985	1990
TFC (MTOE)	20.8	25.5	24.9	26.5	27.9	41.5	54.1
SOLID FUELS	10.0	10.9	10.8	11.8	13.0	18.9	24.3
OIL	9.8	13.1	12.5	12.9	13.1	18.0	21.4
GAS	.0	.0	.0	.0	.0	.2	.3
ELECTRICITY	1.0	1.5	1.6	1.7	1.9	2.9	5.0
OTHER [1]	-	-	-	-	-	1.4	3.1
TFC (PERCENTS)	100.0	100.0	100.0	100.0	100.0	100.0	100.0
SOLID FUELS	48.2	42.6	43.3	44.7	46.5	45.6	44.9
OIL	46.9	51.3	50.0	48.7	46.8	43.3	39.5
GAS	.2	.0	.2	.2	.1	.6	.5
ELECTRICITY	4.8	6.0	6.5	6.4	6.6	7.1	9.2
OTHER [1]	-	-	-	-	-	3.4	5.8

Source: IEA/OECD Energy Balances and IEA Country Submissions (1982).

1. Includes heat production from combined heat and power plants and other non-specified fuels.

TURKEY

5/A3 TOTAL SOLID FUELS BALANCE
(MTOE)

	1973	1978	1979	1980	1981	1985	1990
PRODUCTION	12.0	13.3	13.5	14.4	15.4	22.9	29.3
IMPORTS	-	-	-	-	.4	3.6	7.1
EXPORTS	-	-	-	-	.0	-	-
MARINE BUNKERS	-	-	-	-	-	-	-
STOCK CHANGE	.0	-	.0	-.1	-.1	-	-
PRIMARY REQUIREMENTS	12.0	13.3	13.5	14.3	15.7	26.4	36.3
ELECTRICITY GENERATION	-1.4	-2.0	-2.3	-2.2	-2.6	-7.3	-11.8
GAS MANUFACTURE	-.2	.0	-.1	.0	.0	-.2	-.2
LIQUEFACTION	-	-	-	-	-	-	-
ENERGY SECTOR							
OWN USE + LOSSES [1]	-.4	-.4	-.4	-.2	-.1	-	-
FINAL CONSUMPTION	10.0	10.9	10.8	11.8	13.0	18.9	24.3
INDUSTRY SECTOR	1.0	1.3	1.2	1.4	2.5	7.4	11.5
IRON AND STEEL	.5	.5	.6	.5	1.1	3.6	4.9
CHEMICAL/							
PETROCHEMICAL	.3	.6	.2	.2	.4	.0	.0
OTHER INDUSTRY	.2	.2	.4	.7	1.0	3.8	6.6
TRANSPORTATION SECTOR	.5	.4	.2	.2	.2	.0	.0
OTHER SECTORS	8.5	9.2	9.4	10.3	10.3	11.5	12.7
AGRICULTURE	-	-	-	-	-	-	-
COMMERCIAL AND							
PUBLIC SERVICE	-	-	-	-	.0	-	-
RESIDENTIAL	8.5	9.2	9.4	10.3	10.3	11.5	12.7

Source: IEA/OECD Energy Balances and IEA Country Submissions (1982).

1. Includes statistical difference.

TURKEY

5/B1 SOLID FUELS INDIGENOUS PRODUCTION BY FUEL TYPE
(MTOE)

	1973	1978	1979	1980	1981	1985	1990
TOTAL SOLID FUELS	12.04	13.27	13.50	14.36	15.43	22.88	29.26
HARD COAL	2.83	2.71	2.39	2.17	2.42	2.97	3.26
COKING COAL	na	1.15	1.29	1.09	1.33	2.55	2.81
STEAM COAL	na	1.56	1.10	1.08	1.09	.42	.45
BROWN COAL/LIGNITE	1.72	2.76	3.32	4.09	4.77	14.74	21.22
OTHER SOLID FUELS	7.49	7.80	7.79	8.10	8.24	5.17	4.78

Source: IEA/OECD Coal Statistics, IEA/OECD Energy Balances, IEA Country Submissions (1982).

5/B2 COAL PRODUCTION BY TYPE
(THOUSAND METRIC TONS)

	1973	1978	1979	1980	1981
TOTAL COAL	10375	13630	14975	17191	19859
HARD COAL	4643	4440	3910	3552	3970
COKING COAL	na	1875	2100	1788	2190
STEAM COAL	na	2565	1810	1764	1780
BROWN COAL/LIGNITE	5732	9190	11065	13639	15889
DERIVED PRODUCTS:					
PATENT FUEL	-	-	-	-	-
COKE OVEN COKE	1280	1106	1106	1100	1899
GAS COKE	-	-	175	154	158
BROWN COAL BRIQUETTES	10	20	31	31	-
COKE OVEN GAS(TCAL)	2278	1329	2544	2560	3332
BLAST FURNACE GAS(TCAL)	2718	1450	2906	2930	4101

Source: IEA/OECD Coal Statistics.

TURKEY

5/C1 SOLID FUELS TRADE BY TYPE
(MTOE)

	1973	1978	1979	1980	1981	1985	1990
IMPORTS							
TOTAL SOLID FUELS	-	-	-	-	.40	3.57	7.07
HARD COAL	-	-	-	-	.40	3.57e	7.07e
COKING COAL	-	-	-	-	.40	1.80e	2.37e
STEAM COAL	-	-	-	-	-	1.77e	4.70e
BROWN COAL/LIGNITE	-	-	-	-	-	-	-
OTHER SOLID FUELS	-	-	-	-	-	-	-

Source: IEA/OECD Coal Statistics, IEA/OECD Energy Balances, IEA Country Submissions (1982).

5/C2 COAL IMPORTS BY ORIGIN
(THOUSAND METRIC TONS)

	1978	1979	1980	1981
HARD COAL IMPORTS	-	-	-	650
COKING COAL IMPORTS FROM:	-	-	-	650
AUSTRALIA	-	-	-	-
CANADA	-	-	-	-
GERMANY	-	-	-	134
UNITED KINGDOM	-	-	-	-
UNITED STATES	-	-	-	516
OTHER OECD	-	-	-	-
CZECHOSLOVAKIA	-	-	-	-
POLAND	-	-	-	-
USSR	-	-	-	-
CHINA	-	-	-	-
SOUTH AFRICA	-	-	-	-
OTHER NON-OECD	-	-	-	-
STEAM COAL IMPORTS	-	-	-	-

Source: IEA/OECD Coal Statistics.

TURKEY

5/D1 FUEL INPUT IN ELECTRICITY GENERATION

	1973	1978	1979	1980	1981	1985	1990
TOTAL (MTOE)	3.3	6.0	6.3	6.4	6.9	14.4	23.7
SOLID FUELS	1.4	2.0	2.3	2.2	2.6	7.3	11.8
OIL	1.3	1.8	1.7	1.5	1.4	2.1	2.1
GAS	-	-	-	-	-	-	-
NUCLEAR	-	-	-	-	-	-	-
HYDRO/GEOTHERMAL	.6	2.2	2.4	2.7	3.0	4.9	9.8
OTHER [1]	-	-	-	-	-	-	-
TOTAL (PERCENTS)	100.0	100.0	100.0	100.0	100.0	100.0	100.0
SOLID FUELS	42.7	32.8	35.6	34.9	37.2	51.0	49.8
OIL	38.7	30.4	26.1	23.6	19.9	14.6	8.8
GAS	-	-	-	-	-	-	-
NUCLEAR	-	-	-	-	-	-	-
HYDRO/GEOTHERMAL	18.6	36.8	38.3	41.5	43.0	34.4	41.4
OTHER [1]	-	-	-	-	-	-	-

Source: IEA/OECD Energy Balances and IEA Country Submissions (1982).

1. Solar, wind, etc.

5/D2 ELECTRICTY PRODUCTION BY FUEL TYPE

	1973	1978	1979	1980	1981	1985	1990
TOTAL (TWH)	14.3	21.7	22.5	23.3	24.7	42.0	69.2
SOLID FUELS	4.2	5.7	6.6	6.0	6.1	19.7	32.7
OIL	7.5	6.6	5.7	6.0	5.9	7.8	7.8
GAS	-	-	-	-	-	-	-
NUCLEAR	-	-	-	-	-	-	-
HYDRO/GEOTHERMAL	2.6	9.4	10.3	11.3	12.6	14.4	28.7
OTHER [1]	-	-	-	-	-	-	-
TOTAL (PERCENTS)	100.0	100.0	100.0	100.0	100.0	100.0	100.0
SOLID FUELS	29.2	26.4	29.2	25.6	24.9	47.0	47.3
OIL	52.4	30.5	25.1	25.6	24.0	18.6	11.3
GAS	-	-	-	-	-	-	-
NUCLEAR	-	-	-	-	-	-	-
HYDRO/GEOTHERMAL	18.3	43.1	45.7	48.8	51.1	34.4	41.4
OTHER [1]	-	-	-	-	-	-	-

Source: IEA/OECD Energy Balances and IEA Country Submissions (1982).

1. Solar, wind, etc.

TURKEY

5/D3 PROJECTIONS OF ELECTRICITY GENERATION CAPACITY BY FUEL
(GW)

	Nuclear	Hydro/ Geoth.	Solid Fuels	Oil	Gas	Multi Firing	Total
I 1981							
Operating	-	2.1	1.4	1.6	-	-	5.1
II 1981-1985							
Capacity:							
Under Construction	-	1.8	3.1	0.2	-	-	5.1
Authorised	-	-	-	-	-	-	-
Other Planned	-	-	-	-	-	-	-
Conversion	-	-	-	-	-	-	-
Decommissioning	-	-	-0.1	-0.5	-	-	-0.6
Total Operating 1985	-	3.9	4.4	1.3	-	-	9.6
III 1986-1990							
Capacity:							
Under Construction	-	3.1	1.1	-			4.2
Authorised	-	0.4	0.2	-	-	-	0.6
Other Planned	-	0.6	0.5	-	-	-	1.1
Conversion	-	-	-	-	-	-	-
Decommissioning	-	-	-0.2	-	-	-	-0.2
Total Operating 1990	-	8.0	6.0	1.3	-	-	15.3

Source: IEA/OECD Electricity Statistics and IEA Country Submissions (1982).

TURKEY

5/D4 ENERGY USE IN IRON AND STEEL INDUSTRY BY FUEL

	1973	1978	1979	1980	1981	1985	1990
TOTAL (MTOE)	.7	.8	.9	.8	1.7	4.3	6.2
SOLID FUELS	.5	.5	.6	.5	1.1	3.6	4.9
OIL	.1	.2	.2	.2	.4	.5	.8
GAS	-	-	-	-	-	-	-
ELECTRICITY	.1	.1	.1	.1	.1	.2	.5
HEAT [1]	-	-	-	-	-	-	-
TOTAL (PERCENTS)	100.0	100.0	100.0	100.0	100.0	100.0	100.0
SOLID FUELS	71.6	62.7	65.1	62.8	66.9	82.9	78.7
OIL	21.1	22.3	19.9	21.4	24.7	12.5	13.5
GAS	-	-	-	-	-	-	-
ELECTRICITY	7.3	14.9	14.9	15.8	8.5	4.6	7.7
HEAT [1]	-	-	-	-	-	-	-

Source: IEA/OECD Energy Balances and IEA Country Submissions (1982).

1. Includes only heat from combined heat and power plants.

5/D5 ENERGY USE IN OTHER INDUSTRIES BY FUEL

	1973	1978	1979	1980	1981	1985	1990
TOTAL (MTOE)	3.6	4.9	4.7	5.3	6.7	13.9	19.5
SOLID FUELS	.5	.8	.6	.9	1.4	3.8	6.6
OIL	2.5	3.3	3.2	3.5	4.3	7.9	9.4
GAS	.0	-	.0	.0	-	.1	.2
ELECTRICITY	.6	.9	.9	.9	1.0	1.9	3.2
HEAT [1]	-	-	-	-	-	-	-
TOTAL (PERCENTS)	100.0	100.0	100.0	100.0	100.0	100.0	100.0
SOLID FUELS	13.9	15.5	12.7	16.1	20.3	27.6	34.2
OIL	68.2	67.2	68.1	66.0	64.3	57.4	48.6
GAS	.0	-	.3	.3	-	1.1	1.0
ELECTRICITY	17.9	17.3	19.0	17.7	15.4	14.0	16.2
HEAT [1]	-	-	-	-	-	-	-

Source: IEA/OECD Energy Balances and IEA Country Submissions (1982).

1. Includes only heat from combined heat and power plants.

TURKEY

5/D6 ENERGY USE IN INDUSTRY[1] BY FUEL

	1973	1978	1979	1980	1981	1985	1990
TOTAL (MTOE)	4.3	5.8	5.6	6.2	8.4	18.2	25.7
SOLID FUELS	1.0	1.3	1.2	1.4	2.5	7.4	11.5
OIL	2.6	3.5	3.4	3.7	4.7	8.5	10.3
GAS	.0	-	.0	.0	-	.1	.2
ELECTRICITY	.7	1.0	1.0	1.1	1.2	2.1	3.6
HEAT [2]	-	-	-	-	-	-	-
TOTAL (PERCENTS)	100.0	100.0	100.0	100.0	100.0	100.0	100.0
SOLID FUELS	23.3	22.2	20.7	22.5	29.8	40.7	44.9
OIL	60.5	60.9	60.7	59.9	56.2	46.7	40.1
GAS	.0	-	.2	.3	-	.8	.8
ELECTRICITY	16.2	17.0	18.4	17.4	14.0	11.8	14.2
HEAT[2]	-	-	-	-	-	-	-

Source: IEA/OECD Energy Balances and IEA Country Submissions (1982).

1. Includes non-energy use of petroleum products.
2. Includes only heat from combined heat and power plants.

5/D7 ENERGY USE IN OTHER SECTORS [1] BY FUEL

	1973	1978	1979	1980	1981	1985	1990
TOTAL (MTOE)	12.1	13.6	13.7	14.7	14.9	17.9	22.0
SOLID FUELS	8.5	9.2	9.4	10.3	10.3	11.5	12.7
OIL	3.2	3.9	3.6	3.8	3.9	4.1	4.8
GAS	.0	.0	.0	.0	.0	.1	.1
ELECTRICITY	.3	.5	.6	.6	.7	.8	1.3
OTHER [2]	-	-	-	-	-	1.4	3.1
TOTAL (PERCENTS)	100.0	100.0	100.0	100.0	100.0	100.0	100.0
SOLID FUELS	70.9	67.5	69.0	70.0	69.0	64.2	57.8
OIL	26.4	28.4	26.6	25.7	26.4	23.1	21.6
GAS	.3	.1	.3	.2	.2	.5	.4
ELECTRICITY	2.4	4.0	4.2	4.1	4.4	4.3	5.9
OTHER [2]	-	-	-	-	-	7.9	14.2

Source: IEA/OECD Energy Balances and IEA Country Submissions (1982).

1. Includes use in agricultural, commercial, public services and residential sectors.
2. Includes heat from combined heat and power plants and other non-specified fuels.

UNITED KINGDOM

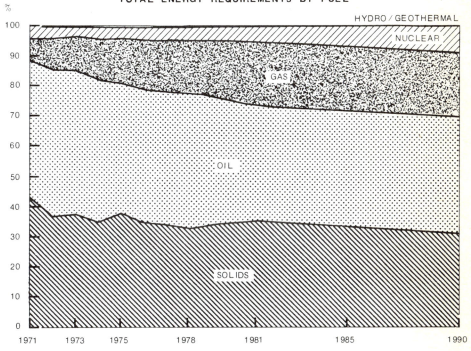

TOTAL ENERGY REQUIREMENTS BY FUEL

UNITED KINGDOM

5/A1 TOTAL PRIMARY ENERGY REQUIREMENTS (TPER) BY FUEL

	1973	1978	1979	1980	1981	1985	1990
TPER (MTOE) [1]	233.2	212.2	221.0	203.2	194.5	203.0	213.0
SOLID FUELS	86.8	69.2	74.7	69.6	67.7	67.0	65.0
OIL	111.6	94.6	93.7	81.8	74.3	78.0	82.0
GAS	25.6	37.6	41.2	41.1	41.6	43.0	45.0
NUCLEAR	7.9	9.5	9.9	9.4	9.5	14.0	19.0
HYDRO/GEOTHERMAL	1.3	1.3	1.4	1.3	1.4	1.0	1.0
OTHER [2]	-	-	-	-	-	-	1.0
TPER (PERCENTS) [1]	100.0	100.0	100.0	100.0	100.0	100.0	100.0
SOLID FUELS	37.2	32.6	33.8	34.2	34.8	33.0	30.5
OIL	47.8	44.6	42.4	40.3	38.2	38.4	38.5
GAS	11.0	17.7	18.6	20.2	21.4	21.2	21.1
NUCLEAR	3.4	4.5	4.5	4.6	4.9	6.9	8.9
HYDRO/GEOTHERMAL	.6	.6	.6	.6	.7	.5	.5
OTHER [2]	-	-	-	-	-	-	.5

Source: IEA/OECD Energy Balances and IEA Country Submissions (1982).

1. Net imports of electricity are only included in totals.
2. Includes other electricity generation (solar, wind, etc.) and traded synthetic fuels.

5/A2 TOTAL FINAL ENERGY CONSUMPTION (TFC) BY FUEL

	1973	1978	1979	1980	1981	1985	1990
TFC (MTOE)	156.0	146.0	152.5	137.6	134.5	144.0	152.0
SOLID FUELS	31.1	19.0	20.2	15.0	15.3	18.0	18.0
OIL	80.8	72.2	73.4	64.3	61.2	66.0	71.0
GAS	24.1	34.5	37.7	38.0	38.2	40.0	42.0
ELECTRICITY	20.0	20.4	21.2	20.2	19.8	20.0	21.0
HEAT [1]	na	na	na	.1	.1	-	-
TFC (PERCENTS)	100.0	100.0	100.0	100.0	100.0	100.0	100.0
SOLID FUELS	20.0	13.0	13.2	10.9	11.4	12.5	11.8
OIL	51.8	49.5	48.1	46.7	45.5	45.8	46.7
GAS	15.4	23.6	24.7	27.6	28.4	27.8	27.6
ELECTRICITY	12.8	13.9	13.9	14.6	14.7	13.9	13.8
HEAT [1]	na	na	na	.1	.1	-	-

Source: IEA/OECD Energy Balances and IEA Country Submissions (1982).

1. Includes only heat production from combined heat and power plants.

UNITED KINGDOM

5/A3 TOTAL SOLID FUELS BALANCE
(MTOE)

	1973	1978	1979	1980	1981	1985	1990
PRODUCTION	87.9	71.1	70.4	74.8	73.2	68.5	66.5
IMPORTS	1.3	1.4	2.7	4.3	2.5	2.5	2.5
EXPORTS	-2.2	-2.0	-1.8	-3.0	-6.2	-4.0	-4.0
MARINE BUNKERS	-	-	-	-	-	-	-
STOCK CHANGE	-.1	-1.4	3.5	-6.6	-1.8	-	-
PRIMARY REQUIREMENTS	86.8	69.2	74.7	69.6	67.7	67.0	65.0
ELECTRICITY GENERATION	-50.5	-48.5	-53.1	-53.1	-51.8	-47.0	-44.0
GAS MANUFACTURE	-.3	.0	.0	.0	-	-	-
LIQUEFACTION	-	-	-	-	-	-	-
ENERGY SECTOR							
OWN USE + LOSSES [1]	-5.0	-1.7	-1.5	-1.4	-.7	-2.0	-3.0
FINAL CONSUMPTION	31.1	19.0	20.2	15.0	15.3	18.0	18.0
INDUSTRY SECTOR	17.1	10.7	11.7	6.8	7.5	11.0	12.0
IRON AND STEEL	8.7	5.7	6.3	3.2	4.4	5.0	5.0
CHEMICAL/							
PETROCHEMICAL	.6	.2	.2	.1	.1	-	-
OTHER INDUSTRY	7.7	4.8	5.2	3.4	3.0	6.0	7.0
TRANSPORTATION SECTOR	.1	.0	.0	.0	.0	-	-
OTHER SECTORS	14.0	8.3	8.4	8.2	7.8	7.0	6.0
AGRICULTURE	.1	.0	.0	.0	.0	-	-
COMMERCIAL AND							
PUBLIC SERVICE	2.2	1.5	1.5	1.0	1.0	-	-
RESIDENTIAL	10.2	6.7	6.9	6.9	6.5	6.0	5.0

Source: IEA/OECD Energy Balances and IEA Country Submissions (1982).

1. Includes statistical difference.

Note: Projections of Solid Fuels Production, Imports and Exports for 1985 and 1990 are based on more recent information from United Kingdom authorities. Therefore, they do not correspond to previous estimates from the 1982 Country Submissions as published in the "Energy Policies and Programmes of IEA Countries, 1982 Review", IEA/OECD Paris 1983.

UNITED KINGDOM

5/B1 SOLID FUELS INDIGENOUS PRODUCTION BY FUEL TYPE
(MTOE)

	1973	1978	1979	1980	1981	1985	1990
TOTAL SOLID FUELS	87.85	71.13	70.41	74.84	73.24	68.50	66.50
HARD COAL	87.67	71.06	70.35	74.81	73.23	68.40	66.40
COKING COAL	na	8.69	6.93	5.78	4.53	4.40e	4.40e
STEAM COAL	na	62.37	63.42	69.03	68.70	64.00e	62.00e
BROWN COAL/LIGNITE	-	-	-	-	-	-	-
OTHER SOLID FUELS	.18	.08	.05	.03	.01	0.10e	0.10e

Source: IEA/OECD Coal Statistics, IEA/OECD Energy Balances, IEA Country Submissions (1982).

Note: See Table 5/A3.

5/B2 COAL PRODUCTION BY TYPE
(THOUSAND METRIC TONS)

	1973	1978	1979	1980	1981
TOTAL COAL	130239	121695	120637	128209	125301
HARD COAL	130239	121695	120637	128209	125301
COKING COAL	na	15110	12055	10050	7883
STEAM COAL	na	106585	108582	118159	117418
BROWN COAL/LIGNITE	-	-	-	-	-
DERIVED PRODUCTS:					
PATENT FUEL	1185	1053	980	926	976
COKE OVEN COKE	17776	12394	12512	10060	9059
GAS COKE	212	-	-	-	-
BROWN COAL BRIQUETTES	-	-	-	-	-
COKE OVEN GAS(TCAL)	30139	19656	20941	15599	14565
BLAST FURNACE GAS(TCAL)	28678	19681	21622	8870	12675

Source: IEA/OECD Coal Statistics.

UNITED KINGDOM

5/C1 SOLID FUELS TRADE BY TYPE
(MTOE)

	1973	1978	1979	1980	1981	1985	1990
IMPORTS							
TOTAL SOLID FUELS	1.27	1.41	2.66	4.33	2.54	2.50	2.50
HARD COAL	1.12	1.35	2.52	4.22	2.47	2.40e	2.40e
COKING COAL	na	.80	1.36	1.38	1.50	1.40e	1.40e
STEAM COAL	na	.56	1.16	2.83	.97	1.00e	1.00e
BROWN COAL/LIGNITE	-	-	-	-	-	-	-
OTHER SOLID FUELS [1]	.15	.06	.14	.11	.07	0.10e	0.10e
EXPORTS							
TOTAL SOLID FUELS	2.23	1.96	1.82	3.00	6.20	4.00	4.00
HARD COAL	1.80	1.30	1.25	2.19	5.24	3.00e	3.00e
COKING COAL	na	.08	.05	.03	.60	.60e	.60e
STEAM COAL	na	1.22	1.20	2.16	4.64	2.40e	2.40e
BROWN COAL/LIGNITE	-	-	-	-	-	-	-
OTHER SOLID FUELS [1]	.43	.66	.57	.81	.96	1.00e	1.00e

Source: IEA/OECD Coal Statistics, IEA/OECD Energy Balances, IEA Country Submissions (1982).

1. Includes derived coal products, e.g. coke.

Note: See Table 5/A3.

UNITED KINGDOM

5/C2 COAL IMPORTS BY ORIGIN
(THOUSAND METRIC TONS)

	1978	1979	1980	1981
HARD COAL IMPORTS	2352	4375	7334	4290
COKING COAL IMPORTS FROM:	1383	2362	2407	2604
AUSTRALIA	380	905	518	831
CANADA	-	-	-	-
GERMANY	207	185	-	-
UNITED KINGDOM	-	-	-	-
UNITED STATES	421	688	1486	1498
OTHER OECD	-	6	-	110
CZECHOSLOVAKIA	-	-	-	-
POLAND	375	578	403	131
USSR	-	-	-	34
CHINA	-	-	-	-
SOUTH AFRICA	-	-	-	-
OTHER NON-OECD	-	-	-	-
STEAM COAL IMPORTS FROM:	969	2013	4927	1686
AUSTRALIA	645	1259	2329	960
CANADA	-	-	-	-
GERMANY	53	44	80	132
UNITED KINGDOM	-	-	1	-
UNITED STATES	1	343	2167	456
OTHER OECD	73	93	78	21
CZECHOSLOVAKIA	-	-	-	-
POLAND	41	80	43	11
USSR	106	65	46	-
CHINA	-	-	-	1
SOUTH AFRICA	26	38	66	80
OTHER NON-OECD	24	91	117	25

Source: IEA/OECD Coal Statistics.

UNITED KINGDOM

5/C3 HARD COAL EXPORTS BY DESTINATION
(THOUSAND METRIC TONS)

	1978	1979	1980	1981
TOTAL EXPORTS	2253	2175	3809	9113
EXPORTS TO:				
AUSTRALIA	-	-	-	-
AUSTRIA	-	-	-	-
BELGIUM	101	100	46	213
CANADA	-	-	-	-
DENMARK	152	181	613	1921
FINLAND	-	-	-	-
FRANCE	922	770	1233	2706
GERMANY	548	679	1472	1586
GREECE	-	-	-	-
ICELAND	-	-	-	-
IRELAND	175	155	210	339
ITALY	53	15	-	67
JAPAN	-	-	-	-
LUXEMBOURG	-	14	38	34
NETHERLANDS	167	56	111	441
NEW ZEALAND	-	-	-	-
NORWAY	77	79	56	64
PORTUGAL	3	2	3	32
SPAIN	13	-	-	90
SWEDEN	34	110	23	157
SWITZERLAND	-	-	-	-
TURKEY	-	-	-	-
UNITED KINGDOM	-	-	-	-
UNITED STATES	-	-	-	-
TOTAL NON-OECD	8	14	4	1463

Source: IEA/OECD Coal Statistics.

UNITED KINGDOM

5/C4 COKING COAL EXPORTS BY DESTINATION
(THOUSAND METRIC TONS)

	1978	1979	1980	1981
TOTAL EXPORTS	132	93	53	1041
EXPORTS TO:				
AUSTRALIA	-	-	-	-
AUSTRIA	-	-	-	-
BELGIUM	22	-	-	174
CANADA	-	-	-	-
DENMARK	-	5	2	-
FINLAND	-	-	-	-
FRANCE	-	-	-	394
GERMANY	45	10	50	55
GREECE	-	-	-	-
ICELAND	-	-	-	-
IRELAND	45	7	1	-
ITALY	-	-	-	48
JAPAN	-	-	-	-
LUXEMBOURG	-	-	-	-
NETHERLANDS	6	-	-	51
NEW ZEALAND	-	-	-	-
NORWAY	-	-	-	5
PORTUGAL	-	-	-	29
SPAIN	10	-	-	49
SWEDEN	4	71	-	31
SWITZERLAND	-	-	-	-
TURKEY	-	-	-	-
UNITED KINGDOM	-	-	-	-
UNITED STATES	-	-	-	-
TOTAL NON-OECD	-	-	-	205

Source: IEA/OECD Coal Statistics.

UNITED KINGDOM

5/C5 STEAM COAL EXPORTS BY DESTINATION
(THOUSAND METRIC TONS)

	1978	1979	1980	1981
TOTAL EXPORTS	2121	2082	3756	8072
EXPORTS TO:				
AUSTRALIA	-	-	-	-
AUSTRIA	-	-	-	-
BELGIUM	79	100	46	39
CANADA	-	-	-	-
DENMARK	152	176	611	1921
FINLAND	-	-	-	-
FRANCE	922	770	1233	2312
GERMANY	503	669	1422	1531
GREECE	-	-	-	-
ICELAND	-	-	-	-
IRELAND	130	148	209	339
ITALY	53	15	-	19
JAPAN	-	-	-	-
LUXEMBOURG	-	14	38	34
NETHERLANDS	161	56	111	390
NEW ZEALAND	-	-	-	-
NORWAY	77	79	56	59
PORTUGAL	3	2	3	3
SPAIN	3	-	-	41
SWEDEN	30	39	23	126
SWITZERLAND	-	-	-	-
TURKEY	-	-	-	-
UNITED KINGDOM	-	-	-	-
UNITED STATES	-	-	-	-
TOTAL NON-OECD	8	14	4	1258

Source: IEA/OECD Coal Statistics.

UNITED KINGDOM

5/D1 FUEL INPUT IN ELECTRICITY GENERATION

	1973	1978	1979	1980	1981	1985	1990
TOTAL (MTOE)	79.7	73.4	77.5	72.7	69.7	67.0	68.0
SOLID FUELS	50.5	48.5	53.1	53.1	51.8	47.0	44.0
OIL	19.1	13.1	12.3	8.2	6.6	5.0	3.0
GAS	1.0	1.0	.8	.6	.4	.0	.0
NUCLEAR	7.9	9.5	9.9	9.4	9.5	14.0	19.0
HYDRO/GEOTHERMAL	1.3	1.3	1.4	1.3	1.4	1.0	1.0
OTHER [1]	-	-	-	-	-	-	1.0
TOTAL (PERCENTS)	100.0	100.0	100.0	100.0	100.0	100.0	100.0
SOLID FUELS	63.3	66.0	68.5	73.1	74.4	70.1	64.7
OIL	23.9	17.8	15.9	11.3	9.4	7.5	4.4
GAS	1.2	1.4	1.0	.8	.6	.0	.0
NUCLEAR	9.9	12.9	12.8	13.0	13.7	20.9	27.9
HYDRO/GEOTHERMAL	1.6	1.8	1.8	1.8	1.9	1.5	1.5
OTHER [1]	-	-	-	-	-	-	1.5

Source: IEA/OECD Energy Balances and IEA Country Submissions (1982).

1. Solar, wind, etc.

5/D2 ELECTRICTY PRODUCTION BY FUEL TYPE

	1973	1978	1979	1980	1981	1985	1990
TOTAL (TWH)	282.0	287.7	299.9	284.9	277.7	268.0	273.0
SOLID FUELS	174.6	188.2	202.9	207.5	206.9	192.0	180.0
OIL	72.2	52.9	50.1	33.1	26.0	15.0	9.0
GAS	2.7	4.2	3.1	2.1	1.5	1.0	1.0
NUCLEAR	28.0	37.2	38.3	37.0	38.0	56.0	79.0
HYDRO/GEOTHERMAL	4.6	5.2	5.5	5.1	5.4	4.0	4.0
OTHER [1]	-	-	-	-	-	-	-
TOTAL (PERCENTS)	100.0	100.0	100.0	100.0	100.0	100.0	100.0
SOLID FUELS	61.9	65.4	67.7	72.8	74.5	71.6	65.9
OIL	25.6	18.4	16.7	11.6	9.4	5.6	3.3
GAS	1.0	1.5	1.0	.7	.5	.4	.4
NUCLEAR	9.9	12.9	12.8	13.0	13.7	20.9	28.9
HYDRO/GEOTHERMAL	1.6	1.8	1.8	1.8	1.9	1.5	1.5
OTHER [1]	-	-	-	-	-	-	-

Source: IEA/OECD Energy Balances and IEA Country Submissions (1982).

1. Solar, wind, etc.

UNITED KINGDOM

5/D3 PROJECTIONS OF ELECTRICITY GENERATION CAPACITY BY FUEL
(GW)

	Nuclear	Hydro/ Geoth.	Solid Fuels	Oil	Gas	Multi Firing	Total
I 1981							
Operating	6.5	2.5	40.0	13.6	-	3.2	70.2
II 1981-1985							
Capacity:							
Under Construction							
Authorised							
Other Planned							
Conversion			na				
Decommissioning							
Total Operating 1985							
III 1986-1990							
Capacity:							
Under Construction							
Authorised							
Other Planned							
Conversion			na				
Decommissioning							
Total Operating 1990							

Source: IEA/OECD Electricity Statistics and IEA Country Submissions (1982)

1. Includes "Other" capacity.

5/D4 ENERGY USE IN IRON AND STEEL INDUSTRY BY FUEL

	1973	1978	1979	1980	1981	1985	1990
TOTAL (MTOE)	15.3	10.4	11.3	6.5	7.5	8.0	8.0
SOLID FUELS	8.7	5.7	6.3	3.2	4.4	5.0	5.0
OIL	4.6	2.5	2.5	1.3	1.3	1.0	1.0
GAS	.9	1.0	1.2	1.0	.9	1.0	1.0
ELECTRICITY	1.1	1.2	1.3	.9	1.0	1.0	1.0
HEAT [1]	-	-	-	-	-	-	-
TOTAL (PERCENTS)	100.0	100.0	100.0	100.0	100.0	100.0	100.0
SOLID FUELS	56.8	54.6	56.0	50.2	57.8	62.5	62.5
OIL	30.1	23.8	22.1	20.5	16.9	12.5	12.5
GAS	5.9	9.7	10.8	16.0	12.5	12.5	12.5
ELECTRICITY	7.3	11.9	11.1	13.4	12.8	12.5	12.5
HEAT [1]	-	-	-	-	-	-	-

Source: IEA/OECD Energy Balances and IEA Country Submissions (1982).

1. Includes only heat from combined heat and power plants.

5/D5 ENERGY USE IN OTHER INDUSTRIES BY FUEL

	1973	1978	1979	1980	1981	1985	1990
TOTAL (MTOE)	56.2	48.2	49.4	42.0	39.3	46.0	50.0
SOLID FUELS	8.3	5.0	5.4	3.5	3.1	6.0	7.0
OIL	31.7	23.7	24.1	19.1	17.8	22.0	24.0
GAS	9.4	12.7	12.9	12.7	12.1	12.0	12.0
ELECTRICITY	6.7	6.8	7.0	6.6	6.3	6.0	7.0
HEAT [1]	-	-	-	-	-	-	-
TOTAL (PERCENTS)	100.0	100.0	100.0	100.0	100.0	100.0	100.0
SOLID FUELS	14.9	10.4	10.9	8.4	7.9	13.0	14.0
OIL	56.4	49.2	48.7	45.4	45.3	47.8	48.0
GAS	16.8	26.3	26.1	30.3	30.9	26.1	24.0
ELECTRICITY	12.0	14.1	14.2	15.8	15.9	13.0	14.0
HEAT [1]	-	-	-	-	-	-	-

Source: IEA/OECD Energy Balances and IEA Country Submissions (1982).

1. Includes only heat from combined heat and power plants.

UNITED KINGDOM

5/D6 ENERGY USE IN INDUSTRY[1] BY FUEL

	1973	1978	1979	1980	1981	1985	1990
TOTAL (MTOE)	71.5	58.6	60.7	48.4	46.8	54.0	58.0
SOLID FUELS	17.1	10.7	11.7	6.8	7.5	11.0	12.0
OIL	36.3	26.2	26.6	20.4	19.0	23.0	25.0
GAS	10.3	13.7	14.1	13.8	13.1	13.0	13.0
ELECTRICITY	7.8	8.0	8.3	7.5	7.2	7.0	8.0
HEAT [2]	-	-	-	-	-	-	-
TOTAL (PERCENTS)	100.0	100.0	100.0	100.0	100.0	100.0	100.0
SOLID FUELS	23.8	18.2	19.3	14.0	15.9	20.4	20.7
OIL	50.8	44.7	43.7	42.1	40.7	42.6	43.1
GAS	14.4	23.3	23.3	28.4	27.9	24.1	22.4
ELECTRICITY	11.0	13.7	13.6	15.5	15.4	13.0	13.8
HEAT [2]	-	-	-	-	-	-	-

Source: IEA/OECD Energy Balances and IEA Country Submissions (1982).

1. Includes non-energy use of petroleum products.
2. Includes only heat from combined heat and power plants.

5/D7 ENERGY USE IN OTHER SECTORS [1] BY FUEL

	1973	1978	1979	1980	1981	1985	1990
TOTAL (MTOE)	53.6	54.6	58.2	55.4	55.1	56.0	57.0
SOLID FUELS	14.0	8.3	8.4	8.2	7.8	7.0	6.0
OIL	13.9	13.4	13.5	10.4	9.8	9.0	9.0
GAS	13.8	20.8	23.6	24.3	25.1	27.0	29.0
ELECTRICITY	12.0	12.1	12.7	12.4	12.3	13.0	13.0
HEAT [2]	na	na	na	.1	.1	-	-
TOTAL (PERCENTS)	100.0	100.0	100.0	100.0	100.0	100.0	100.0
SOLID FUELS	26.1	15.1	14.5	14.8	14.1	12.5	10.5
OIL	25.9	24.6	23.3	18.8	17.9	16.1	15.8
GAS	25.7	38.1	40.5	43.8	45.6	48.2	50.9
ELECTRICITY	22.3	22.1	21.8	22.3	22.3	23.2	22.8
HEAT [2]	na	na	na	.2	.1	-	-

Source: IEA/OECD Energy Balances and IEA Country Submissions (1982).

1. Includes use in agricultural, commercial, public services and residential sectors.
2. Includes only heat from combined heat and power plants.

OECD SALES AGENTS
DÉPOSITAIRES DES PUBLICATIONS DE L'OCDE

ARGENTINA – ARGENTINE
Carlos Hirsch S.R.L., Florida 165, 4° Piso (Galería Guemes)
1333 BUENOS AIRES, Tel. 33.1787.2391 y 30.7122
AUSTRALIA – AUSTRALIE
Australia and New Zealand Book Company Pty, Ltd.,
10 Aquatic Drive, Frenchs Forest, N.S.W. 2086
P.O. Box 459, BROOKVALE, N.S.W. 2100
AUSTRIA – AUTRICHE
OECD Publications and Information Center
4 Simrockstrasse 5300 BONN. Tel. (0228) 21.60.45
Local Agent/Agent local :
Gerold and Co., Graben 31, WIEN 1. Tel. 52.22.35
BELGIUM – BELGIQUE
CCLS – LCLS
19, rue Plantin, 1070 BRUXELLES. Tel. 02.521.04.73
BRAZIL – BRÉSIL
Mestre Jou S.A., Rua Guaipa 518,
Caixa Postal 24090, 05089 SAO PAULO 10. Tel. 261.1920
Rua Senador Dantas 19 s/205-6, RIO DE JANEIRO GB.
Tel. 232.07.32
CANADA
Renouf Publishing Company Limited,
2182 St. Catherine Street West,
MONTRÉAL, Que. H3H 1M7. Tel. (514)937.3519
OTTAWA, Ont. K1P 5A6, 61 Sparks Street
DENMARK – DANEMARK
Munksgaard Export and Subscription Service
35, Nørre Søgade
DK 1370 KØBENHAVN K. Tel. +45.1.12.85.70
FINLAND – FINLANDE
Akateeminen Kirjakauppa
Keskuskatu 1, 00100 HELSINKI 10. Tel. 65.11.22
FRANCE
Bureau des Publications de l'OCDE,
2 rue André-Pascal, 75775 PARIS CEDEX 16. Tel. (1) 524.81.67
Principal correspondant :
13602 AIX-EN-PROVENCE : Librairie de l'Université.
Tel. 26.18.08
GERMANY – ALLEMAGNE
OECD Publications and Information Center
4 Simrockstrasse 5300 BONN Tel. (0228) 21.60.45
GREECE – GRÈCE
Librairie Kauffmann, 28 rue du Stade,
ATHÈNES 132. Tel. 322.21.60
HONG-KONG
Government Information Services,
Publications/Sales Section, Baskerville House,
2/F., 22 Ice House Street
ICELAND – ISLANDE
Snaebjörn Jönsson and Co., h.f.,
Hafnarstraeti 4 and 9, P.O.B. 1131, REYKJAVIK.
Tel. 13133/14281/11936
INDIA – INDE
Oxford Book and Stationery Co. :
NEW DELHI-1, Scindia House. Tel. 45896
CALCUTTA 700016, 17 Park Street. Tel. 240832
INDONESIA – INDONÉSIE
PDIN-LIPI, P.O. Box 3065/JKT., JAKARTA, Tel. 583467
IRELAND – IRLANDE
TDC Publishers – Library Suppliers
12 North Frederick Street, DUBLIN 1 Tel. 744835-749677
ITALY – ITALIE
Libreria Commissionaria Sansoni :
Via Lamarmora 45, 50121 FIRENZE. Tel. 579751/584468
Via Bartolini 29, 20155 MILANO. Tel. 365083
Sub-depositari :
Ugo Tassi
Via A. Farnese 28, 00192 ROMA. Tel. 310590
Editrice e Libreria Herder,
Piazza Montecitorio 120, 00186 ROMA. Tel. 6794628
Costantino Ercolano, Via Generale Orsini 46, 80132 NAPOLI. Tel.
405210
Libreria Hoepli, Via Hoepli 5, 20121 MILANO. Tel. 865446
Libreria Scientifica, Dott. Lucio de Biasio "Aeiou"
Via Meravigli 16, 20123 MILANO Tel. 807679
Libreria Zanichelli
Piazza Galvani 1/A, 40124 Bologna Tel. 237389
Libreria Lattes, Via Garibaldi 3, 10122 TORINO. Tel. 519274
La diffusione delle edizioni OCSE è inoltre assicurata dalle migliori
librerie nelle città più importanti.
JAPAN – JAPON
OECD Publications and Information Center,
Landic Akasaka Bldg., 2-3-4 Akasaka,
Minato-ku, TOKYO 107 Tel. 586.2016
KOREA – CORÉE
Pan Korea Book Corporation,
P.O. Box n° 101 Kwangwhamun, SÉOUL. Tel. 72.7369

LEBANON – LIBAN
Documenta Scientifica/Redico,
Edison Building, Bliss Street, P.O. Box 5641, BEIRUT.
Tel. 354429 – 344425
MALAYSIA – MALAISIE
and/et SINGAPORE - SINGAPOUR
University of Malaya Co-operative Bookshop Ltd.
P.O. Box 1127, Jalan Pantai Baru
KUALA LUMPUR. Tel. 51425, 54058, 54361
THE NETHERLANDS – PAYS-BAS
Staatsuitgeverij
Verzendboekhandel Chr. Plantijnstraat 1
Postbus 20014
2500 EA S-GRAVENHAGE. Tel. nr. 070.789911
Voor bestellingen: Tel. 070.789208
NEW ZEALAND – NOUVELLE-ZÉLANDE
Publications Section,
Government Printing Office Bookshops:
AUCKLAND: Retail Bookshop: 25 Rutland Street,
Mail Orders: 85 Beach Road, Private Bag C.P.O.
HAMILTON: Retail Ward Street,
Mail Orders, P.O. Box 857
WELLINGTON: Retail: Mulgrave Street (Head Office),
Cubacade World Trade Centre
Mail Orders: Private Bag
CHRISTCHURCH: Retail: 159 Hereford Street,
Mail Orders: Private Bag
DUNEDIN: Retail: Princes Street
Mail Order: P.O. Box 1104
NORWAY – NORVÈGE
J.G. TANUM A/S Karl Johansgate 43
P.O. Box 1177 Sentrum OSLO 1. Tel. (02) 80.12.60
PAKISTAN
Mirza Book Agency, 65 Shahrah Quaid-E-Azam, LAHORE 3.
Tel. 66839
PHILIPPINES
National Book Store, Inc.
Library Services Division, P.O. Box 1934, MANILA.
Tel. Nos. 49.43.06 to 09, 40.53.45, 49.45.12
PORTUGAL
Livraria Portugal, Rua do Carmo 70-74,
1117 LISBOA CODEX. Tel. 360582/3
SPAIN – ESPAGNE
Mundi-Prensa Libros, S.A.
Castelló 37, Apartado 1223, MADRID-1. Tel. 275.46.55
Libreria Bosch, Ronda Universidad 11, BARCELONA 7.
Tel. 317.53.08, 317.53.58
SWEDEN – SUÈDE
AB CE Fritzes Kungl Hovbokhandel,
Box 16 356, S 103 27 STH, Regeringsgatan 12,
DS STOCKHOLM. Tel. 08/23.89.00
SWITZERLAND – SUISSE
OECD Publications and Information Center
4 Simrockstrasse 5300 BONN. Tel. (0228) 21.60.45
Local Agents/Agents locaux
Librairie Payot, 6 rue Grenus, 1211 GENÈVE 11. Tel. 022.31.89.50
TAIWAN – FORMOSE
Good Faith Worldwide Int'l Co., Ltd.
9th floor, No. 118, Sec. 2
Chung Hsiao E. Road
TAIPEI. Tel. 391.7396/391.7397
THAILAND – THAILANDE
Suksit Siam Co., Ltd., 1715 Rama IV Rd,
Samyan, BANGKOK 5. Tel. 2511630
TURKEY – TURQUIE
Kültur Yayinlari Is-Türk Ltd. Sti.
Atatürk Bulvari No : 77/B
KIZILAY/ANKARA. Tel. 17 02 66
Dolmabahce Cad. No : 29
BESIKTAS/ISTANBUL. Tel. 60 71 88
UNITED KINGDOM – ROYAUME-UNI
H.M. Stationery Office, P.O.B. 276,
LONDON SW8   5DT. Tel. (01) 622.3316, or
49 High Holborn, LONDON WC1V 6 HB (personal callers)
Branches at: EDINBURGH, BIRMINGHAM, BRISTOL,
MANCHESTER, BELFAST.
UNITED STATES OF AMERICA – ÉTATS-UNIS
OECD Publications and Information Center, Suite 1207,
1750 Pennsylvania Ave., N.W. WASHINGTON, D.C.20006 – 4582
Tel. (202) 724.1857
VENEZUELA
Libreria del Este, Avda. F. Miranda 52, Edificio Galipan,
CARACAS 106. Tel. 32.23.01/33.26.04/31.58.38
YUGOSLAVIA – YOUGOSLAVIE
Jugoslovenska Knjiga, Terazije 27, P.O.B. 36, BEOGRAD.
Tel. 621.992

Les commandes provenant de pays où l'OCDE n'a pas encore désigné de dépositaire peuvent être adressées à :
OCDE, Bureau des Publications, 2, rue André-Pascal, 75775 PARIS CEDEX 16.

Orders and inquiries from countries where sales agents have not yet been appointed may be sent to:
OECD, Publications Office, 2 rue André-Pascal, 75775 PARIS CEDEX 16.

66628-6-1983

OECD PUBLICATIONS, 2, rue André-Pascal, 75775 PARIS CEDEX 16 - No. 42705 1983
PRINTED IN FRANCE
(61 83 07 1) ISBN 92-64-12489-6